Large-sample confidence interval for $p_1 - p_2$	$\hat{p}_1 - \hat{p}_2 \pm z\sigma_{\hat{p}_1 - \hat{p}_2}$
Large-sample test of $H_0: \mu_1 - \mu_2 = 0$	$z = \dfrac{\bar{y}_1 - \bar{y}_2}{\sigma_{\bar{y}_1 - \bar{y}_2}}$
Large-sample test of $H_0: p_1 - p_2 = 0$	$z = \dfrac{\hat{p}_1 - \hat{p}_2}{\sigma_{\hat{p}_1 - \hat{p}_2}}$
Small-sample confidence interval for μ	$\bar{y} \pm t\dfrac{s}{\sqrt{n}}$
Small-sample confidence interval for $\mu_1 - \mu_2$	$\bar{y}_1 - \bar{y}_2 \pm ts\sqrt{\dfrac{1}{n_1} + \dfrac{1}{n_2}}$
Small-sample test of $H_0: \mu = \mu_0$	$t = \dfrac{\bar{y} - \mu_0}{s/\sqrt{n}}$
Small-sample test of $H_0: \mu_1 - \mu_2 = 0$	$t = \dfrac{\bar{y}_1 - \bar{y}_2}{s\sqrt{\dfrac{1}{n_1} + \dfrac{1}{n_2}}}$
Least squares line	$\hat{y} = \hat{\beta}_0 + \hat{\beta}_1 x$
Slope of least squares line	$\hat{\beta}_1 = \dfrac{S_{xy}}{S_{xx}}$
Intercept of least squares line	$\hat{\beta}_0 = \bar{y} - \hat{\beta}_1 \bar{x}$
Test of $H_0: \beta_1 = 0$	$t = \dfrac{\hat{\beta}_1}{\sqrt{s^2/S_{xx}}}$
Population correlation coefficient	ρ
Sample correlation coefficient	$\hat{\rho} = \dfrac{S_{xy}}{\sqrt{S_{xx}S_{yy}}}$
Test of $H_0: \rho = 0$	$t = \hat{\rho}\sqrt{\dfrac{n-2}{1-\hat{\rho}^2}}$
Paired-difference test	$t = \dfrac{\bar{d}}{s_d/\sqrt{n}}$
Test of $H_0: \sigma_1^2 = \sigma_2^2$	$F = \dfrac{s_1^2}{s_2^2}$
Chi-square test of independence	$\chi^2 = \sum \dfrac{(O-E)^2}{E}$

third edition

Understanding Statistics

DATE DUE		
OCT 16 1987		OCT 12 2001
FEB 19 1988	SEP 26 1996	
OCT -5 1989	OCT -7 1996	OCT 18 2001
APR 24 1992		JAN 24 2003
APR 2 1993	SEP 18 1997	MAR 14 2008
DEC -9 1994	JAN 20 1998	
JAN 23 1995	JUN -3 1998	
	NOV 23 1998	
JUL 3 - 1995	SEP 23 1999	
JAN 29 1996	SEP -7 2000	
FEB 12 1996		

c.1 Ch

```
QA 276.12 M46 1980 c.1
MENDENHALL, WILLIAM.
UNDERSTANDING STATISTICS  3D
ED.
```

FRASER VALLEY COLLEGE
2 6025 00253 0159

third edition

Understanding Statistics

WILLIAM MENDENHALL
University of Florida

LYMAN OTT
Merrell Research Center

DUXBURY PRESS
North Scituate, Massachusetts

Understanding Statistics, Third Edition, was produced by the following people:

Copy Editor: Carol Beal
Interior Designer: David Ford
Cover Designer: David Ford

Duxbury Press
A Division of Wadsworth, Inc.

© 1980 by Wadsworth, Inc., Belmont, California 94002. All rights reserved. No part of this book may be reproduced, stored in a retrieval system, or transcribed, in any form or by any means, electronic, mechanical, photocopying, recording, or otherwise, without the prior written permission of the publisher, Duxbury Press, a division of Wadsworth, Inc., Belmont, California.

Library of Congress Cataloging in Publication Data
Mendenhall, William.
   Understanding statistics.

   Includes bibliographies and index.
   1. Statistics.   I. Ott, Lyman, joint author.
II. Title.
QA276.12.M46   1980   519.5'4   79-20914
ISBN 0-87872-241-6

Printed in the United States of America
1 2 3 4 5 6 7 8 9 — 84 83 82 81 80

Contents

PREFACE ix

1 WHAT IS STATISTICS? 1

1.1 What Is Statistics? 1
1.2 Why Study Statistics? 4
1.3 Some Current Applications of Statistics 5
1.4 What Do Statisticians Do? 8
1.5 A Note to the Student 9

2 HOW TO PHRASE AN INFERENCE: GRAPHICAL METHODS 13

2.1 Why Describe a Set of Measurements? 13
2.2 Circle Charts (Optional) 14
2.3 Bar Charts (Optional) 15
2.4 Frequency Histogram 19
2.5 Frequency Polygon (Optional) 27
2.6 Comments Concerning Histograms 28
    Summary 31

3 HOW TO PHRASE AN INFERENCE: NUMERICAL METHODS 37

3.1 Introduction 37
3.2 Measures of Central Tendency 38
3.3 Measures of Variability 47
3.4 Shortcut Method for Calculating the Variance and Standard Deviation 60
3.5 How to Guess the Standard Deviation of Sample Data 64
    Summary 67

4 PROBABILITY AND PROBABILITY DISTRIBUTIONS 73

- 4.1 Probability and Inference 73
- 4.2 What Is Probability? 74
- 4.3 Additivity of Probabilities 76
- 4.4 Conditional Probability and Independence 78
- 4.5 Random Variables 80
- 4.6 The Binomial Probability Distribution 81
- 4.7 The Normal Probability Distribution 91
- Summary 101

5 SAMPLING DISTRIBUTIONS 111

- 5.1 Introduction 111
- 5.2 Random Sampling 111
- 5.3 The Central Limit Theorem and the Sampling Distribution for a Sample Mean 117
- 5.4 The Sampling Distribution of a Sample Proportion 126
- Summary 131

6 MAKING INFERENCES: ESTIMATION 137

- 6.1 Introduction 137
- 6.2 Point Estimation of a Population Mean 141
- 6.3 Interval Estimation of a Population Mean 146
- 6.4 Point Estimation of the Binomial Parameter p 150
- 6.5 Interval Estimation of the Binomial Parameter p 155
- Summary 157

7 MAKING INFERENCES: TESTING HYPOTHESES 165

- 7.1 Introduction 165
- 7.2 Testing an Hypothesis About a Population Mean 167
- 7.3 Testing an Hypothesis About a Binomial Parameter 176
- 7.4 The Level of Significance of a Statistical Test 179
- Summary 182

8 COMPARISONS 189

- 8.1 Introduction 189
- 8.2 The Sampling Distribution of the Difference Between Two Sample Statistics 190
- 8.3 Comparing Two Population Means 194
- 8.4 Comparing Two Binomial Proportions 200
- Summary 208

9 INFERENCES BASED ON SMALL SAMPLES 217

- 9.1 Introduction 217
- 9.2 A Small-Sample Test of an Hypothesis About the Population Mean μ 217
- 9.3 A Small-Sample Confidence Interval for μ 227
- 9.4 A Small-Sample Test of a Difference in Means 229
- 9.5 A Small-Sample Confidence Interval for $(\mu_1 - \mu_2)$ 234
- Summary 237

10 REGRESSION AND CORRELATION 247

- 10.1 Introduction 247
- 10.2 Scatter Diagrams and the Freehand Regression Line 249
- 10.3 Method of Least Squares 255
- 10.4 Inferences Concerning the Slope of the Least Squares Regression Line 260
- 10.5 The Coefficient of Linear Correlation 266
- 10.6 Multiple Regression (Optional) 270
- Summary 272

11 THE DESIGN OF AN EXPERIMENT: GETTING MORE INFORMATION FOR YOUR MONEY 279

- 11.1 Introduction 279
- 11.2 The Paired-Difference Experiment: An Example of a Designed Experiment 280
- 11.3 Choosing the Sample Size to Estimate a Population Mean μ or a Population Proportion p 288
- Summary 294

12 TESTING THE EQUALITY OF POPULATION VARIANCES 303

- 12.1 Introduction 303
- 12.2 A Test of an Hypothesis Concerning Two Population Variances 304
- Summary 310

13 ANALYSIS OF VARIANCE 319

- 13.1 Introduction 319
- 13.2 The Logic Behind an Analysis of Variance 319
- 13.3 A Test of an Hypothesis Concerning More Than Two Population Means: An Example of an Analysis of Variance 321
- Summary 328

14 CONTINGENCY TABLES 341

14.1 Introduction 341
14.2 A Test for Determining Whether Two Methods of Classifying Observed Events Are Independent 342
Summary 348

15 NONPARAMETRIC STATISTICS 361

15.1 Introduction 361
15.2 A Sample Comparative Test: The Sign Test 362
15.3 Wilcoxon's Signed-Rank Test 369
15.4 Wilcoxon's Rank-Sum Test 373
15.5 Spearman's Rank Correlation Coefficient 378
Summary 383

16 LYING WITH STATISTICS 389

16.1 Lying in General 389
16.2 Graphical Distortions 391
16.3 Biased Samples (Loading the Dice) 392
16.4 The Honest Inference (What Is the Sample Size?) 394
16.5 Distorting the Truth (and Being Honest) 395
16.6 Just Plain Lying 395

17 A SUMMARY 399

Appendix I: Useful Statistical Tests and Confidence Intervals 403
Appendix II: Statistical Tables 410
Glossary of Common Statistical Terms 435
Selected Answers 439
Index 455

Preface

This text is designed for a one-quarter or one-semester introductory course in statistics. It could be used in those colleges and universities that teach a general course appropriate for students majoring, or intending to major, in many different areas. The approach, examples, and exercises provide a basic knowledge of statistical concepts that will be useful in business and in the biological, social, or physical sciences.

Certainly one of the primary objectives of an introductory statistics course is to develop in the student an appreciation of the role that statistics plays in society. This third edition of *Understanding Statistics* retains this objective but places an increased emphasis on the analysis and interpretation of experimental data. Entry into the topics of statistical estimation and testing of hypotheses is prefaced by new chapters on probability and sampling distributions. The net effect is to present a course that attempts to develop an introductory level of statistical competence as well as a general understanding and appreciation of the contribution of statistics in business and the social, physical, and biological sciences.

The level of difficulty of a noncalculus introductory statistics course is frequently set by the amount of probability theory that it contains and by the extent to which this theory is woven into the major topic of the course, statistical inference. The new chapter on probability contained in this edition is much stronger than the chapter of earlier editions, but it lacks the difficulty of many treatments in that there is no attempt to develop an ability to solve probability problems. Rather, the chapter presents the basic properties and concepts of probability needed for an understanding of sampling distributions, statistical independence, and so on—that is, topics that will be encountered in later chapters. The two most important probability distributions, the binomial and the normal, are used to illustrate the practical applications of probability distributions and to set the stage for the new chapter (chapter 5) on sampling distributions. The result is an introduction to statistical inference that is of moderate difficulty but one that contains much useful and practical information.

To summarize, the third edition is an integrated and connected introductory presentation of statistics, aimed at developing competence in the use of statistical methods and an understanding of their contributions in business and the sciences. Major changes and additions to the third edition include the following:
— A new chapter (chapter 4) on probability. Elementary concepts of probability are introduced. Properties and uses of the binomial and normal probability distributions are discussed.
— A new chapter (chapter 5) on sampling and sampling distributions.

Preface

— A new chapter (chapter 15) on nonparametric statistics.
— Many new and interesting exercises. The number of exercises has been increased from 268 in the second edition to 478.
— A new section (section 10.6) that introduces multiple regression.
— A reversal in the ordering of chapters 7 and 8 from the second edition, which reflects the greater difficulty students have in understanding the basic concepts of a statistical test. In the third edition, discussion of estimation (chapter 6) precedes a discussion of hypothesis testing (chapter 7). Both chapters have been rewritten to reflect the increased emphasis on the analysis and interpretation of data.
— A new section (section 7.4) on the level of significance of a statistical test.
— *Detailed* summaries of estimation and test procedures that are set off clearly from the text discussion.
— Computer output in several chapters to illustrate solutions to standard problems using statistical software packages.
— Defined words printed in color in the margin and underlined within the text for easy reference of the reader.
— Fundamental ideas of statistics emphasized by the use of color within the text.
— An expanded and more comprehensive index.

Many students and instructors who used previous editions of this text have contacted us with suggestions, and these are much appreciated. The authors also want to acknowledge the reviewers for this edition: Steven Balkin, University of Illinois at Chicago Circle; Giorgio Gnugnoli, Randolph-Macon College; Frank J. Lambiase, Bridgewater State College; Hubert Lilliefors, George Washington University; Roy Saunders, Northern Illinois University; James G. Seiler, San Diego City College; Bill Stines, North Carolina State University; Dennis D. Wackerly, University of Florida; James G. Ware, University of Tennessee at Chattanooga.

The success of this book through the years is in many ways a team effort, and we gratefully thank the following individuals for their assistance: Carol Sexton, Susan Reiland, and Janice Warmke. A special note of gratitude goes to Barbara Beaver for her diligent preparation of the Solutions Manual and Robert Beaver for his preparation of the Study Guide.

The production staff of Duxbury Press has been of invaluable assistance to us, and a special expression of gratitude is extended to Carol Beal for her editing prowess. And, last but not least, we acknowledge the patience, encouragement, and assistance of our families in this project.

William Mendenhall
Lyman Ott

third
edition

Understanding Statistics

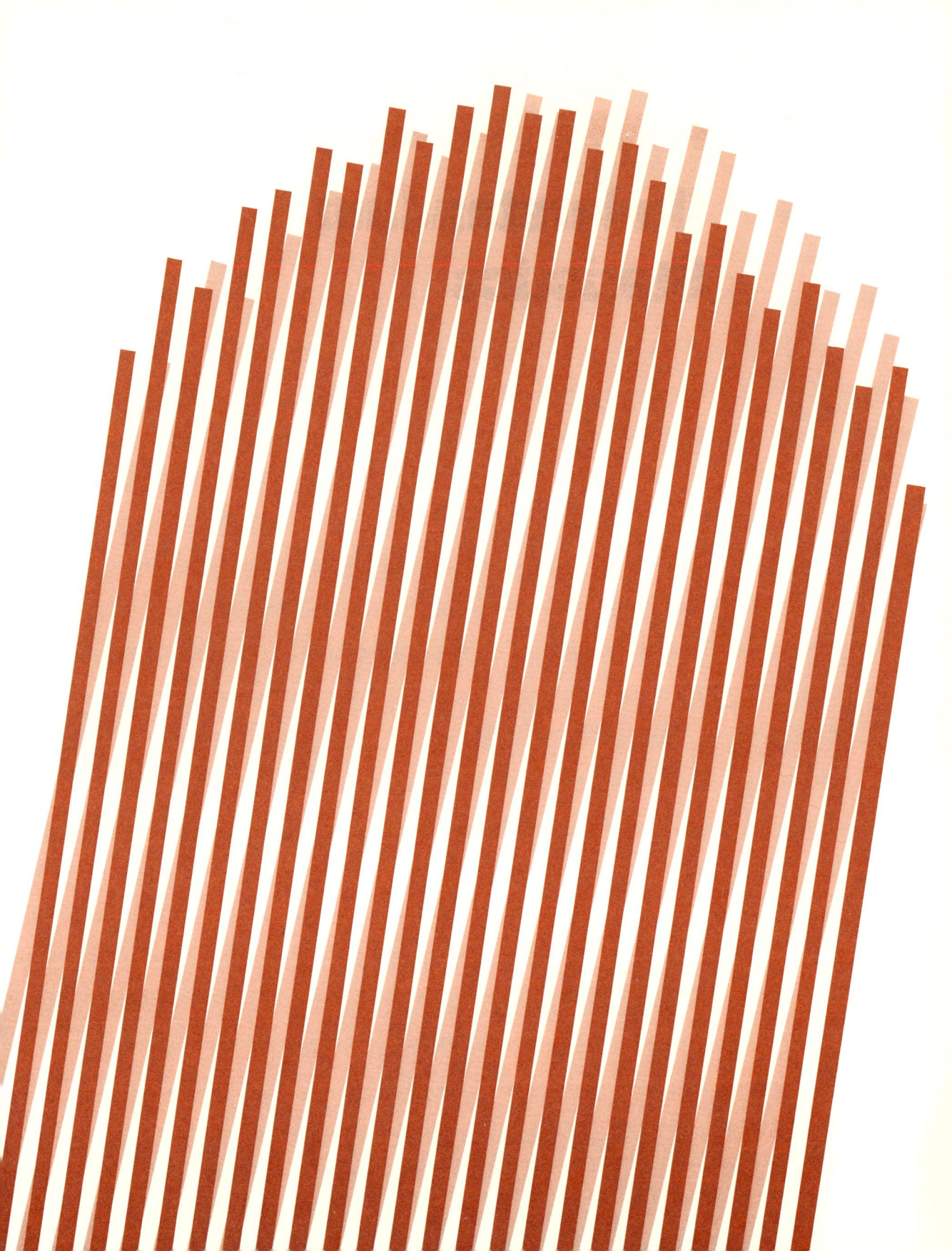

1 What Is Statistics?

CHAPTER OUTLINE

1.1 What Is Statistics?
1.2 Why Study Statistics?
1.3 Some Current Applications of Statistics
1.4 What Do Statisticians Do?
1.5 A Note to the Student

1.1 WHAT IS STATISTICS?

What is statistics? Is it the addition of numbers? Is it graphs, batting averages, percentages of passes completed, percentages of unemployment, and, in general, numerical descriptions of society and nature? Or is statistics a modern and scientific method for penetrating the unknown? The best way to answer this question is to consider a few examples of its application.

Suppose that a manufacturer of light bulbs produces roughly a half million bulbs per day. Concerned about customer reaction to its product, the firm wishes to determine the fraction of bulbs produced on a given day that are defective. It can solve the problem in two ways. All the half million bulbs could be inserted into sockets and tested, but the cost of this solution would be substantial and could greatly increase the price per bulb. A second method for determining the fraction that are defective is to select 1000 bulbs from the half million produced and test each one. The fraction of bulbs defective in the 1000 tested could be used to estimate the fraction defective in the entire day's production. We will show in later chapters that the fraction defective in the bulbs tested will probably be quite close to the fraction defective for the entire half million bulbs. Also, we will be able to tell you by how much you might expect this estimate to differ from the fraction of defective bulbs produced on a given day.

A second and similar example of statistics is brought to mind by the frequent use of the Gallup poll, the Harris poll, and other public opinion polls. How can these pollsters presume to know the opinions of more than 100 million Americans? They certainly cannot reach their conclusions by contacting every voter in the United States. Rather, as we have suggested in the light bulb example, they sample the opinions of a small number of voters, perhaps as few as 1000, to estimate the reaction of every voter in the country. The amazing result of this process is that the fraction of those people contacted who hold a particular opinion will match very closely the fraction of voters holding that opinion in the complete population. Most students find this assertion difficult to believe; convincing supportive evidence will be supplied in subsequent chapters.

A third example of a statistical problem is taken from the field of medicine. Suppose a research physician wishes to investigate the effect of a new drug on the stimulation of a patient's heart. Note that the physician is really interested in the effect of the drug on all future heart patients who might be treated by the drug. Fifty heart patients are selected and each treated with the drug. The increase in the pulse rate is recorded for each. After observing the effect of the drug on the 50 patients, the physician may infer that the drug will have a similar effect on all heart patients in the future.

characteristics

measurement

Let us now try to identify from the preceding examples the characteristics common to inferential statistical problems. First, each example involved making an observation or measurement which cannot be predicted with certainty in advance. Indeed, results of repeated observations are likely to bob about in a haphazard or random manner. For example, we cannot say in advance whether a particular light bulb selected from production will work, whether a randomly selected voter will vote for a particular political candidate, or what a new heart patient's pulse will be after receiving the drug.

sampling

Second, each example involved sampling. A group of light bulbs was selected from the day's production, a group of people was taken from the entire voting population of the United States, and a group of 50 heart patients was obtained from the total of all heart patients.

collection of data

Third, each example involved the collection of data or measurements, one measurement corresponding to each element of the sample (group). We realize that observations on the elements of the sample may be quantitative in nature (when we record age, income, heart rate, and so on) or qualitative in nature (when we record political affiliation, sex, preferences, and so on). But even these qualitative observations can be viewed as measurements if we assign a number to each qualitative category. For example, when a light bulb is tested we will associate a 1 with a defective bulb and a 0 with a good bulb. Thus the entire daily output of bulbs can be associated with a set of numbers. Each number in the set will be a 0 or a 1. The total number of defective light bulbs in the sample is the sum of these measurements. Similarly, in a sample of voter intentions, we can assign the measurement 1 to each person who supports a specified candidate and 0 to each person who does not.

common objective

Finally, each example exhibits a common objective. That is, **the purpose of sampling is to obtain information that can be used to make an inference about a much larger set of measurements called the population.** For the manufacturer who wishes to make a statistical inference about (estimate) the fraction of defectives in a day's production, the population of interest is the set of ones and zeros—roughly a half million in number—corresponding to the defective and nondefective bulbs produced that day. Similarly, the population for the voter problem is the set of ones and zeros corresponding to the 100 million or more voters in the United States. Each voter is assigned a 1 if the voter intends to vote for the political candidate and a 0 if not. The objective of sampling is to estimate the fraction of eligible voters who favor the candidate—that is, the fraction of ones in the population.

The population associated with the drug–heart rate experiment is the set of pulse rates for all heart patients who might be treated by the drug. The sample of 50 pulse rates for the group of heart patients is presumably representative of this population. The objective of the experiment is to estimate the average pulse rate that might be induced by the drug. Thus the experimenter wishes to make a statistical inference about a large set of pulse measurements that could be acquired from patients treated by the drug at some time in the future. Note again that populations are collections of measurements and are not collections of people (which is the usual connotation of the term). Note also that populations may exist in fact or may be conceptual in nature. The populations for the first two examples exist even though we do not actually possess the complete collection of ones and zeros corresponding to defectives and nondefectives, or the entire set of ones and zeros corresponding to voters favoring or opposing the political candidate. In contrast, the population of pulse rates measured on all heart patients who could be administered the drug, now or in the future, is referred to as an imaginary or conceptual population.

Putting the four characteristics together—random observations, sampling, numerical data, and a common inferential objective—we might define **statistics** to be a theory of information. Information is obtained by experimentation or, equivalently, by sampling; it is employed to make an inference about a larger set of measurements, existing or conceptual, called a population.

Relevant definitions for this section are as follows:

DEFINITION 1.1

A population is the set of all measurements of interest to the sample collector.

DEFINITION 1.2

A sample is any subset of measurements selected from the population.

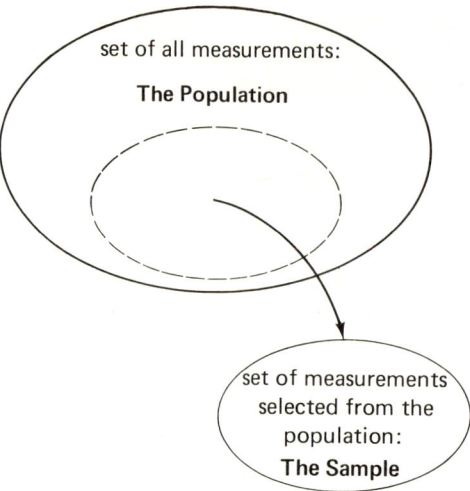

Figure 1.1   Population and Sample

DEFINITION 1.3

objective of statistics   The objective of statistics is to make an inference about a population based on information contained in a sample.

1.2   WHY STUDY STATISTICS?

We can think of two good reasons for taking an introductory course in statistics. First, every person is exposed to manufacturers' claims for products, to the results of sociological, consumer, and political polls, and to the published achievements of scientific research. Many of these results are inferences based on sampling. Some of the inferences are valid; others are invalid. Some are based on samples of adequate size; others are not. Yet all these published results bear the ring of truth. Some people say that statistics can be made to support almost anything (particularly statisticians). Others say it is easy to lie with statistics. Both statements are true. It is easy, purposely or unwittingly, to distort the truth by using statistics when presenting the results of sampling to the uninformed. Thus one reason for studying statistics is that you need to know how to evaluate published numerical facts, when to believe them, when to place tongue in cheek, and when to reject them.

first reason

second reason

A second reason for studying statistics is that your profession or employment may require you to interpret the results of sampling (surveys or experimentation) or to employ statistical methods of analysis to make inferences in your work. For example, practicing physicians receive large amounts of advertising describing the benefits of new drugs. These advertisements frequently display the numerical results of experiments that compare a new drug with an older one. Do such data really imply that the new drug is more effective, or is the observed difference in results due simply to random variation in the experimental measurements?

Recent trends in the conduct of court trials indicate an increasing use of probability and statistical inference in evaluating the quality of evidence. The use of statistics in the social, biological, and physical sciences is essential because all these sciences make use of observations of natural phenomena, through sample surveys or experimentation, to develop and test new theories. Statistical methods are employed in business when sample data are used to forecast sales and profit. In addition, they are used in engineering and manufacturing to monitor product quality. The sampling of accounts is a new and useful tool to assist accountants in conducting audits. Thus statistics plays an important role in almost all areas of science, business, and industry; persons employed in these areas need to know the basic concepts, strengths, and limitations of statistics.

1.3 SOME CURRENT APPLICATIONS OF STATISTICS

The Environmental Effects of the SST

The recent development of supersonic jet transports (SST) by Russia, Great Britain, and France has created some unexpected environmental problems. In addition to the problems of fuel consumption, noise, and space required for landing, it is feared that extensive use of the SST may increase the intensity of ultraviolet rays emitted by the sun. This, in turn, may increase the incidence of skin cancer in humans and have a profound effect on plant growth.

Ultraviolet radiation from the sun (which causes sunburn) is present at all times, on cloudy as well as sunny days, but the intensity of the radiation is reduced by a cloud cover of various substances between the earth and the sun. One of the major protective ultraviolet ray barriers is a thick layer of ozone that blankets the earth at high altitudes. Because SSTs fly at high altitudes, there is concern that they will disturb this protective ozone blanket. Many questions arise which must be answered by research. If ultraviolet rays are passed through ozone, how will the intensity of the rays be affected by changes in the ozone density? How is the intensity of ultraviolet radiation related to the incidence of skin cancer and plant growth? How will the use of a specific number of SSTs affect the density of ozone in the atmosphere and, consequently, the intensity of the sun's ultraviolet rays? These and many other questions must be answered by experimentation (sampling) and statistical analysis of the experimental data.

Determining the Effectiveness of a New Drug Product

The development and testing of the Salk vaccine for protection against poliomyelitis (polio) provide an excellent example of how statistics can be used in solving practical problems. Most parents and children growing up in the years before 1954 can recall the panic brought on by the outbreak of cases of polio during the summer months. Although relatively few children fell victim to the disease each year, the pattern of outbreak of polio was unpredictable and caused great concern because of the possibility of paralysis or death. The fact that very few of today's youth have even heard of polio demonstrates the great success of the vaccine and the testing program that preceded its release on the market.

It is standard practice in establishing the effectiveness of a particular drug product to conduct an experiment (often called a *clinical trial*) with human subjects. For some clinical trials assignments of subjects are made at random, with half receiving the drug product and the other half receiving a nonfunctioning saline solution (called a *placebo*). One statistical problem concerns the determination of the total number of subjects to be included in the clinical trial. This problem was particularly important in the testing of the Salk vaccine because data from previous years suggested that the incidence rate might be less than 50 cases for every 100,000 children. Hence a large number of subjects had to be included in the clinical trial in order to detect a difference in the incidence rates for those treated with the vaccine and those receiving the placebo.

With the assistance of statisticians it was decided that a total of 400,000 children should be included in the Salk clinical trial begun in 1954, with half of them randomly assigned the vaccine and the remaining children assigned the placebo. No other clinical trial had ever been attempted on such a large group of subjects. Through a public school inoculation program, the 400,000 subjects were treated and then observed over the summer to determine the number of children contracting polio. Although less than 200 cases of polio were reported for the 400,000 subjects in the clinical trial, more than three times as many cases appeared in the group receiving the placebo. These results together with some statistical calculations were sufficient to indicate the effectiveness of the Salk polio vaccine. However, these conclusions would not have been possible if the statisticians and scientists had not planned for and conducted such a large clinical trial.

The development of the Salk vaccine is not an isolated example of the use of statistics in the testing and developing of drug products. In recent years the Food and Drug Administration has placed stringent requirements on pharmaceutical firms to establish the effectiveness of proposed new drug products. Thus statistics has played an important role in the development and testing of birth control pills, rubella vaccines, chemotherapeutic agents in the treatment of cancer, and many other preparations.

The Food Crisis: A Search for Additional Food Supplies

Hardly a day goes by that we are not reminded of the millions of people throughout the world who are suffering from malnutrition or dying from starvation. Recent droughts in various countries throughout the world have drastically curtailed the world food supply. Consequently, new food sources must be sought in order to meet the rising food demands of an ever-increasing world population.

One possible new source lies in untapped food supplies in the seas, but this view is not shared by all. Some experts, such as French Diplomat Michel Lenuyeauz-Commène, have said that the seas are already overfished and we can expect to exhaust this food source in 20 years. While this is certainly a pessimistic view, others hold a brighter outlook. It has been estimated that the seas currently supply approximately 13% of our total animal-protein intake per year. Ronald F. Smith of the National Oceanic and Atmospheric Administration estimates that the seas could be used to feed almost half the world's population. One way to increase the world protein intake from the seas is to develop sea farming. Aquaculture, as it is called, involves farming of various fish varieties from spawn until maturity and growth of microorganisms (plankton) which serve as food for fish.

These developments will make use of experimentation to determine the best procedures for harvesting ocean food resources. Statisticians will assist in the design of experiments and surveys and in the interpretation of research data. Thus statistics will be used to help determine the effectiveness of such programs in terms of the increase in fish and plankton yields and to evaluate the economic feasibility of aquaculture.

The Energy Crisis: A Search for New Sources and a Search for Oil

The oil crisis of 1973–1974 brought to America's attention a problem that will be with us for decades, a shortage of energy. Thus in the face of rapidly rising annual increases in the demand for energy, the United States is confronted by a supply that cannot meet current demands. This led to the "energy rush" of 1974 and subsequent problems since then.

Possible sources of energy needed to supply the present and future requirements of the United States include the development of vast coal and oil shale reserves, the development of nuclear reactors, the search for new oil and natural gas reserves, and the possible use of solar energy. In which of these resources should we, the American public, invest the capital necessary for development? Which source will yield a given amount of energy at minimum cost? What unfavorable impact will each have on the environment or the quality of life? Which might yield dangerous side effects? These questions and others must be answered by experimentation. Statisticians will assist in designing experiments and in interpreting experimental data.

1.4 WHAT DO STATISTICIANS DO?

What do statisticians do; or, equivalently, with what is the field of statistics concerned? Statisticians, both in consulting and research, devote their time to two major areas. The first concerns the **acquisition of the sample data**. Sample surveys and experiments cost money and yield information, usually numbers on sheets of paper. By varying the survey or experimental procedure—where you select the data and how many observations you take from each source—you can vary the cost and quantity of information in the experiment. Rather simple modifications in the data selection procedure can reduce the cost of the sample drastically. Thus statisticians study various methods for designing sample surveys and experiments (selecting the sample) and attempt to find the method that will yield a specified amount of information at minimum cost.

acquisition of sample data

The second task facing statisticians is the **selection of** the appropriate **method of inference** for a given sample survey or experimental design. Some of these methods are good, some are bad, and some seem to be good for most occasions. It is the statistician's job to choose the appropriate method for a given situation.

selection of method of inference

The foregoing discussion leads to the most important contribution of statistics to science and business. Anyone can devise a method to make inferences based on the sample data. **The major contribution of statistics is in evaluating the "goodness" of the inference.** When predicting, we seek an upper limit to the error of prediction. In reaching a decision concerning a characteristic of the population, we ask the probability of reaching a correct conclusion.

To summarize, statisticians first design surveys and experiments to minimize the cost of obtaining a specified quantity of information. Second, they seek the best method for analyzing the data and making an inference for a given sampling situation. Finally, statisticians can measure the goodness of an inference.

Exercises

How a Statistician Can Be of Assistance in Scientific Research

1. helping to plan the experiment
2. choosing the appropriate method of inference
3. giving a measure of the goodness of an inference

1.5 A NOTE TO THE STUDENT

We think with words and concepts. Thus statistics, a theory of information, requires the memorization of new terms and concepts (as does the study of a foreign language). Commit definitions, theorems, and concepts to memory.

Second, focus on the broader aspects of statistics. What is statistics? How does it work? What are some of the more important applications? Do not let details obscure these broader characteristics of the subject. The teaching objective of this text is to identify and amplify these broader concepts of statistics.

EXERCISES

1.1. Selecting the proper diet for shrimp or other sea animals is an important aspect of sea farming. A researcher wishes to estimate the mean weight of shrimp maintained on a specific diet for a period of six months. One hundred shrimp are randomly selected from an artificial pond and each is weighed.
(a) Identify the population of measurements that is of interest to the researcher.
(b) Identify the sample.
(c) What characteristics of the population would be of interest to the researcher?
(d) If the sample measurements are used to make inferences about certain characteristics of the population, why would a measure of the reliability of the inferences be important?

1.2. Radioactive waste disposal as well as the production of radioactive material in some mining operations is creating a serious pollution problem in some areas of the United States. State health officials decided to investigate the radioactivity levels in one suspect area. Two hundred points were randomly selected in the area and the level of radioactivity was measured at each point. Answer questions (a), (b), (c), and (d) in exercise 1.1 for this sampling situation.

10 What Is Statistics?

 1.3. A social researcher in a particular city wishes to obtain information on the number of children in households that receive welfare support. A random sample of 400 households was selected from the welfare rolls of the city. A check on welfare recipient data provided the number of children in each household. Answer questions (a), (b), (c), and (d) of exercise 1.1 for this sample survey.

EXPERIENCES WITH REAL DATA

Search issues of your local newspaper to locate the results of a recent Harris or Gallup survey.

1. Identify the items that will be observed in order to obtain the sample measurements.

2. Identify the measurement made on each item.

3. Clearly identify the population associated with the survey.

4. What characteristic(s) of the population is (are) of interest to the pollster?

5. Does the article explain how the sample was selected?

6. Does the article include the number of measurements in the sample?

7. What type of inference is made concerning the population characteristics?

8. Does the article tell you how much faith you can place in the inference about the population characteristic?

2 How to Phrase an Inference: Graphical Methods

CHAPTER OUTLINE

2.1 Why Describe a Set of Measurements?
2.2 Circle Charts (Optional)
2.3 Bar Charts (Optional)
2.4 Frequency Histogram
2.5 Frequency Polygon (Optional)
2.6 Comments Concerning Histograms

2.1 WHY DESCRIBE A SET OF MEASUREMENTS?

We stated in chapter 1 that the objective of statistics is to make an inference about a population based on information contained in a sample. Since a population is usually a large set of measurements, it is necessary to find some means of condensing and describing these measurements. Only after completing this description can we begin to talk about a population, or, in other words, to phrase an inference about it.

It is also necessary to condense information about large sets of economic or sociological data. For example, production figures for all oil wells in the United States in a given year need to be described in condensed form before they can be published. The data collected by the U.S. Bureau of the Census require similar condensation and description. Although census, sociological, and economic data might ultimately be employed to make statistical inferences, simple descriptions of these large quantities of data are necessary to make them immediately available for public consumption.

Descriptions of most things are difficult. Try, for example, describing the person sitting next to you so precisely that a stranger could select the individual from a group of others having similar physical characteristics. It is not an easy task. Fingerprints, voiceprints, and photographs, all pictorial descriptions, are the most precise methods of human identification. The description of a set of measurements is also a difficult task but, like the description of a person, it can be accomplished more easily by using graphic or pictorial methods. Thus we could agree with the old axiom, "A picture is worth a thousand words," and we might add, "or more."

Pictorial or graphical description is as old as recorded history. Cave drawings convey to us scattered bits of information about the life of prehistoric people. Similarly, vast quantities of knowledge about the ancient lives and cultures of the Babylonians, Egyptians, Greeks, and Romans are brought to life by means of drawings and sculpture. Art has been used to convey a picture of various life-styles, history, and culture in all ages. Not surprisingly, it is also of value in describing a set of measurements.

Sets of measurements are described in two ways, graphically and numerically. Graphical description is the objective of chapter 2. Numerical descriptive measures will follow in chapter 3. Four of the most common graphical methods for describing data—circle charts, bar charts, frequency histograms, and frequency polygons—are presented in this chapter. Coverage of three of these—circle charts, bar charts, and frequency polygons—is optional, but a study of the relative frequency histogram is essential to an understanding of statistical inference.

2.2 CIRCLE CHARTS (OPTIONAL)

circle chart

A circle chart, or pie chart, is often used to show how a number of objects are apportioned to a group of categories. For example, table 2.1 provides a summary of the employment status of 1000 adult males and shows the number of men falling in each of the three categories of employment. From the table we see that 122, or 12.2%, had no jobs; 536, or 53.6%, had one job; and 342, or 34.2%, held more than one job. The percentages of men falling into each of the three employment categories are shown graphically in figure 2.1.

Table 2.1 Employment Data

No Job	One Job	More Than One Job
122	536	342

Figure 2.1 was constructed by apportioning the 360° of the circle to match the percentage of each employment category. Thus 12.2% of 360° [or (.122)(360) = 43.9°] was assigned to the group holding no job. Similarly, 53.6% (or 193°) was assigned to those holding one job. The remainder of the pie was allocated to those holding more than one job. The pie itself was then constructed by marking the angles with a protractor. (If a protractor is not available, you can approximate the portions by eye.)

Bar Charts (Optional)

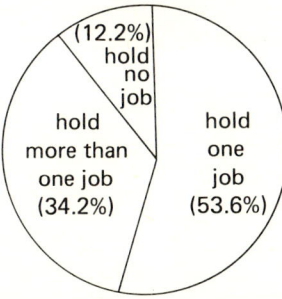

Figure 2.1 Employment Pie Chart

Figure 2.2 depicts the location of known oil reserves in the Western Hemisphere. Note that the largest oil reserves, 46.1%, are in Mexico, 29.7% are in the United States, 17.4% in South America, and 6.8% in Canada. The circle chart gives a rapid description of the relative portion of known oil reserves in these four locations.

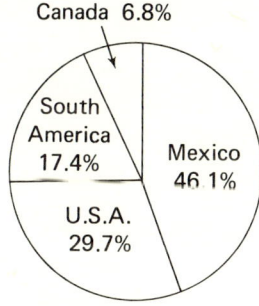

Figure 2.2 Location of Oil Reserves in the Western Hemisphere

2.3 BAR CHARTS (OPTIONAL)

bar chart

The bar chart is another graphical method for showing the proportions of a number of objects that fall in a group of categories. We can illustrate its use with the following example: Hospital officials were interested in studying the financial status of people visiting a particular outpatient clinic. Since it was impossible to run a lengthy check on each patient entering the clinic, a sample of 75 patients was included in the study. It was decided to classify patients into one of three economic groups: patients unable to purchase any prescription drugs were classified as group I, patients able to purchase some prescription drugs were classified as group II, and those able to purchase any and all

Table 2.2 Classification of Patients by Economic Status

Economic Group	Number of Patients per Group (Group Frequency)	Percentage of Patients per Group
group I	30	40.0
group II	21	28.0
group III	24	32.0
Total	75	100.0

necessary prescription drugs were classified as group III. The results of this study are tabulated in table 2.2. We could certainly represent the percentages given in table 2.2 by using a circle chart, but we can also display the same data by using a bar chart, or bar graph. To construct a bar chart for this example, each group is located and labeled on the horizontal axis. Rectangles are then constructed over each group to a height equal to the number of patients in that group. The number in each group is called the **group frequency**, or simply the frequency. See figure 2.3.

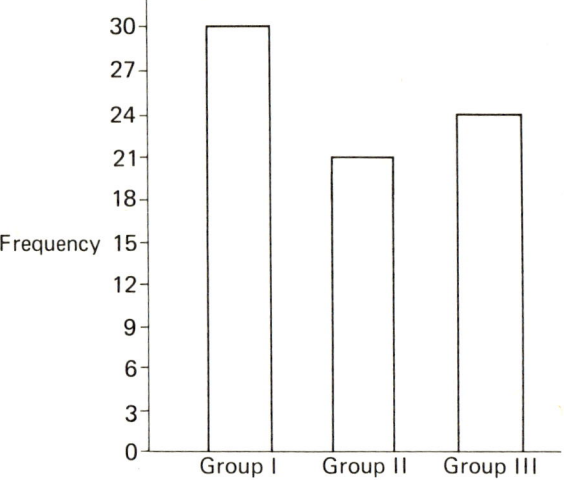

Figure 2.3 Bar Chart for Table 2.2

Bar charts are used extensively to display economic data. For example, universities throughout the country have had steadily increasing operating budgets over the last 20 years. The bar chart in figure 2.4 illustrates the increases in total operating budgets from the years 1950–1951 to 1975–1976 for the Florida state university system.

Bar Charts (Optional)

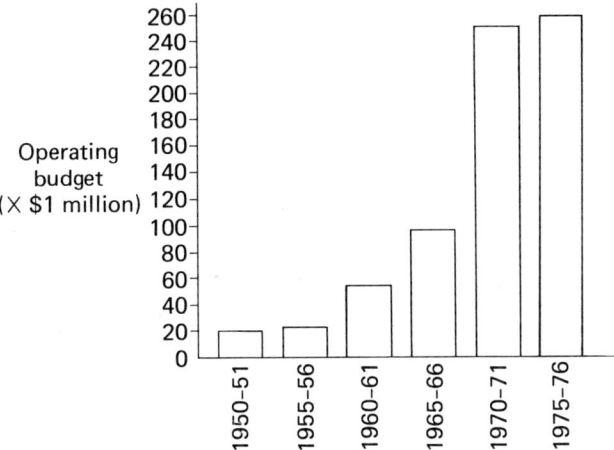

Figure 2.4 Cost of Higher Education in Florida

Bar charts are also used extensively to display percentage data. The bar chart in figure 2.5 offers a comparison of the percentages of owner-occupied housing units for blacks and whites in metropolitan and nonmetropolitan areas.

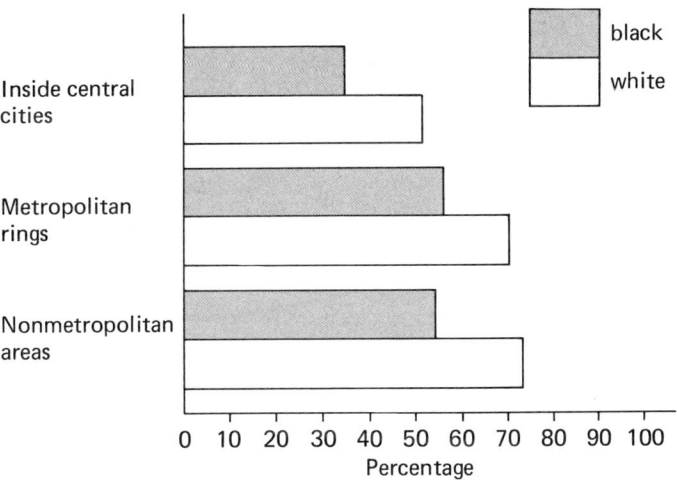

Figure 2.5 Percentage of Owner-occupied Housing Units, by Metropolitan-Nonmetropolitan Residence. *Source:* Bureau of the Census, *Current Population Reports, Special Studies,* series P-23, no. 54, "The Social and Economic Status of the Black Population in the United States, 1974" (Washington, D.C.: Government Printing Office, 1975), p. 133.

How to Phrase an Inference: Graphical Methods

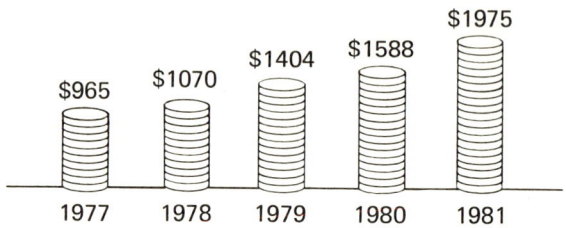

Figure 2.6 Maximum Annual Worker Contribution for Social Security, 1977–1981

A modification of the bar chart that is commonly employed to display monetary expenditures is shown in figure 2.6. Columns of coins replace the bars. This figure presents a picture of the growing annual contribution that workers must make for Social Security for the years 1977–1981. The maximum annual contribution for a given year is proportional to the height of the stack of coins.

EXERCISES

(Optional exercises are starred.)

*2.1. The accompanying data give a breakdown of the total United States oil consumption, in percentages, by the various purposes for which the oil is used. Thus the percentages show how the United States uses oil. Describe these data by using a circle chart.

Use	Percentage of Total
gasoline	43.1
industrial fuel oil	12.0
heating oil	17.0
jet fuel	6.9
diesel fuel	5.4
petrochemical	3.6
other	12.0
Total	100.0

*2.2. Refer to exercise 2.1. Use the consumption figures to construct a bar graph.

*2.3. The U.S. Bureau of the Census publishes the *Statistical Abstract of the United States*. This reference was used to obtain the listing in the accompanying table of four racial-ethnic population sizes in 1975. Use these data to construct a pie chart for the male population by racial-ethnic groups.

Four Racial-Ethnic Populations in the United States in 1975, by Sex (× 1000)

Racial-Ethnic Grouping	Male		Female	
	Number	Percentage	Number	Percentage
Indian	389	36.1	404	37.3
Japanese	271	25.1	320	29.5
Chinese	229	21.2	206	19.0
Filipino	189	17.5	154	14.2
Total	1078	99.9	1084	100.0

Source: U.S. Bureau of the Census, *Statistical Abstract of the United States: 1975* (Washington, D.C.), p. 30.

*2.4. Refer to exercise 2.3. Construct a bar chart for the female population by racial-ethnic groups.

*2.5. Refer to exercise 2.3. Combine the male and female data for each racial-ethnic group. Construct a bar chart for the combined population.

*2.6. Refer to exercise 2.3. Combine the male and female data for each racial-ethnic group. Construct a circle chart for the combined population. (Hint: We cannot add the percentages for males and females in each racial-ethnic group. A new percentage must be computed based on the total combined population.)

2.4 FREQUENCY HISTOGRAM

The frequency histogram offers a third graphical method for describing a set of measurements. We will illustrate the frequency histogram for the following situation.

Most of us are so accustomed to drinking water treated in a municipal water treatment plant that we rarely concern ourselves with the quality of the water we drink. If you were to visit a water treatment plant, you would observe that many different measurements are obtained throughout the day to monitor the quality of water sent into the system. For example, 1000 1-milliliter samples of water are taken from the water flow and sent through a filtering process. If, after filtering, one or more colonies of coliform bacteria appear, the water is unacceptable. However, since the testing process is so delicate, the appearance of a colony could mean there was a laboratory error in the analysis (something as trivial as coughing over the water). Thus more samples would have to be obtained to verify the findings.

In addition to checking for the presence of coliform bacteria, measurements are obtained on the chlorine residual in the water leaving the treatment plant. A reading of 2.2 parts per million (ppm) could be an acceptable value for a particular community. Similarly, readings on the hardness, turbidity, acidity, and color also are made in monitoring the quality of water sent throughout the system. And for those communities that add fluoride to their water as a tooth decay preventative (especially for children), the level of fluoride must also be monitored in the system.

The regulations of the board of health in a particular state specify that the fluoride level must not exceed 1.5 ppm. The 25 measurements in table 2.3 represent the fluoride levels for a sample of 25 days. Although fluoride levels are measured more than once per day, these data represent the early morning readings for the 25 days sampled.

Table 2.3 Fluoride Levels (ppm) for a Sample of 25 Days

.75	.86	.84	.85	.97
.94	.89	.84	.83	.89
.88	.78	.77	.76	.82
.72	.92	1.05	.94	.83
.81	.85	.97	.93	.79

We note from table 2.3 that the recorded fluoride levels range from .72 ppm to 1.05 ppm. Although we might examine the table closely, it would be difficult to describe precisely how the measurements are distributed along this range. For example, are most of the fluoride measurements less than .90 ppm or are most greater than .90 ppm?

range

class

class width

To answer these and other questions, we will partition the range of the data—the difference between the largest and the smallest measurements—into a number of subintervals or classes of equal width. Then by tallying the number of measurements falling in each class, we can see how the data tend to be distributed over their scale of measurement. For example, suppose we choose 7 classes for the fluoride data. Then a suitable subinterval or class width can be found by dividing the range by the number of classes, 7. Thus

$$\text{class width} = \frac{\text{range}}{\text{number of classes}} = \frac{1.05 - .72}{7} = \frac{.33}{7} \approx .05$$

lower boundary

We choose .705 as the lower boundary of the first class, a point slightly below the smallest fluoride measurement (.72). We obtain the other boundaries by adding multiples of the class width .05. Beginning with .705, the 7 classes are

.705 to .755
.755 to .805
.805 to .855
.855 to .905

Frequency Histogram

.905 to .955
.955 to 1.005
1.005 to 1.055

Note that each class has a width of .05 and that the 7 classes span the range of the fluoride measurements. Further, note that by choosing the lower class boundary at .705 (rather than .70 or .71), we made it impossible for a measurement to fall on a class boundary.

Since we are interested in knowing how the fluoride readings are distributed among the 7 classes, we examine each of the 25 measurements in table 2.3 and tally the number of measurements falling in each class. The number of measurements falling into a given class is called the **class frequency**. These frequencies are shown in table 2.4. Notice that the sum of the frequencies equals the total number of measurements n, and this will always be true.

class frequency

Table 2.4 Frequency Table for the Fluoride Data

Class	Class Boundaries	Class Frequency	Relative Frequency
1	.705– .755	2	$2/25$
2	.755– .805	4	$4/25$
3	.805– .855	8	$8/25$
4	.855– .905	4	$4/25$
5	.905– .955	4	$4/25$
6	.955–1.005	2	$2/25$
7	1.005–1.055	1	$1/25$
Total		$n = 25$	1

relative frequency

The **relative frequency** of a class is defined as the frequency of the class divided by the total number n of measurements (total frequency). Thus if we let f_i equal the frequency for class i, then

$$\text{relative frequency} = \frac{f_i}{n} \quad \text{for class } i$$

For example, the relative frequency for class 6 in table 2.4 can be found as follows: The total number of measurements is $n = 25$, and the frequency for class 6 is $f_6 = 2$. Hence

$$\text{relative frequency for class 6} = \frac{2}{25}$$

frequency histogram

A frequency table can be presented as a graph called a **frequency histogram**. We mark the class boundaries (.705, .755, and so on) along the horizontal axis. Frequencies (2, 4, 8, and so on) are labeled along the vertical axis. Rectangles are then constructed over each subinterval, with the height of the rectangle equal to the class frequency. The frequency histogram for the data in table 2.4 is given in figure 2.7.

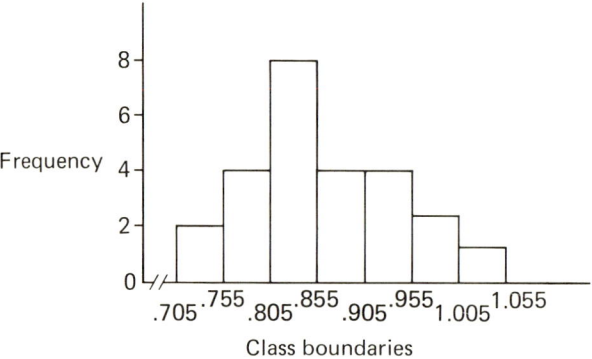

Figure 2.7 Frequency Histogram for the Fluoride Data

relative frequency histogram

Sometimes the results of a frequency table are presented graphically by using a relative frequency histogram. The only difference between the frequency histogram and the relative frequency histogram is that the vertical axis in the relative frequency histogram is scaled for relative frequency rather than for frequency. The relative frequency histogram for table 2.4 is presented in figure 2.8. Very little distinction is made between these two histograms since they become the same figures if drawn to the same scale. We frequently label either one as simply a histogram.

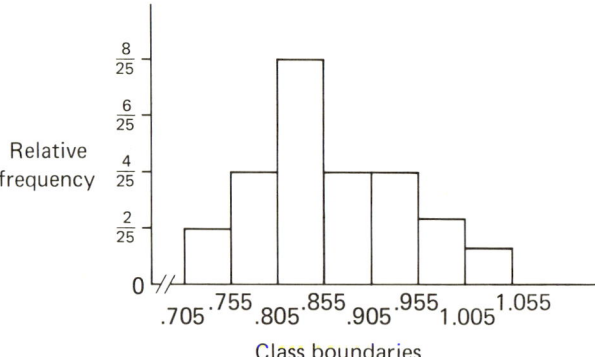

Figure 2.8 Relative Frequency Histogram for the Fluoride Data

Frequency Histogram

A relative frequency histogram describing a set of 1000 recorded fluoride readings is shown in figure 2.9. Note that we have used more intervals to obtain a better description of the data. Also note that the relative frequency histogram begins to approach a smooth curve when both the number of measurements and the number of class intervals are increased.

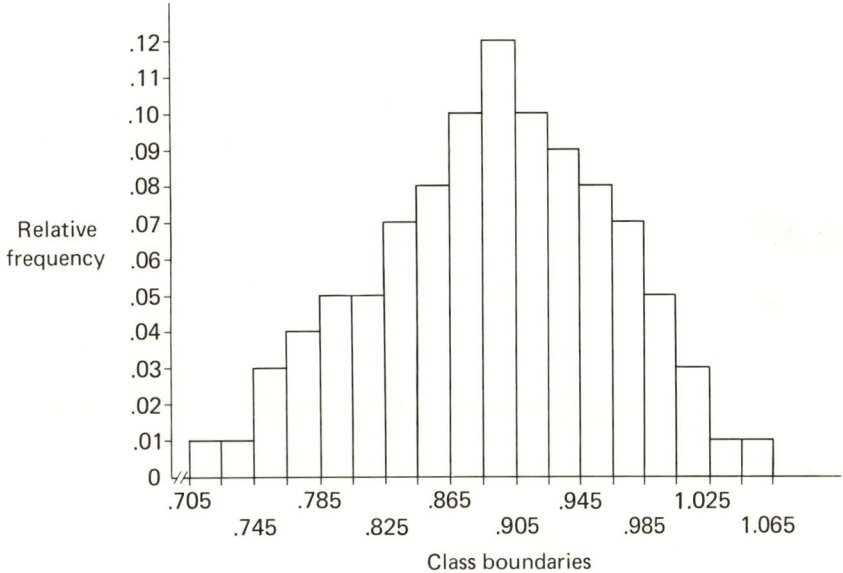

Figure 2.9 Histogram for 1000 Fluoride Readings

number of classes

How do you decide on the number of classes to be used to construct a histogram for a set of data? There is no exact answer to this question, but we can obtain a partial answer if we consider what the histogram is supposed to accomplish. Ideally, we want to choose the number of classes that will provide the best graphical description of the way the data are distributed over the scale of measurement. Particularly, we would like to see the rise and fall of the relative frequencies over the interval of measurement and the approximate location of the maximum relative frequency. These characteristics, which are visible in the histograms of figures 2.8 and 2.9, would be lost if the number of classes were either very small or very large. For this reason most histograms are constructed by using between 5 and 20 classes, the smaller number being reserved for small sets of data. More than 7 class intervals would probably be too many to properly characterize the 25 measurements of table 2.3, but 18 classes were successfully used to characterize the 1000 measurements in fig-

EXERCISES

2.7. The degree of job satisfaction among employees in any job classification is difficult to quantify. Attempts have been made to develop questionnaires (sometimes called *instruments*) composed of specific questions related to a variable of interest. Each respondent is asked to quantify his or her answer to each question on a scale (perhaps from 0 to 5). In this way a respondent obtains a total score (often called *index score*) for the entire questionnaire. It is hoped that different numerical scores will differentiate the variable of interest, such as the degree of job satisfaction among employees. One such instrument was used by Kimball P. Marshall to measure the degree of job satisfaction among a sample of 219 nurses. (A high index score indicates a high degree of job satisfaction.) The data are summarized in the accompanying frequency table. Construct a frequency histogram for these data.

Degree of Job Satisfaction among 219 Nurses

Index Score	f	Index Score	f
67.5–71.5	4	39.5–43.5	7
63.5–67.5	13	35.5–39.5	7
59.5–63.5	24	31.5–35.5	3
55.5–59.5	43	27.5–31.5	1
51.5–55.5	38	23.5–27.5	1
47.5–51.5	43	19.5–23.5	2
43.5–47.5	33		
Total			219

Source: Kimball P. Marshall, "A Study of Job Satisfaction among 219 Nurses in the Southeast," unpublished paper; by permission.

2.8. Refer to the data of exercise 2.7. Compute the relative frequency for each class and construct a relative frequency histogram. Note that the shape of the relative frequency histogram should be the same as that of the frequency histogram of exercise 2.7.

 2.9. Many metropolitan areas throughout the country have experienced staggering increases in size over the past decade. It has been estimated that

Frequency Histogram

nearly 70% of the entire U.S. population lives in 264 metropolitan areas, and by the year 2000 this percentage could go as high as 83%. To further emphasize the crowded conditions in these areas, the average number of people per square mile for the 264 metropolitan areas is 400. The accompanying table lists the 50 most crowded and the 50 least crowded metropolitan areas from the original list of 264.

America's 50 Most Crowded Metropolitan Areas . . . And the 50 Least Crowded

Metropolitan Area	Number of People per Square Mile	Metropolitan Area	Number of People per Square Mile
1. Jersey City	12,963	1. Reno	19
2. New York	7,206	2. Laredo	22
3. Paterson-Clifton-Passaic	2,400	3. Richland-Kennewick, Wash.	31
4. Boston	2,351	4. Great Falls, Mont.	31
5. Meriden, Conn.	2,332	5. Billings	33
6. Nassau-Suffolk, N.Y.	2,096	6. Yakima, Wash.	34
7. Newark, N.J.	2,039	7. Las Vegas	35
8. Bridgeport	2,029	8. Duluth-Superior	36
9. Chicago	1,877	9. Tucson	38
10. New Brunswick-Perth Amboy-Sayreville, N.J.	1,871	10. Bakersfield, Calif.	40
11. Anaheim-Santa Ana-Garden Grove	1,816	11. Riverside-San Bernardino-Ontario	42
12. Los Angeles-Long Beach	1,728	12. Fargo-Moorhead	43
13. Stamford	1,706	13. Abilene, Tex.	45
14. New Britain	1,670	14. Eugene-Springfield, Oreg.	47
15. Norwalk	1,449	15. Fort Smith	47
16. Cleveland	1,359	16. San Angelo, Tex.	47
17. Philadelphia	1,356	17. Pueblo, Colo.	49
18. Trenton	1,333	18. Texarkana	56
19. San Francisco-Oakland	1,254	19. St. Cloud, Minn.	62
20. Lowell, Mass.	1,219	20. Alexandria, La.	66
21. Providence-Warwick-Pawtucket	1,212	21. Albuquerque	68
22. Detroit	1,132	22. Provo-Orem	68
23. New Haven-West Haven	1,109	23. Fresno	69
24. Brockton, Mass.	1,098	24. Midland, Tex.	70
25. Honolulu	1,056	25. Fayetteville-Springdale, Ark.	71
26. Washington	1,034	26. Salinas-Seaside-Monterey, Calif.	75
27. Long Branch-Asbury Park, N.J.	965	27. Killeen-Temple, Tex.	76
28. Milwaukee	964	28. Wichita Falls, Tex.	76
29. Baltimore	917	29. Amarillo	80
30. Bristol, Conn.	885	30. Salt Lake City-Ogden	82
31. Springfield-Chicopee-Holyoke	856	31. Tallahassee	86
32. Buffalo	849	32. Tuscaloosa	87
33. Lawrence-Haverhill, Mass.	848	33. Colorado Springs	88
34. Waterbury, Conn.	844	34. Bloomington-Normal, Ill.	89

(continued)

America's 50 Most Crowded Metropolitan Areas . . . And the 50 Least Crowded
(Continued)

Metropolitan Area	Number of People per Square Mile	Metropolitan Area	Number of People per Square Mile
35. San Jose	819	35. Sherman-Denison, Tex.	89
36. Fall River, Mass.	807	36. Williamsport, Pa.	93
37. Pittsburgh	788	37. Florence, Ala.	94
38. New Bedford, Mass.	783	38. Tulsa, Okla.	97
39. Akron	752	39. Santa Barbara-Santa Maria-Lompoc	97
40. Hartford	698	40. Lynchburg, Va.	97
41. Gary-Hammond-East Chicago	675	41. Salem, Oreg.	98
42. Worcester, Mass.	667	42. Pine Bluff, Ark.	98
43. Cincinnati	644	43. Bryan-College Station, Tex.	99
44. Louisville	623	44. Lawton, Okla.	100
45. Miami	621	45. Odessa, Tex.	101
46. Lewiston-Auburn, Maine	604	46. Topeka	102
47. Fitchburg-Leominster, Mass.	581	47. Sioux City	103
48. Nashua, N.H.	560	48. Wilmington, N.C.	103
49. Norfolk-Virginia Beach-Portsmouth	548	49. Tyler, Tex.	104
50. New Orleans	532	50. Phoenix	106

You will note immediately that Jersey City and New York have population densities (number of people per square mile) that far exceed the densities of the remaining 48 cities in the list of the 50 most crowded metropolitan areas. Since it would be difficult to include Jersey City and New York on the same graph with the remaining 48 because of the extremely high densities in these two cities, graph only the remaining 48. Use a frequency histogram to describe the population densities for the 48 cities from Paterson-Clifton-Passaic to New Orleans. Begin the first subinterval at 531.5 and construct each subinterval with a width of 190.

2.10. Construct a frequency histogram for the 50 least crowded cities among the original 264 cities. Use approximately 10 subintervals with an interval width of 9. Begin the first interval at 18.5.

2.11. Refer to exercise 2.9. Construct a relative frequency histogram for the most crowded metropolitan area data (excluding Jersey City and New York). Use the same subintervals as in exercise 2.9.

2.12. Refer to exercise 2.9. Construct a relative frequency histogram for the 50 least crowded metropolitan areas, using the same subintervals as in exercise 2.10. Note that the shape of the relative frequency histogram is identical to that for the frequency histogram of exercise 2.10.

2.13. The length of time an outpatient must wait for treatment is a variable that plays an important role in the design of outpatient clinics. The waiting times (in minutes) for 50 patients at a pediatric clinic are as follows:

35	22	63	6	49	19	15	83	46	19
16	31	24	29	36	68	42	57	64	8
23	47	21	51	7	40	19	46	16	32
108	33	55	32	22	36	25	27	37	58
39	10	42	28	72	13	51	45	77	16

Construct a relative frequency histogram for these data.

2.5 FREQUENCY POLYGON (OPTIONAL)

frequency polygon

Class relative frequencies can also be portrayed in the form of a frequency polygon. The only difference between a frequency polygon and a frequency histogram is that the frequency associated with each class is indicated by a dot placed over the midpoint of the class interval. The dots are then joined by straight lines. The frequency polygon for the fluoride data of table 2.4 is presented in figure 2.10. Note that we actually used two additional classes, one at either end. This device provides a neater graph, with endpoints that fall off to zero on the horizontal axis.

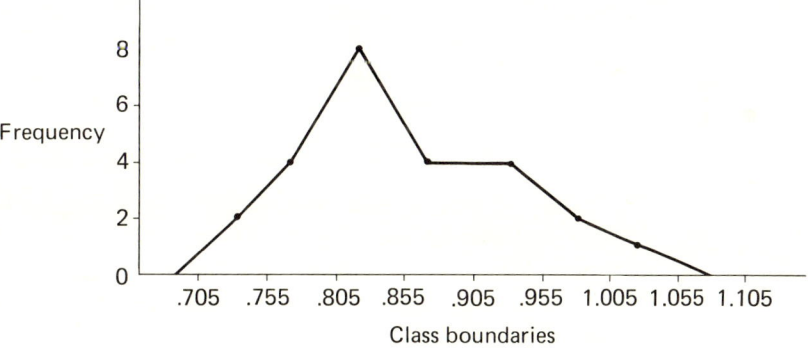

Figure 2.10 Frequency Polygon for the Fluoride Data

Frequency polygons are often used in news articles when it is necessary to summarize frequency data. For example, the data summarized in figure 2.11 were published in a news article concerning the number and intensity of tropical storms. Some of these results are presented in figure 2.11, which provides information on the frequency of hurricanes and tropical storms by month from data collected over a 73-year period.

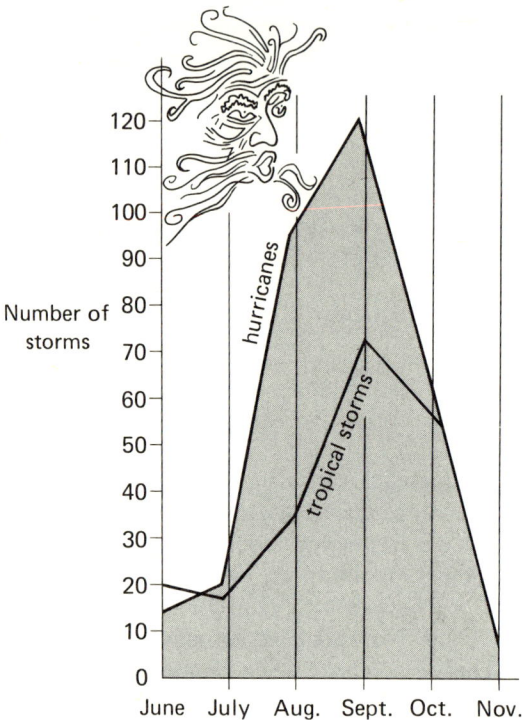

Figure 2.11 Monthly Variation of Hurricanes and Tropical Storms (Taken from the *St. Petersburg Times*)

2.6 COMMENTS CONCERNING HISTOGRAMS

The relative frequency histogram is the most important graphical descriptive method in statistical inference. It is important to note several things before we continue. **First, the area under the frequency histogram over a particular interval is proportional to the fraction of the total number of measurements falling in that interval.**

For example, consider the relative frequency histogram for the sample of 25 fluoride readings (figure 2.8). The fraction of the 25 measurements less than or equal to .905 is $18/25$. You will note that $18/25$ of the total area under the histogram (shaded) lies to the left of .905. See figure 2.12.

The second pertinent point is that the area under the histogram over any particular interval is proportional to the probability of selecting a measurement from that interval if a single observation is selected from the total number.

For example, if 25 cards were labeled with the respective fluoride readings

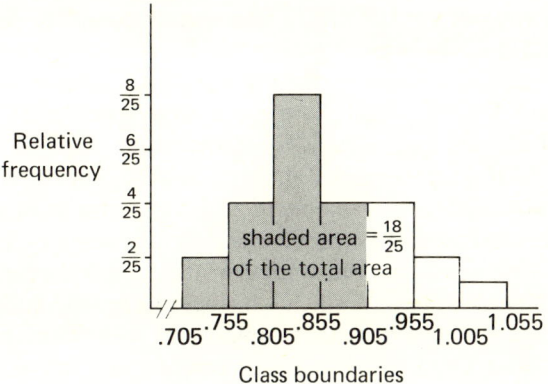

Figure 2.12 Relative Frequency Histogram for the Fluoride Data

in table 2.3, then shuffled, the probability of choosing a card with a fluoride reading between .805 and .855 ppm is $8/25$, because 8 out of the 25 readings fall within this interval. From figure 2.13 we note that the shaded area under the relative frequency histogram over the interval .805 to .855 is equal to $8/25$ of the total area.

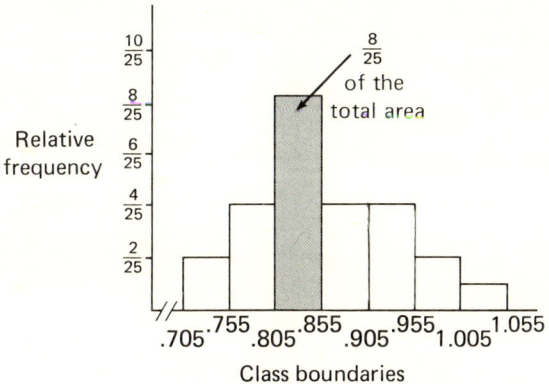

Figure 2.13 Relative Frequency Histogram Showing the Fraction of Fluoride Readings in the Interval .805 to .855

A third important point to note is that we can construct a frequency histogram for any set of measurements, any sample, or any population; but our primary objective is to describe or make inferences about a population.

We will never actually have in hand all the measurements for a population, and hence we will not be able to construct a complete population frequency histogram. However, we can image that if one could be constructed, it would possess an outline similar to that obtained for a sample from that population.

Since populations usually contain a large number of measurements, the number of classes can usually be made rather large so that the population frequency histogram becomes almost a smooth curve. The 25 fluoride readings were obtained from a sample of 25 days. As the number of measurements in a sample increases, we can select smaller class intervals. The resulting histogram will then become more regular and tend to become a smooth curve. Figure 2.14 shows three frequency histograms for the fluoride measurements. The first is a histogram for a sample of 25 measurements; the second is a histogram for a sample of 100 measurements; and the third is a possible histogram for the entire population. Note that the scale of relative frequency (the vertical scale) will change from one figure to another. Also observe that it may be necessary to change the endpoints of the class intervals [as was done in (a) and (b)] so that no measurement falls on the boundary between two classes.

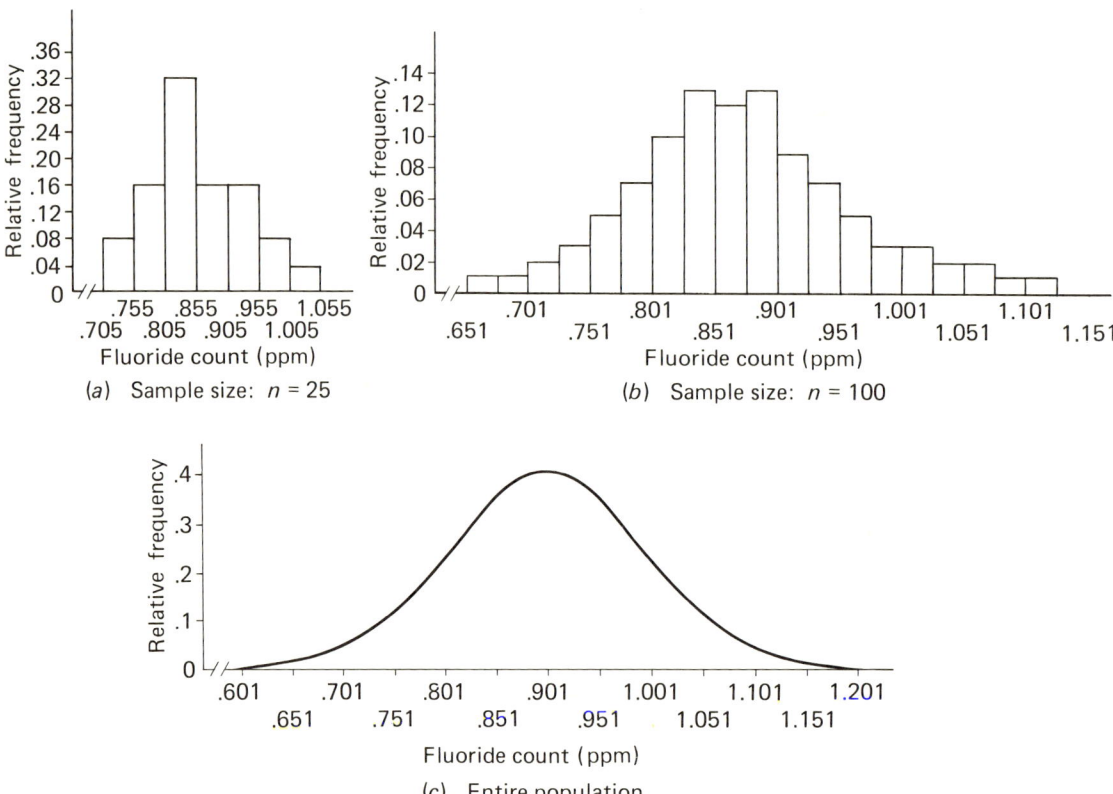

Figure 2.14 Relative Frequency Histograms for the Fluoride Data

The frequency histogram is an excellent way to describe a population of measurements because the area under the histogram for the population tells us the fraction of the total number of measurements falling in given intervals. We can also see from the population frequency histogram the largest and the smallest measurements.

SUMMARY

We study how to describe a set of measurements for two reasons. First, the objective of statistics is to make inferences about a population based on information contained in a sample. Since populations are sets of measurements, existing or conceptual, we need some method for talking about the population or, equivalently, for describing a set of measurements. Second, graphical description of data is very useful because it is easily comprehended by both the novice and the scientist. For example, there is a need for the condensation and description of large quantities of economic or sociological data that are collected annually by various government bureaus or even every ten years by the U.S. Bureau of the Census. Graphical descriptions of these data are easily understood by the layperson and the scientist engaged in economic or sociological research.

Data can be described either graphically or numerically. Graphical methods such as circle charts, bar charts, histograms, and frequency polygons are presented in this chapter. Numerical descriptive measures are discussed in chapter 3.

Note the role of data description in statistical inference; it is impossible to make an inference about anything, measurements or what have you, unless you are able to describe the object of your interest. But graphical description has limitations when used for making inferences. These will be made apparent in the next chapter.

REFERENCES

O'Toole, A. L. *Elementary Practical Statistics*. New York: Macmillan, 1964.
Ott, L.; Mendenhall, W.; and Larson, R. *Statistics: A Tool for the Social Sciences*. 2d ed. N. Scituate, Mass.: Duxbury Press, 1978.

SUPPLEMENTARY EXERCISES

(Optional exercises are starred.)

 *2.14. A measure of the growth in the expenditures of the U.S. Department of Health, Education, and Welfare is provided by data contained in an article in *U.S. News & World Report*, 11 June 1979. The percentages shown in the accompanying table indicate the share of all federal outlays spent by HEW. Construct a bar graph depicting the growth in this percentage over the period from 1953 through 1979.

Year	1953	1958	1963	1968	1973	1979
Percentage of All Federal Expenditures	2	3	18	23	33	37

*2.15. Consumers have been increasing their personal debts (through credit card and time purchases) during the past 29 years. Total consumer debt, expressed as a percentage of total personal income after taxes, is shown in the accompanying table for the period from 1950 through 1978. Use a bar chart to display these data graphically.

Year	1950	1955	1960	1965	1970	1975	1978
Total Consumer Debt as a Percentage of Total Personal Income	37	53	65	75	73	73	83

*2.16. The oil reserves in the Western Hemisphere are estimated to be as shown in the accompanying table. Display these data in a circle (pie) chart.

Location	Barrels (in billions)
United States	38.7
South America	22.6
Canada	8.8
Mexico	60.0

*2.17. Display the data of exercise 2.16 in a bar chart.

*2.18. The 1970–1971 operating costs by school for the universities of the state university system in Florida are given in the accompanying table. Present these figures in a circle chart.

School	Operating Costs (× $1 million)
University of North Florida	1.5
Florida International	1.5
Florida Technological Institute	8.5
University of West Florida	9.3

School	Operating Costs (× $1 million)
Florida A and M	11.3
Florida Atlantic	13.6
University of South Florida	32.1
Florida State University	55.7
University of Florida	107.9

*2.19. Present the data of exercise 2.18 in a bar chart.

 2.20. A study was conducted to evaluate the absorption characteristics of an antibiotic preparation, chloramphenicol. A subject was given a .5-gram oral dose of the preparation. Urine collections were then made over the next 12 hours, and the number of milligrams of chloramphenicol excreted was recorded at each collection. The data are given in the accompanying table. Use a frequency histogram to display graphically the chloramphenicol excretion pattern for the subject.

Urine Collection Period (hours after medication)	Chloramphenicol Excreted (milligrams)
0–2	94
2–4	107
4–6	70
6–8	62
8 10	31
10–12	28

2.21. Construct a frequency polygon for the data in exercise 2.20.

*2.22. Urine collections for the subject in exercise 2.20 continued to be taken over a 48-hour period after he received the .5-gram oral dose of chloramphenicol. These data appear in the accompanying table. Construct a bar chart to display these results.

Urine Collection Period (hours after medication)	Chloramphenicol Excreted (milligrams)
0–2	94
2–4	107
4–6	70
6–8	62
8–12	59
12–24	47
24–48	11

34   How to Phrase an Inference: Graphical Methods

*2.23. Present an explanation of the difference between a frequency histogram and a bar chart.

2.24. An investigator was interested in studying the sedative effect on rats of different doses of a drug. A small cage was constructed with several electric eyes focused at different angles. Attached to each electric eye was a counter that monitored the number of times a rat broke any of the light beams in a 15-minute period. Twenty-five rats were injected with a specified dose of the drug, and each one was observed in the cage for a 15-minute period. These data are recorded in the accompanying table. Construct a frequency histogram for these data using five or more class intervals.

Rat	Number of Times a Light Beam Was Broken in the 15-Minute Period	Rat	Number of Times a Light Beam Was Broken in the 15-Minute Period
1	107	14	128
2	99	15	106
3	171	16	177
4	116	17	144
5	101	18	102
6	109	19	196
7	199	20	191
8	142	21	169
9	118	22	182
10	173	23	148
11	155	24	130
12	184	25	159
13	132		

2.25. A questionnaire circulated in late 1978 asked 94 economic forecasters to estimate the probability of decline in the gross national product from the third quarter to the fourth quarter of 1979. The results of the survey are summarized in the accompanying table. Construct a relative frequency histogram for these results.

Estimated Probability of a Decline in GNP	Frequency
.01–.10	38
.11–.20	18
.21–.30	15
.31–.40	15
.41–.50	8

2.26. A study was conducted among smokers to examine symptoms of

Experiences with Real Data

breathlessness and wheeze. A total of 1827 subjects exhibiting breathlessness and wheeze were classified by age. Use the data in the accompanying table to construct a relative frequency histogram.

Age	Frequency	Age	Frequency
20–24	9	45–49	269
25–29	23	50–54	404
30–34	54	55–59	406
35–39	121	60–64	372
40–44	169		

*2.27. Summarize the data from exercise 2.26 into the following age categories: 20–29, 30–39, 40–49, and 50 and over. Construct a circle chart depicting the percentage of subjects in each of these categories.

EXPERIENCES WITH REAL DATA

Select a sample of 50 measurements from a population of interest to you and construct a relative frequency distribution for the data. The data could be observations on a variable measured in a chemistry or physics experiment, they could be measurements of highway traffic on a given route during a fixed period of time, or they could be measurements on the number of people waiting in a queue (ticket counter, supermarket checkout counter, etc.) at particular times. Carefully define the population before you select your sample and make certain that each sample measurement is selected from the target population.

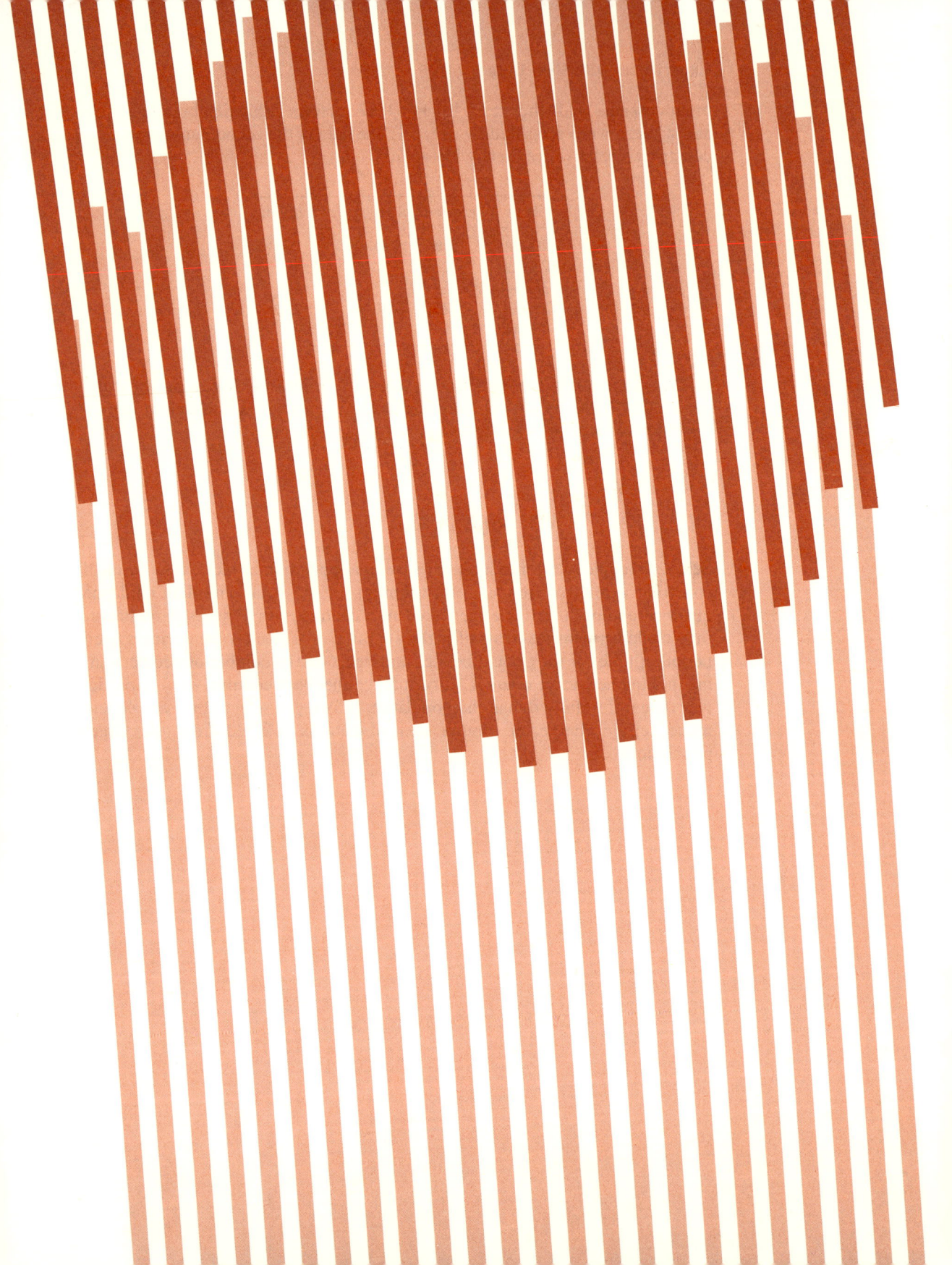

3 How to Phrase an Inference: Numerical Methods

CHAPTER OUTLINE

3.1 Introduction
3.2 Measures of Central Tendency
3.3 Measures of Variability
3.4 Shortcut Method for Calculating the Variance and Standard Deviation
3.5 How to Guess the Standard Deviation of Sample Data

3.1 INTRODUCTION

Numerical descriptions, like graphical ones, are commonly used to convey a mental image of physical objects or phenomena. Consequently, we seek one, two, or more numbers—called <u>numerical descriptive measures</u>—to create a mental picture of a set of data.

numerical descriptive measures

There are two good reasons for this search. First, we frequently wish to discuss sets of measurements with others, and it is inconvenient to carry frequency histograms about in our pockets. Discussion is much easier if we can project a picture of the frequency distribution to the minds of our listeners by means of one or two descriptive numbers. Second, the frequency distribution is an excellent method for characterizing a population, but it possesses severe limitations when used to make inferences. An irregular frequency histogram of a sample will be similar to the corresponding distribution for the population. But how similar? How can we measure the "goodness" of our inference? How can we measure the degree of dissimilarity between the histogram for the sample and the histogram for the population?

The sample frequency histogram can be used to make an inference concerning the shape of a population frequency distribution, but it is difficult to determine how good that inference is. In contrast, numerical descriptive measures of the population can be estimated by using the sample measurements. We can say, with a measured degree of uncertainty, how close the estimate will be to the population descriptive measure.

Since the same numerical descriptive measure could be computed for either a sample or a population, it is desirable to distinguish between these two applications. Numerical descriptive measures computed from a sample are called statistics. Those that describe a population are called parameters.

DEFINITION 3.1

statistics

Numerical descriptive measures computed from a sample are called statistics.

DEFINITION 3.2

parameters

Numerical descriptive measures of a population are called parameters.

The two most important types of parameters are those that locate the center of the distribution and those that describe its spread. They are called, respectively, *measures of central tendency* and *measures of variability*. We will show that two numbers, one locating the center of a distribution and one quantifying the amount of variability or spread, provide good descriptions of the frequency distributions for most sets of measurements. As you might suspect, we will frequently use a descriptive measure of the sample to estimate the value of the corresponding parameter of the population.

Measures of central tendency (averages, medians, and modes) and their definitions, interpretations, and applications will be presented in section 3.2. Measures of variability and their calculation and interpretation occupy the remainder of the chapter. Thus chapter 3 provides the final touches to the first step in our study of statistical inference—namely, finding a way to describe a set of measurements. We will use these descriptive measures in later chapters to make inferences about populations based on sample measurements.

3.2 MEASURES OF CENTRAL TENDENCY

Numerical descriptive measures that locate the center of a distribution of measurements are called measures of central tendency. The most common of these are the arithmetic mean, the median, and the mode.

The Arithmetic Mean

Perhaps the most widely used measure of central tendency is the arithmetic mean (or "average") of a set of measurements.

DEFINITION 3.3

arithmetic mean

The arithmetic mean of a set of measurements is the sum of the measurements divided by the number of measurements in the set.

The arithmetic mean—or, simply, the mean—is used extensively in many fields of science and business. You have undoubtedly observed phrases such as the mean income for persons living in ghetto areas, the mean tensile strength of a cable, the mean velocity of the first stage of a missile, the mean increase in the cost of living index over the past six months, and the mean closing price of a group of stocks.

Example 3.1

The length of survival (in years) after discovery of a rare type of cancer was recorded for each of six cancer patients. The data are shown in table 3.1. Calculate the mean survival time for this sample of six patients.

Table 3.1 Length of Survival After Discovery of Cancer

Patient	Length of Survival (years)
1	1.7
2	3.2
3	2.1
4	4.6
5	1.4
6	2.8

Solution

$$\text{mean} = \frac{\text{sum of the measurements}}{6}$$
$$= \frac{1.7 + 3.2 + 2.1 + 4.6 + 1.4 + 2.8}{6} = \frac{15.8}{6} = 2.63$$

Note that the mean, 2.63, falls somewhere near the "middle" of the set of 6 survival time measurements.

How to Phrase an Inference: Numerical Methods

The means for both a sample and a population are defined and computed in the same way, since both sets are sets of measurements, but we use different symbols for each. The symbol $\bar{y}$ (y bar) will be used to denote the mean of a sample and the symbol μ (Greek letter mu) will denote the mean of a population.

Note that we will rarely possess all the measurements in the population, and consequently we will rarely know the value of μ. But that leads us back to the objective of statistics. We will sample the population and use the sample mean $\bar{y}$ to estimate the value of μ. We will show you how to measure the goodness (accuracy) of this estimate in chapter 6.

$\bar{y}$ **is the sample mean.**

μ **is the population mean.**

It is convenient here to introduce some notation, which we will use in the computational formulas encountered in this and later chapters. First, let the letter y represent a measurement. If we refer specifically to a sample, y represents a measurement in that set. In the case of multiple measurements, a subscript is used to denote a particular measurement in the set. If we consider the six measurements from example 3.1,

$$1.7 \quad 3.2 \quad 2.1 \quad 4.6 \quad 1.4 \quad 2.8$$

we let y_1 denote the first observation ($y_1 = 1.7$). In the same manner we let $y_2 = 3.2, \ldots, y_6 = 2.8$.

To indicate the sum of our measurements, we use the Greek symbol Σ (sigma). Thus Σy indicates the sum of the measurements that we denoted by the symbol y. Using the data of example 3.1, we have

$$\Sigma y = y_1 + y_2 + \cdots + y_6$$
$$\Sigma y = 1.7 + 3.2 + \cdots + 2.8 = 15.8$$
$$\bar{y} = \frac{15.8}{6} = 2.63$$

We can expand this example to include any number of measurements. If we have a sample of n measurements, which we denote by $y_1, y_2, \ldots, y_n$, the **sample mean** is given by the following formula:

sample mean

$$\bar{y} = \frac{\Sigma y}{n} = \frac{y_1 + y_2 + \cdots + y_n}{n}$$

The Median

The median is a second measure of central tendency. It is computed in the same way for either a sample or a population.

Measures of Central Tendency

median, odd number of measurements

DEFINITION 3.4

The median for an odd number of measurements is the middle measurement when the measurements are arranged in order of magnitude (size).

Example 3.2

A sample of 7 students was given a reading achievement test. Find the median for these test scores:

$$95 \quad 86 \quad 78 \quad 90 \quad 62 \quad 73 \quad 89$$

Solution

We must first arrange the scores in order of magnitude.

$$62 \quad 73 \quad 78 \quad 86 \quad 89 \quad 90 \quad 95$$

Since we have an odd number (7) of measurements, the median is then the middle score—that is, 86.

median, even number of measurements

DEFINITION 3.5

The median for an even number of measurements is the mean of the two middle observations when the measurements are arranged in order of magnitude.

Example 3.3

Suppose 3 more students out of a class of 30 took the achievement test of example 3.2 and scored 73, 75, and 91, respectively. Determine the sample mean and median for the combined 10 test scores.

Solution

The sample mean is given by

$$\bar{y} = \frac{\Sigma y}{n}$$

$$= \frac{62 + 73 + 73 + 75 + 78 + 86 + 89 + 90 + 91 + 95}{10}$$
$$= 81.2$$

Since we have an even number of observations, the sample median is the mean of the two middle scores when the scores are arranged in order of magnitude. The scores, arranged in order of magnitude, are 62, 73, 73, 75, 78, 86, 89, 90, 91, and 95. The two middle scores are 78 and 86; hence the median is given by

$$\text{median} = \frac{78 + 86}{2} = 82.$$

The median seems to be the preferred measure of central tendency for describing economic, sociological, and educational data. Newspaper reports and magazines frequently refer to the median wage increase won by unions, the median income of families in the United States, the median age of persons receiving Social Security benefits, and the gap between the median income for men and the median income for women.

Why is the median so popular in the social sciences? The answer is that many of the frequency distributions of measurements in the social sciences are **skewed** (tail off rapidly to the right or left). Because the mean is greatly affected by extremely large (or small) observations and the median is not, the median is preferred in locating the center of skewed distributions. For example, suppose you have five measurements, 1, 2, 3, 4, and 20. The mean is 6 and the median is 3. Notice how the mean is affected by the largest measurement. Also, notice that it does not appear to fall near the "center" of the five measurements. Figure 3.1 shows the location of the mean and the median for a distribution skewed to the right.

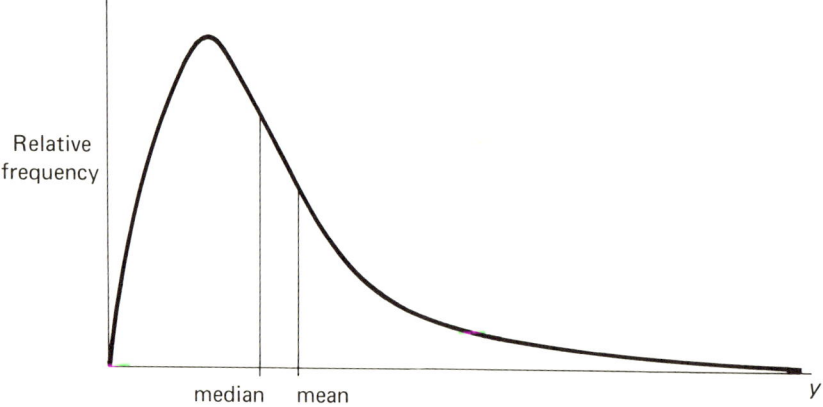

Figure 3.1 Relationship Between the Mean and the Median for a Distribution Skewed to the Right

Measures of Central Tendency

As we continue, keep in mind these three important characteristics of the median:

1. The mean and the median are not necessarily equal in value.
2. The median has most relevance when used to describe a large set of data. Thus published median ages, median wage increases, and median incomes are usually based on large amounts of data.
3. For a large set of measurements the median is a number such that approximately half the measurements lie below it and half above it.

The Mode

A third commonly used measure of central tendency is the mode of a set of measurements.

DEFINITION 3.6

The mode of a set of measurements is the measurement that occurs most often in the set.

The mode is the least common of the three measures of central tendency considered in this text, but it is very useful in business planning for identifying those products or product sizes that are in greatest demand. A shirt or dress manufacturer, for example, is interested in customer sizes most frequently purchased. Frequent reference is also made to the mode of a set of measure-

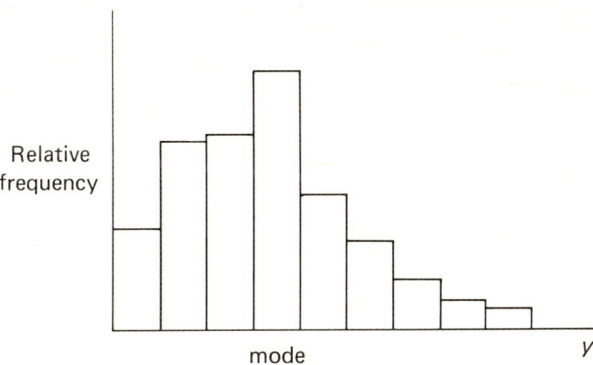

Figure 3.2 Location of the Mode for a Relative Frequency Histogram

ments in advertising campaigns. For example, we hear that more people choose brand *W* than any other aspirin, more doctors smoke "Lungs" cigarettes than any other brand, or that a two-child family is the most preferred family size among young married couples in the United States. When the data are described by a relative frequency histogram, the mode is the midpoint of the class interval for which the relative frequency is the largest. (See figure 3.2 p. 43.)

Example 3.4

A research team consisting of an anthropologist and a sociologist analyzed the family structure of a large Hutterite community. One task involved counting the number of children in each family. The sample data given in table 3.2 represent a portion of the total number of Hutterite families. Determine the mode for the sample of 25 measurements listed in the table.

Table 3.2 Number of Children in 25 Hutterite Families

7	10	8	11	9
9	9	8	9	8
9	9	9	8	9
8	8	9	10	11
10	7	10	9	7

Solution

First, let us arrange the measurements in order of magnitude, ranging from the smallest to the largest:

7, 7, 7, 8, 8, 8, 8, 8, 8, 9, 9, 9, 9, 9, 9, 9, 9, 9, 9, 10, 10, 10, 10, 11, 11

It is clear from this listing that the sample mode for the number of children per Hutterite family is 9.

Sometimes the frequency distribution for a set of measurements possesses more than one peak. For example, in a particular athletic shoe worn by both adults and youths, two widely differing sizes might be in great demand. In such cases the sets are said to be bimodal, trimodal, and so on. A variation of this situation occurs when all observations appear the same number of times. In this circumstance the mode gives no information for locating the center of the distribution, and we say that the frequency distribution possesses no mode.

We have discussed three measures of central tendency—the mean, the median, and the mode—and noted that one measure might be better than the other two in a given situation. We will concern ourselves almost exclusively with the sample mean as a measure of central tendency for the remainder of this text. The reason for this choice is that the sample mean is most widely employed in statistical inference, and the study of inference is our ultimate objective. Recall that we wish to make inferences about the population from which the sample is drawn. A sample mean is not only descriptive of the sample observations but, more importantly, it can easily be used to estimate the population mean with some degree of accuracy.

EXERCISES

3.1. Class exercise: Suppose you were an environmental scientist and you were required to measure the dissolved oxygen content (which is a measure of pollution) at a particular point in a lake. Why would a single chemical determination be unsatisfactory?

The answer to this question can be found by recalling your own experiences. If you have ever constructed something, measuring and cutting pieces to be assembled, you may have noticed that they frequently did not fit. This problem is often caused by errors in measuring the component parts. Similar random measurement errors occur in almost all experimentation (if the measuring instruments are accurate enough to detect the variation).

Most errors caused by inaccurate measuring instruments are reduced (not eliminated) by using not one measurement but the mean of several measurements to characterize the true value of the quantity being measured.

To illustrate, have 10 people in your class measure some object in the room (for instance, the length of the room). Notice the variation in the recorded measurements. Calculate the mean for the 10 sample measurements ($n = 10$). Note that the mean falls near the center of the set of measurements and that it tends to offset overly large measurements against small ones.

If we were to measure an object repeatedly, millions and millions of times, a population of measurements would be generated. It is likely that the mean of this population, μ, would coincide with the true length of the object. Viewed in this manner, the 10 classroom measurements represent a sample, and sample mean $\bar{y}$ *estimates* μ. In later chapters we will learn how to evaluate the error of this estimate—that is, the difference between the estimate $\bar{y}$ and the true mean μ.

3.2. Most wristwatches are subject to some degree of error. Some read fast and others slow. Consequently, if you wish an accurate estimate of the time, it would be desirable to combine the time readings from several watches in order to "average out" the errors. Try this experiment. At a particular point in time, have each person in your class (or a small subset of them) record the time on his or her wristwatch. Calculate the average of these time readings. Note that the average gives a time that counterbalances the errors caused by watches that read fast with those that read slow.

 3.3. A recent newspaper article stated that the median wage for workers in a particular trade union is $8.48 per hour. Interpret this statistic.

3.4. In order for a health clinic to be capable of handling the desired patient load, designers need to know something about the demand for services in the area where the clinic is to be located. This would include information on the patient arrival rate as well as the length of time to treat a patient. Both the arrival rate and the treatment time will vary in a random manner. The arrival rate varies because of the random occurrence of outbreaks of flu and other common illnesses; the time of treatment varies because treatment time will depend on a particular patient's illness. Thus the demand for physician and nurse time will vary in a random manner. By studying the frequency distributions of patient arrival rates and treatment times, the designer can specify the number of doctors, nurses, technicians, orderlies, and physical equipment needed to meet the demand. These numbers will affect the length of time a patient must wait in the clinic before receiving attention.

To answer some of the questions about clinic treatment times, a designer acquired data from an established clinic in a locale that possessed similar characteristics to the proposed new clinic location. The treatment times for 50 patients, randomly selected from the clinic's records, are as follows:

21	20	31	24	15	21	24	18	33	8
26	17	27	29	24	14	29	41	15	11
13	28	22	16	12	15	11	16	18	17
29	16	24	21	19	7	16	12	45	24
21	12	10	13	20	35	32	22	12	10

(a) What is the average treatment time for the sample of 50 patients? [You can verify that the sum of the 50 measurements ($n = 50$) is 1016.] Interpret this statistic.

(b) Find the median treatment time. Interpret this statistic.

(c) Construct a relative frequency histogram for the data. Use as class intervals 5.5–10.5, 10.5–15.5, 15.5–20.5, and so on. Mark the mean and median on the horizontal axis. Note that both measures do a reasonably good job of locating the center of the distribution of treatment times.

(d) Must the sample mean and sample median be equal? Explain.

 3.5. The following news clipping on the human life span demonstrates the manner in which statistics are presented via the news media. Interpret the numerical descriptive measures contained in the article.

U. S. Life Span 71.4

According to a United Nations report, improved health services have helped to add about 7 months to the life span of the average American. The average American can now expect to live 71.4 years.

However, this longer life span for Americans is still two years less than the 73.3 years the average Japanese can expect to live. Japan is at the top of the list.

The figures in the report show that all the most developed regions list an average life expectancy of 71.2 years. The less developed regions list an average life expectancy of 53.9 years.

According to the report, the longer life expectancy in developed countries has slowed down. However, women on the average live 5 to 7 years longer than men and are further widening the gap.

The leading causes of death at an advanced age in the developed nations, the report stated, are cancer and diseases of the circulatory and nervous systems.

The report also predicted that by the end of the year there would be one billion more people on the earth than there were 15 years ago. All but one-sixth of the increased population was in the less developed countries, which also account for almost three-fourths of the world's people.

3.3 MEASURES OF VARIABILITY

The importance of data variation is exemplified in a joke that is often directed at statistics and statisticians. "Have you ever heard the story about the statistician who couldn't swim and drowned in a river with an average depth of three feet?" While we admit to some discomfort every time we hear the joke, it does stress the importance of data variation. The mean (or any other measure of central tendency) only tells part of the story—or, equivalently, only partially describes a distribution of measurements.

A machine manufacturing size-9 shoes "on the average" would not be considered satisfactory if the actual sizes varied from $8\frac{3}{8}$ to $9\frac{1}{2}$. A machine producing 1-inch nails "on the average" would not be very satisfactory if the actual lengths randomly varied from $\frac{1}{2}$ to $1\frac{1}{2}$ inches. Indeed, variation of product quality is probably of far greater importance to a manufacturer than corresponding measures of central tendency.

Keep in mind that the objective of numerical description is to obtain a set of measures, one or more, that can be used to create a mental reconstruction of the frequency distribution of the data and that a measure of central tendency only performs one function—locating the center of the distribution. The examples above amply illustrate the need for numerical measures of data variation or spread.

The Range

The simplest measure of data variation is the range.

DEFINITION 3.7

range

The range of a set of measurements is the difference between the largest and smallest measurements.

The range is used extensively as a measure of variability in summaries of data that are made available to the general public. We read, for instance, that the range of salaries for psychologists with the rank of assistant professor is $4000, the range in temperature in Miami throughout the year is 50°, and the range in personal property taxes for a given state is $2900. The range is also widely used to describe the variability in the quality of an industrial product when small samples are selected periodically from an operating production line.

Example 3.5

Compute the range of the $n = 5$ measurements shown here:

$$3.2 \quad 7.6 \quad 5.7 \quad 6.6 \quad 4.7$$

Solution

The range is the difference between the largest and the smallest measurements in a set. Hence

$$\text{range} = 7.6 - 3.2 = 4.4$$

Although simple to define and calculate, the range is not a sensitive measure of variability except for very small samples. In figure 3.3 the two relative

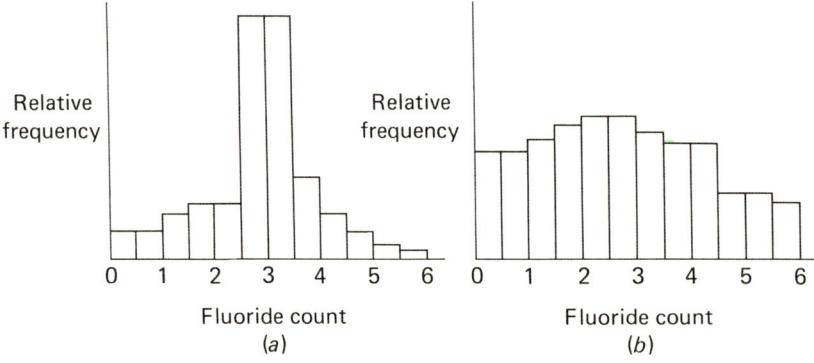

Figure 3.3 Two Distributions of Fluoride Readings with the Same Range

Measures of Variability

frequency distributions of fluoride readings have the same range, 6, but the data for the two distributions differ greatly in variability. The data for figure 3.3(a) are much less variable than those for figure 3.3(b). In figure 3.3(a) most of the measurements are very close to the mean, while in figure 3.3(b) the measurements are spread evenly throughout the range.

Percentiles

Percentiles can be used to characterize the spread or variation of a (usually large) set of data, and they also can be used to give the relative standing of one measurement in relation to the others in the set.

DEFINITION 3.8

pth percentile

For a large set of measurements arranged in order of magnitude, the pth percentile is the value such that $P\%$ of the measurements are less than that value and $(100 - P)\%$ are greater.

For example, the 80th percentile of a large set of measurements on a variable y is the value y such that 80% of the measurements fall below it and 20% lie above it. See figure 3.4.

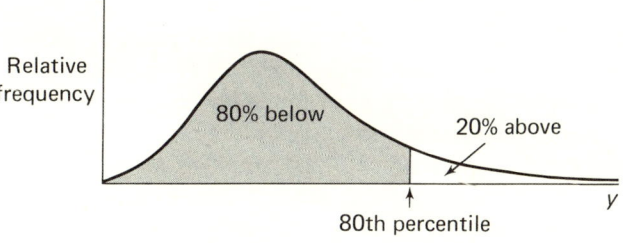

Figure 3.4 80th Percentile

Percentiles are frequently used to describe the results of achievement tests and the final ranking of people taking examinations. Such phrases as "John S. scored in the 90th percentile on the Scholastic Aptitude Test (SAT) this year" or "Sue J. scored in the 85th percentile on a national speed-reading examination for 10th graders" relate a person to the others who took the same examination.

Note that the 50th percentile of a set of measurements is the median. See figure 3.5. The 25th and 75th percentiles are called quartiles.

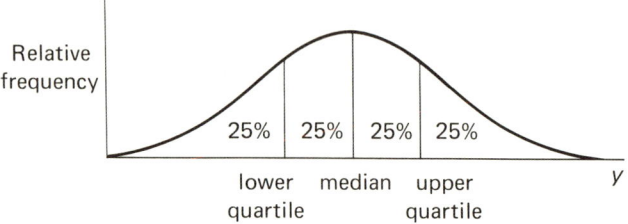

Figure 3.5 Lower Quartile, Median, and Upper Quartile

DEFINITION 3.9

lower quartile

The lower quartile of a large set of data is the 25th percentile. Thus 25% of the measurements fall below the lower quartile and 75% fall above it.

DEFINITION 3.10

upper quartile

The upper quartile of a large set of data is the 75th percentile. Thus 25% of the measurements fall above the upper quartile and 75% fall below it.

Specification of the median, the lower quartile, and the upper quartile provides a fairly good description of a set of measurements except that it gives no notion of the maximum or minimum measurements that one might expect in the set. Thus using percentiles requires a total of five (including the maximum and minimum) numerical descriptive measures to create a mental image of a frequency distribution. Since a smaller number of numerical descriptive measures would be easier to interpret, we seek a more sensitive measure to describe the variation of a set of measurements.

The Variance

dot diagram

The variance of a set of measurements utilizes the deviations of the measurements from their mean. To illustrate, suppose we have a set of 5 measurements, $y_1 = 6$, $y_2 = 7$, $y_3 = 5$, $y_4 = 3$, and $y_5 = 4$. These are shown on the dot diagram in figure 3.6. (Dot diagrams are used to depict very small sets of measurements.) Each measurement is located by a dot above the horizontal axis of the diagram.

Measures of Variability

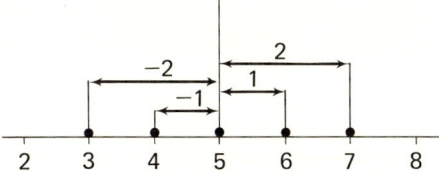

Figure 3.6 Dot Diagram

We use the mean,

$$\bar{y} = \frac{\Sigma y}{n} = \frac{25}{5} = 5.0$$

deviations

to locate the center of the set. We then construct horizontal lines on figure 3.6 to represent the distances (deviations) of the measurements from their mean. The larger the deviations, the greater will be the variation of the set of measurements. The deviations of the measurements are computed by using the formula $(y - \bar{y})$. In our example the deviation of y_1 from the mean is

$$(y_1 - \bar{y}) = 6 - 5 = 1.0$$

The 5 measurements in figure 3.6 and their deviations from the mean are shown in columns (a) and (b), respectively, of table 3.3.

Table 3.3 Deviations of the 5 Measurements from Their Mean

(a) Measurement, y	(b) Deviation, $(y - \bar{y})$	(c) Squared Deviation, $(y - \bar{y})^2$
6	1	1
7	2	4
5	0	0
3	−2	4
4	−1	1
$\Sigma y = 25$	$\Sigma (y - \bar{y}) = 0$	$\Sigma (y - \bar{y})^2 = 10$

Many different measures of variability could be constructed using the deviations of the measurements from their mean. A first thought would be to use their average, but this will always equal zero. This is because the negative and positive deviations balance one another, so that their sum (and hence mean) equals zero [see column (b), table 3.3]. A second possibility would be to ignore the minus signs and compute the average of the absolute (positive) values of the deviations. This quantity would be easy to compute but difficult to interpret. A third possibility, and one that we will employ, makes use of the squared deviations of the measurements from their mean. Recall that the deviation of a measurement y from the sample mean $\bar{y}$ is expressed as $(y - \bar{y})$. The square of a deviation, then, is represented as $(y - \bar{y})^2$.

DEFINITION 3.11

variance

The <u>variance</u> of a set of n measurements is the sum of the squared deviations of the measurements from their mean divided by $(n - 1)$.

Because we will use information from a sample to make inferences about the population from which it was selected, it becomes convenient to draw a distinction between the variance of a set of sample measurements and the variance of a population. We use the symbol s^2 to represent the sample variance. Thus

$$s^2 = \frac{\Sigma(y - \bar{y})^2}{n - 1}$$

The corresponding population variance is denoted by σ^2 (σ is the Greek lower case letter sigma).

s^2 **is the sample variance.**
σ^2 **is the population variance.**

Example 3.6

Calculate the variance for the $n = 5$ measurements, 6, 7, 5, 3, and 4, given in table 3.3.

Solution

Using the information in column (c) of table 3.3, we find the sample variance to be

$$s^2 = \frac{\Sigma(y - \bar{y})^2}{n - 1} = \frac{10}{4} = 2.5$$

What can we say about the spread of a set of measurements with a variance of 2.5? The greater the variability or spread in a set of measurements, the larger is the variance; but how large is large? **Although we can compare variances of sets of measurements to compare variability, it is difficult to interpret the variance for a single set of measurements.**

Measures of Variability

We now consider a measure of variability useful not only for comparison purposes but also for describing a single set of measurements.

The Standard Deviation

DEFINITION 3.12

standard deviation

The standard deviation of a set of measurements is the positive square root of the variance.

The sample standard deviation is denoted by s and the corresponding population standard deviation by the symbol σ.

s **is the sample standard deviation.**
σ **is the population standard deviation.**

We now state a rule that gives practical significance to the standard deviation as a measure of variability. The rule applies only to data that have mound-shaped frequency distributions—the most frequently encountered distributions in the analysis of data. No theory exists to "prove" that the rule will always hold, but it is a fact that it provides a realistic characterization of many sets of data.

EMPIRICAL RULE

For a frequency distribution that is mound-shaped, the interval

$$(\bar{y} - s) \text{ to } (\bar{y} + s)$$

68% contains approximately 68% of the measurements;

$$(\bar{y} - 2s) \text{ to } (\bar{y} + 2s)$$

95% contains approximately 95% of the measurements;

$$(\bar{y} - 3s) \text{ to } (\bar{y} + 3s)$$

all or nearly all contains all or nearly all of the measurements.

mound-shaped frequency distribution

An example of a mound-shaped frequency distribution appears in figure 3.7. The relative frequencies are largest near the center of the distribution and tend to decrease as you move toward the distribution tails. **The exact shape of the**

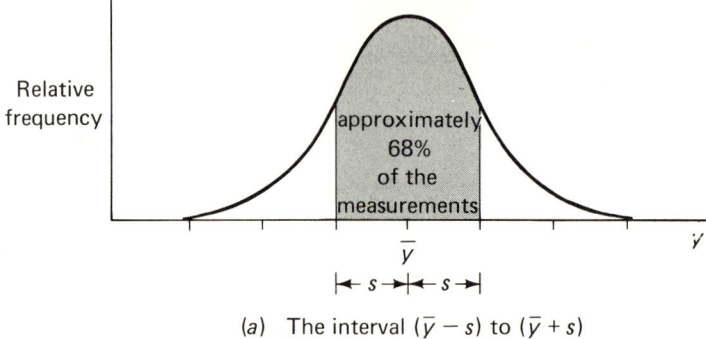

(a) The interval $(\bar{y} - s)$ to $(\bar{y} + s)$

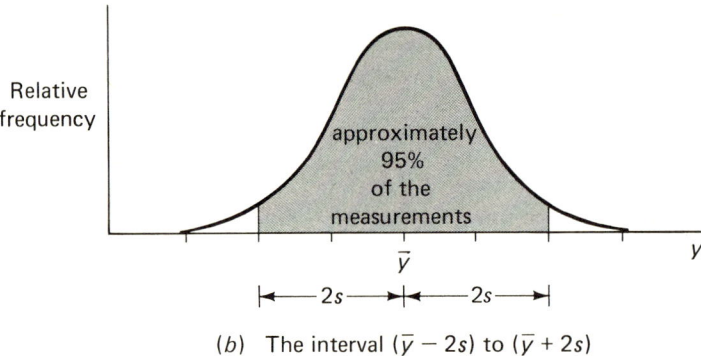

(b) The interval $(\bar{y} - 2s)$ to $(\bar{y} + 2s)$

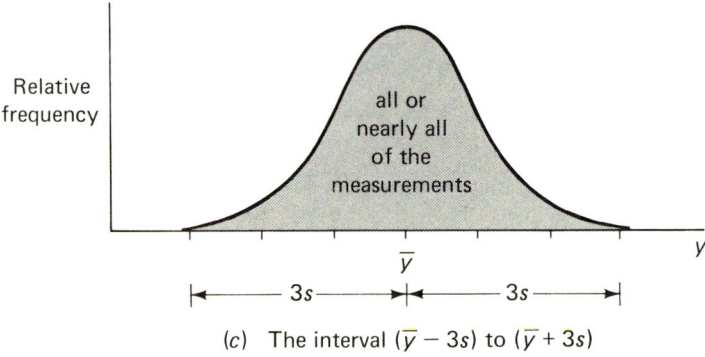

(c) The interval $(\bar{y} - 3s)$ to $(\bar{y} + 3s)$

Figure 3.7 Graphical Representation of the Empirical Rule

Measures of Variability

distribution is unimportant, because the rule will adequately describe the variability for mound-shaped distributions of data encountered in real life. Because it is a rule of thumb, a rule that has been observed to work in practice, it has been called the Empirical Rule.

Example 3.7

Students in a college economics class were interested in examining price increases for a certain make of car over a specified period of time. To do this, they obtained price increases on $n = 24$ different models over a six-month period. The increases (in dollars) are presented in table 3.4.

Table 3.4 Price Increases (dollars)

100	121	130	129
150	116	120	117
154	125	110	119
130	115	125	123
90	109	100	120
92	112	115	118

The sample mean and standard deviation for this set of measurements are $\bar{y} = 118.33$ and $s = 15.01$, respectively. Previous experience working with price increases indicates we can assume that the set of 24 price increases is mound-shaped. (To verify this, we could construct a relative frequency distribution for the data.) Describe the variability of the sample by using the Empirical Rule.

Solution

According to the Empirical Rule, approximately 68% of the measurements lie in the interval $(\bar{y} - s)$ to $(\bar{y} + s)$. Substituting the values given for the sample mean and standard deviation, we obtain $(118.33 - 15.01)$ to $(118.33 + 15.01)$, that is, 103.32 to 133.34, for the interval. From table 3.4 we find that 18 of the 24 (75%) price increases lie in this interval. Note that the 68% specified by the Empirical Rule provides a rough approximation to the actual percentage, 75%, found in the interval $(\bar{y} \pm s)$.

The Empirical Rule also states that approximately 95% of the measurements lie in the interval $(\bar{y} - 2s)$ to $(\bar{y} + 2s)$. When we substitute the values for $\bar{y}$ and s, we obtain the interval $[118.33 - 2(15.01)]$ to $[118.33 + 2(15.01)]$, or 88.31 to 148.35. We find that 22 of the 24 (92%) price increases of table 3.4 lie in this interval.

Similarly, the interval $(\bar{y} - 3s)$ to $(\bar{y} + 3s)$ should contain all or nearly all of the measurements. Adding and subtracting $3s = 3(15.01) = 45.03$ from $\bar{y} = 118.33$, we have the interval 73.30 to 163.36. As can be seen from table 3.4, all of the measurements do lie in this interval.

Keep in mind that the Empirical Rule is not intended to give exact percentages in the specified intervals. Rather, it gives approximate percentages, which will be surprisingly accurate for the intervals $(\bar{y} \pm 2s)$ and $(\bar{y} \pm 3s)$ for most sets of data.

Example 3.8

To remain competitive, manufacturers must be concerned with the efficiency of their operation. One corporation conducted a study to determine the average length of time it takes for an item to be completed on an assembly line. A sample of 50 items ($n = 50$) was timed. The mean and standard deviation (in hours) for the 50 measurements were $\bar{y} = 4.8$ and $s = .42$. A relative frequency histogram for the sample measurements indicated that the distribution is mound-shaped. Describe the 50 assembly line completion times by using the Empirical Rule.

Solution

The Empirical Rule tells us that approximately 68% of the measurements lie in the interval $(\bar{y} - s)$ to $(\bar{y} + s)$, that is, the interval 4.38 to 5.22. The interval $(\bar{y} - 2s)$ to $(\bar{y} + 2s)$, or 3.96 to 5.64, should contain approximately 95% of the measurements. All or nearly all of the measurements should be in the interval $(\bar{y} - 3s)$ to $(\bar{y} + 3s)$, or the interval 3.54 to 6.06. These results are summarized in table 3.5.

Table 3.5 Empirical Rule Results for Example 3.8

k	$\bar{y} \pm ks$	Approximate Percentage in Interval
1	4.38 to 5.22	68
2	3.96 to 5.64	95
3	3.54 to 6.06	almost all

To bolster your confidence in the Empirical Rule, let us see how well it describes the five frequency distributions of figure 3.8. We calculated the mean and standard deviation for each of the five data sets (not given) and

Measures of Variability

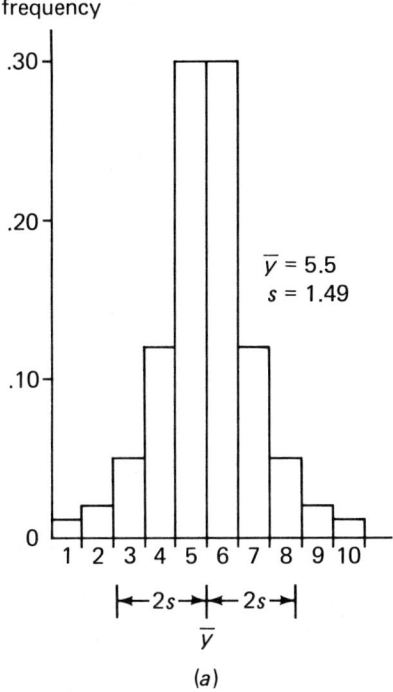

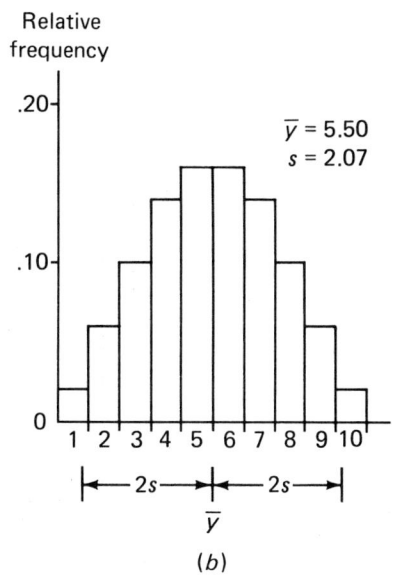

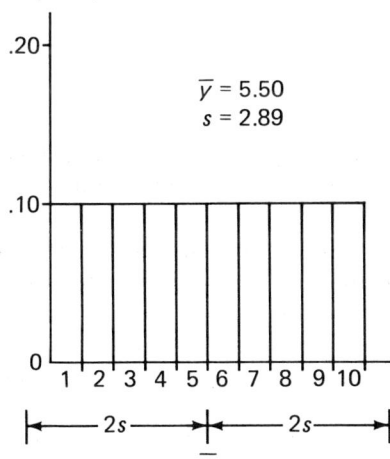

Figure 3.8 A Demonstration of the Utility of the Empirical Rule

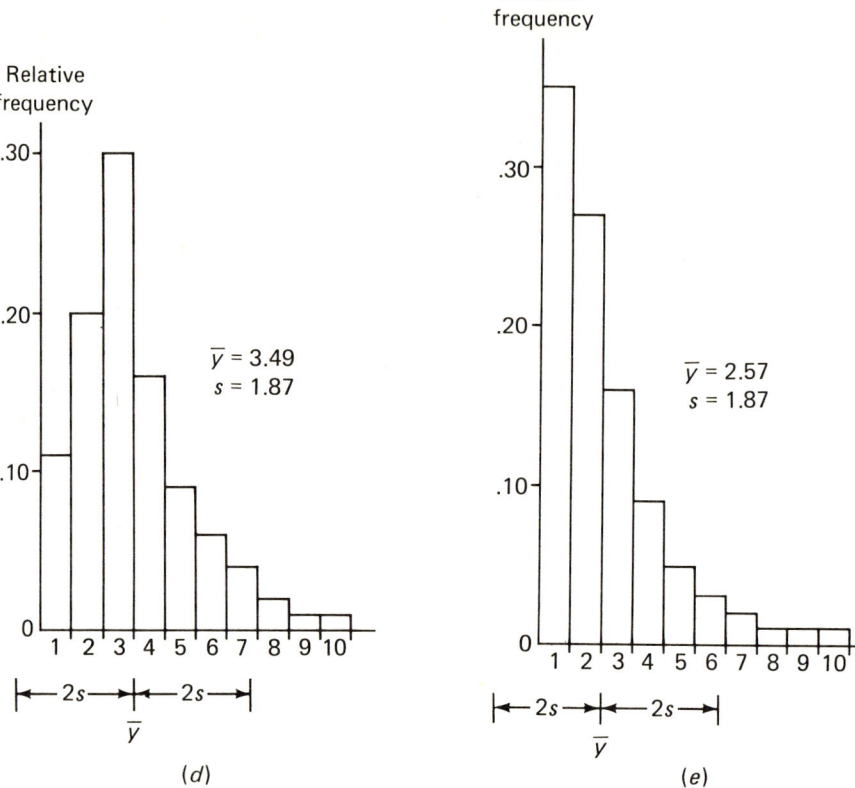

Figure 3.8 (Continued)

these are shown next to each frequency distribution. Figure 3.8(a) shows the frequency distribution for measurements made on a variable that can take values $y = 0, 1, 2, \ldots, 10$. The mean and standard deviation, $\bar{y} = 5.50$ and $s = 1.49$, for this symmetric mound-shaped distribution were used to calculate the interval ($\bar{y} \pm 2s$), which is marked below the horizontal axis of the graph. We found 94% of the measurements falling in this interval, that is, lying within two standard deviations of the mean. You will note that this percentage is very close to the 95% specified in the Empirical Rule. We also calculated the percentage of measurements lying within one standard deviation of the mean. We found this percentage to be 60%, a figure that is not too far from the 68% specified by the Empirical Rule. Consequently, we think the Empirical Rule provides an adequate description for figure 3.8(a).

Figure 3.8(b) shows another mound-shaped frequency distribution, but one that is less peaked than the distribution of figure 3.8(a). The mean and standard deviation for this distribution, shown to the right of the figure, are 5.50

Measures of Variability

and 2.07, respectively. The percentages of measurements lying within one and two standard deviations of the mean are 64% and 96%, respectively. Once again, these percentages agree very well with the Empirical Rule.

Now let us look at three other distributions. The distribution in figure 3.8(c) is perfectly flat, while the distributions of figures 3.8(d) and (e) are nonsymmetric and skewed to the right. The percentages of measurements lying within two standard deviations of the mean are 100%, 96%, and 95%., respectively, for these three distributions. All these percentages are reasonably close to the 95% specified by the Empirical Rule. The percentages lying within one standard deviation of the mean (60%, 75%, and 87%, respectively) show some disagreement with the 68% of the Empirical Rule.

To summarize, you can see that the Empirical Rule accurately forecasts the percentage of measurements falling within two standard deviations of the mean for all five distributions of figure 3.8, even for the distributions that are flat, figure 3.8(c), or highly skewed to the right, figure 3.8(e). The Empirical Rule is less accurate in forecasting the percentages within *one* standard deviation of the mean, but the forecast, 68%, compares reasonably well for the three distributions that might be called mound-shaped, figures 3.8(a), (b), and (d).

We have discussed four measures of variability in this section: the range, percentiles, the variance, and the standard deviation. Although each of these measures is useful in data description, the variance and the standard deviation provide us with information to

variance and standard deviation

compare variability

interpret variability

1. compare variability between sets of measurements
2. interpret the variability of a single set of measurements by using the Empirical Rule (recall that the Empirical Rule only applies to data that have a mound-shaped frequency distribution)

In the remainder of this text we will use the standard deviation almost exclusively to measure the variability of a single set of measurements.

EXERCISES

3.6. Calculate $\bar{y}$ and s for the following sample of five measurements: 3, 9, 4, 1, 3.

3.7. The test scores for a nationally administered college achievement test have a mound-shaped distribution with a mean of 520 and a standard deviation of 110.
(a) Use the Empirical Rule to assist you in describing this distribution of scores.
(b) The median for the scores is reported to be 510 and the 98th percentile is 795. Interpret these two statistics.
(c) The range of the scores is 702. Interpret this statistic.

How to Phrase an Inference: Numerical Methods

3.8. Students in a chemistry class were assigned the task of determining the purity of a chemical substance. Three hundred class members independently analyzed the substance. The mean and the standard deviation for the $n = 300$ measurements were 78.1% and 1.3%, respectively. Joe Smith missed his laboratory class but analyzed the substance the next day. His analysis gave a purity reading of 83.4%. Is there reason to suspect his analytical result? Why?

3.9. In manufacturing oxygen tents it is very important for the actual percentage of oxygen generated at a particular time to be close to the amount specified by a physician and indicated on the oxygen control valve. To investigate a particular manufacturer's oxygen tents, 50 tents were selected and all the control valves were adjusted to the same oxygen input setting. The atmosphere within each tent was then sampled, and the difference between the actual percentage of oxygen and the valve setting was recorded for each. If the mean and the standard deviation for the sample were 1.3% and .6%, respectively, describe the distribution of the 50 readings.

3.10. Trichinosis, a disease derived from improperly cooked pork products (main source: pork sausage) may be on the increase. Two hundred eighty-four (284) cases of trichinosis, including one fatality, were reported to the Communicable Disease Center in 1975. This number was 2½ times higher than the mean number of cases reported during the previous five years, and it represents the highest annual incidence since 1961.* From your knowledge about data variation, do you think these data confirm an increase in the per capita rate of incidence of trichinosis? Explain. (Note: This exercise is only intended to stimulate discussion. More clear-cut decisions will be derived from more complete sets of data in later chapters.)

3.4 SHORTCUT METHOD FOR CALCULATING THE VARIANCE AND STANDARD DEVIATION

Calculation of the variance and standard deviation can be tedious if we use the definition and calculate each individual deviation, $(y - \bar{y})$. This method also leads to rounding error caused by rounding the values of $(y - \bar{y})$. However, there is a shortcut method that saves time and leads to more accurate calculations. We will use a sample of five measurements, 5, 7, 1, 2, 4, to introduce this shortcut method for calculating the sum of squares of deviations.

We need both the sum and the sum of squares of the y values for the shortcut formula. You will recall that Σy was used to indicate the sum of the y values. Thus

$$\Sigma y = 5 + 7 + 1 + 2 + 4 = 19$$

*Source: *Veterinary Medicine Newsletter*, Florida Cooperative Extension Service, August 1977.

Shortcut Method for Calculating the Variance and Standard Deviation

Similarly, we use the symbol Σy^2 to indicate the sum of the squares of the y values. Thus

$$\Sigma y^2 = (5)^2 + (7)^2 + (1)^2 + (2)^2 + (4)^2$$
$$= 25 + 49 + 1 + 4 + 16 = 95$$

These quantities, Σy and Σy^2, which are given as totals in table 3.6, are substituted into the formula (which follows) to find $\Sigma (y - \bar{y})^2$. As you will see, this formula gives us an easy way to calculate the sum of squares of deviations, the quantity that appears in the formula for s^2.

Table 3.6 Shortcut Calculations for $\Sigma (y - \bar{y})^2$

	y	y^2
	5	25
	7	49
	1	1
	2	4
	4	16
Total	19	95

Shortcut Formula for Calculating $\Sigma(y - \bar{y})^2$

$$\Sigma (y - \bar{y})^2 = \Sigma y^2 - \frac{(\Sigma y^2)}{n}$$

(Proof of this formula is omitted.)

Substituting the 5 measurements ($n = 5$) of table 3.6 into the shortcut formula, we arrive at the following:

$$\Sigma (y - \bar{y})^2 = 95 - \frac{(19)^2}{5}$$
$$= 95 - \frac{361}{5}$$
$$= 95 - 72.2 = 22.8$$

To check the validity of this result, we can compute, as before, the sum of the squares of the individual deviations; see table 3.7. The sample mean is $\bar{y}$

= 3.8. We see that $\Sigma(y - \bar{y})^2$ is the same when using either computational procedure.

Table 3.7 Calculating $\Sigma (y - \bar{y})^2$

y	$(y - \bar{y})$	$(y - \bar{y})^2$
5	1.2	1.44
7	3.2	10.24
1	−2.8	7.84
2	−1.8	3.24
4	0.2	0.04
Total $\Sigma y = 19$	$\Sigma (y - \bar{y}) = 0$	$\Sigma (y - \bar{y})^2 = 22.8$

The variance and the standard deviation are then found as before. Thus

$$s^2 = \frac{\Sigma (y - \bar{y})^2}{n - 1} = \frac{22.8}{4} = 5.7$$

and using table 6 of the Appendix,

$$s = \sqrt{s^2} = \sqrt{5.7} = 2.39$$

EXERCISES

3.11. Always use the shortcut method to calculate s^2 and s. To give you some practice, consider a small set of 5 measurements, say 5, 3, 1, 2, 2.
(a) So that you can see the variation in the measurements, construct a dot diagram similar to figure 3.6.
(b) Use Σy and Σy^2 to calculate $\Sigma (y - \bar{y})^2$.
(c) Calculate s^2 and s.
(d) Because the number of measurements in the sample is so small, the frequency distribution for the sample measurements is not mound-shaped. Nevertheless, note that the interval $(\bar{y} \pm 2s)$ contains all the measurements. (Construct this interval on the dot diagram for the data so that you can see the location of the points within the interval.)

3.12. Repeat the instructions of exercise 3.11 for the 6 measurements 1, 0, 3, 1, 2, 2.

3.13. Repeat the instructions of exercise 3.11 for the 10 measurements 4, 1, 3, 5, 2, 3, 1, 3, 0, 2.

Shortcut Method for Calculating the Variance and Standard Deviation

 3.14. The treatment times for patients at a health clinic (see exercise 3.4) are as follows:

```
21  20  31  24  15  21  24  18  33   8
26  17  27  29  24  14  29  41  15  11
13  28  22  16  12  15  11  16  18  17
29  16  24  21  19   7  16  12  45  24
21  12  10  13  20  35  32  22  12  10
```

(a) Use the shortcut formula to calculate s^2 and s. You can verify that $\Sigma y = 1{,}016$ and $\Sigma y^2 = 24{,}080$ for the 50 treatment times (or you can accept our calculations).

(b) To increase your confidence in the applicability of the Empirical Rule, construct the intervals $(\bar{y} \pm s)$, $(\bar{y} \pm 2s)$, and $(\bar{y} \pm 3s)$, and count the number of treatment times falling in each of the three intervals. From these frequencies calculate the corresponding percentage of measurements falling in the three intervals. Does the Empirical Rule give a reasonable approximation to the relative frequencies you have observed?

3.15. Refer to exercise 3.2. Calculate $\bar{y}$, s^2, and s, and follow the instructions of exercise 3.14. Once again you will see that $\bar{y}$ and s describe this distribution of data and that the Empirical Rule gives an effective interpretation to s.

3.16. To assist in estimating the amount of lumber in a tract of timber, an owner decided to count the number of trees with diameters exceeding 12 inches in randomly selected 50 × 50-foot squares. Seventy 50 × 50 squares were randomly selected from the tract and the number of trees (with diameters in excess of 12 inches) were counted for each. The data were as follows:

```
 7   8   6   4   9  11   9   9   9  10
 9   8  11   5   8   5   8   8   7   8
 3   5   8   7  10   7   8   9   8  11
10   8   9   8   9   9   7   8  13   8
 9   6   7   9   9   7   9   5   6   5
 6   9   8   8   4   4   7   7   8   9
10   2   7  10   8  10   6   7   7   8
```

(a) Construct a relative frequency histogram to describe these data.
(b) Calculate the sample mean $\bar{y}$ as an estimate of μ, the mean number of timber trees with diameters exceeding 12 inches for all 50 × 50-foot squares in the tract.
(c) Calculate s for the data. Construct the intervals $(\bar{y} \pm s)$, $(\bar{y} \pm 2s)$, and $(\bar{y} \pm 3s)$. Count the percentage of squares falling in each of the three intervals, and compare these percentages with the corresponding percentages given by the Empirical Rule.

Exercises 3.11 and 3.14 to 3.16 are intended to aid you in learning how to calculate a standard deviation and to use the Empirical Rule to interpret it. In practice, we are interested in the population relative frequency distribution, not the sample. We will subsequently use $\bar{y}$ and s to estimate the population parameters μ and σ, and we will employ them to describe the population frequency distribution. The accuracy of these estimates will be discussed in later chapters.

3.5 HOW TO GUESS THE STANDARD DEVIATION OF SAMPLE DATA

Many times sample data are presented without accompanying numerical descriptive measures such as the mean and standard deviation. Although the sample mean can be computed very easily, the calculations required for obtaining s (even with the shortcut formula of section 3.4) can be difficult and time-consuming. In situations where we are interested in obtaining a rough approximation to the actual sample standard deviation, without going through the tedious calculations, we can use the following formula:

$$s \approx \frac{\text{range}}{4}$$

check calculations

(read, "s is approximately equal to the range divided by 4"). One additional reason for using this approximation is to check our calculations of s when using the shortcut formula. Arithmetic mistakes or errors in reading a square root table can easily occur, so we suggest you use the range approximation as a check even when the actual calculation of s is required.

In this chapter we presented a rule, the Empirical Rule, which was useful in interpreting the variability of a mound-shaped distribution. The Empirical Rule states that approximately 95% of the measurements in a set will be within

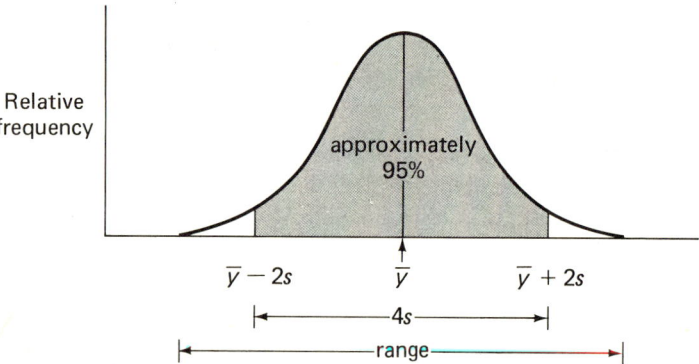

Figure 3.9 Approximation to s Using the Range

How to Guess the Standard Deviation of Sample Data

two standard deviations of their mean, or, using notation for a sample approximately 95% of the measurements will be in the interval $(\bar{y} - 2s)$ to $(\bar{y} + 2s)$. See figure 3.9.

We note also from figure 3.9 that the interval $(\bar{y} - 2s)$ to $(\bar{y} + 2s)$ has a length equal to $4s$. Since the range of the measurements is approximately $4s$, one-fourth of the range will provide an approximate value for s.

Approximation to s Using the Range

$$s \approx \frac{\text{range}}{4}$$

Note: This approximation will work best for mound-shaped distributions.

Example 3.9

Fifteen students with similar socioeconomic backgrounds were compared on an IQ test at the end of their freshman year. Their scores are presented in table 3.8. Use the range of the observations to approximate s. Calculate s by using the shortcut formula and compare this result with the approximate value.

Table 3.8 IQ Scores

116	118	126
129	114	130
129	122	128
132	131	125
123	134	126

Solution

We approximate the value of s by using the range of the measurements divided by 4. From table 3.8 we see the range is $(134 - 114) = 20$. Our approximation is then

$$\text{approximate value of } s = \frac{\text{range}}{4} = \frac{20}{4} = 5.0$$

Using the shortcut formula for $\Sigma (y - \bar{y})^2$ to calculate the exact value of s, we need to compute Σy^2 and Σy.

$$\Sigma y^2 = (116)^2 + (129)^2 + \cdots + (126)^2 = 236{,}873$$
$$\Sigma y = 116 + 129 + \cdots + 126 = 1{,}883$$

Hence

$$\Sigma (y - \bar{y})^2 = \Sigma y^2 - \frac{(\Sigma y)^2}{n}$$
$$= 236{,}873 - \frac{(1{,}883)^2}{15}$$
$$= 236{,}873 - 236{,}379.267 = 493.733$$

The sample variance s^2 is then

$$s^2 = \frac{\Sigma (y - \bar{y})^2}{n - 1} = \frac{493.733}{14} = 35.267$$

and the standard deviation (see table 6 in the Appendix and use the square root for 35.3, the nearest tabulated value to 35.267) is

$$s = \sqrt{35.267} = 5.94$$

Although the approximate value of s, 5.0, differs somewhat from the actual value, 5.94, it still provides a check on our calculations. For instance, we know that our calculation of 5.94 is at least reasonable. If we had computed s to be 59.4, it would certainly not agree with our check. We would advise running this simple check every time you compute a standard deviation.

EXERCISES

3.17. Use the 10 measurements 3, 1, 0, 1, 3, 4, 2, 3, 2, 3.
(a) Use the range approximation to guess the value of s.
(b) Calculate s (use the shortcut formula) and compare this computed value of s with your approximation in part (a).

3.18. Use the 20 measurements 2, 5, 3, 4, 0, 1, 4, 2, 5, 7, 4, 4, 3, 5, 4, 6, 5, 8, 1, 3.
(a) Use the range approximation to guess the value of s.
(b) Calculate s (use the shortcut formula) and compare this computed value of s with your approximation in part (a).

3.19. Refer to the data in exercise 3.13. Calculate the range of the measurements and then calculate an approximate value of s by using the range. Compare this result with the value of s calculated in exercise 3.13 (or with the exercise answer given in the back of the text). The range estimate is close enough to detect gross errors in calculating s [like forgetting to divide by $(n-1)$ or forgetting to take the square root].

3.20. Find the range for the treatment time data of exercise 3.14 and then obtain the range approximation to s. Compare this result with the value of s calculated in exercise 3.14. Note that the approximation provides a good method for detecting gross errors in calculation.

3.21. Find the range of the observations taken for the time-recording experiment of exercise 3.2. Calculate the range approximation to s and compare this result with the value of s calculated in exercise 3.15.

3.22. An experiment was conducted to investigate the effect of root temperature on the growth of soybeans. Ten soybean plants were subjected to a standard soil condition and a root temperature of 25°. Fourteen days after seed germination the weights of the exposed tops were obtained. The weights, in grams, for the 10 plants were as follows:

$$.23, \quad .27, \quad .41, \quad .25, \quad .37, \quad .51, \quad .29, \quad .34, \quad .33, \quad .48$$

The first step in the analysis of these data would require the computation of the sample mean and standard deviation.
(a) Calculate the sample mean.
(b) Approximate the value of s for these data by using the range.
(c) Calculate s and compare this value with your approximation in part (b). (You will find that $\Sigma y = 3.48$ and $\Sigma y^2 = 1.2924$ for these data.)

SUMMARY

The objective of statistics is to make an inference about a population based on information contained in a sample. Since making a statistical inference implies describing the population, we seek a way to describe a set of measurements.

graphical descriptive methods

Graphical descriptive methods, presented in chapter 2, are very effective for describing a set of measurements, but they are unsatisfactory for statistical inference. The difficulty is that, although we can say that a sample frequency histogram will be similar to the population frequency histogram, we have no way of knowing how similar they might be. In other words, we have no way to measure the goodness of the statistical inference.

numerical descriptive measures

Numerical descriptive measures are numbers that provide a mental image of the frequency distribution for a set of measurements. The most important of these are the mean and standard deviation, which measure the center and the spread, respectively, of a frequency distribution. The standard deviation of a set of mound-shaped measurements is a meaningful measure of variability when interpreted by the Empirical Rule. Numerical descriptive measures of the sample and population are called, respectively, statistics and parameters.

Numerical descriptive measures are suitable for developing both descriptions and inferences. Thus we can use a numerical descriptive measure of the sample, say the sample mean, to estimate a parameter such as the population mean. The great advantage of the numerical descriptive measure in making inferences is that we can give a quantitative measure of the goodness of the inference. In particular, we can show that the sample mean will lie within a specified distance of the population mean with some predetermined probability.

Now that we know how to phrase an inference, that is, how to talk about a population of interest to us, we turn to the problem of making the inference. As suggested in an earlier discussion, all types of inferences based on partial information, such as sample data, are subject to a degree of uncertainty. If we decide that a population possesses a certain characteristic, we may be correct, but there is always some element of doubt in our minds. **Thus we will find that probability, which is a measure of uncertainty, plays a major role both in making an inference and in measuring how good it is.**

REFERENCES

Barr, A. J.; Goodnight, J. H.; Sall, J. P.; and Helwig, J. T. *A User's Guide to SAS 76*. Raleigh, N.C.: SAS Institute, Inc., 1976.

Dixon, W. J., and Brown, M. B., eds. *Biomedical Computer Programs*, Rev. ed. Los Angeles: University of California Press, 1978.

Freund, J. E. *Statistics: A First Course*. 2d ed. Englewood Cliffs, N.J.: Prentice-Hall, 1976.

Huntsberger, D. V., and Billingsley, P. *Elements of Statistical Inference*. 4th ed. Boston: Allyn and Bacon, 1977.

Mendenhall, W. *Introduction to Probability and Statistics*. 5th ed. N. Scituate, Mass.: Duxbury Press, 1979. Chapter 3.

Nie, N.; Hull, C. H.; Jenkins, J. G.; Steinbrenner, K.; and Bent, D. H. *Statistical Package for the Social Sciences*. 2d ed. New York: McGraw-Hill, 1975.

SUPPLEMENTARY EXERCISES

3.23. Calculate $\bar{y}$ and s for the $n = 5$ measurements 6, 1, 4, 5, 2.

3.24. For the $n = 7$ measurements 2, 1, 1, 0, 4, 5, 2, find $\bar{y}$ and s.

3.25. Suppose you were told that the lower quartile for the annual income of unskilled workers in the United States is $3200. What does this mean?

3.26. The mean contribution per person for a college alumni fund drive was $100.50, with a standard deviation of $38.23. What fraction of the contributions could be expected to lie in the interval $62.27 to $138.73? Explain.

Supplementary Exercises

3.27. The following data represent a sample of $n = 10$ measurements on the tensile strength (in pounds per square inch) of sutured wounds in 10 days after repair:

$$63, \ 56, \ 32, \ 56, \ 48, \ 45, \ 45, \ 96, \ 57, \ 71$$

(a) Determine the sample mean, median, and mode.
(b) Calculate s.
(c) Do the values of $\bar{y}$ and s help you to describe this set of data?

3.28. The percentage of city telephone subscribers who use unlisted numbers is increasing. The distribution of the percentage of unlisted numbers in cities possesses a mean and a standard deviation that are near 14% and 6%, respectively. If you were to pick a city at random, is it likely that the percentage of unlisted numbers would exceed 20%? Explain. Within what limits would you expect the percentage to fall?

3.29. Experiments are conducted on small animals as a prelude to experimentation with larger animals or humans. The object of an experiment conducted by Dr. Garth Resch, University of Florida, was to investigate the effect of propranolol in reducing hypertension in rats. The hypertension was induced by cold exposure. Two groups of rats were used, an untreated (or control) group and another that received the drug dosage. The extent of hypertension in a given rat was monitored by recording its blood pressure.

To illustrate, the blood pressures of 5 rats in the control group, which were exposed for six weeks to a 5°C temperature, were 152, 179, 182, 176, and 149. Corresponding blood pressures for 7 rats treated with propranolol were 113, 111, 143, 151, 109, 111, and 168. The first step in the analysis of the data requires the computation of the sample means and variances. (How we decide whether the drug was effective in reducing hypertension will come later.)
(a) Calculate $\bar{y}$ and s^2 for the $n = 5$ rats in the control group. (For this sample $\Sigma \ y = 838$ and $\Sigma \ y^2 = 141{,}446$.)
(b) Calculate $\bar{y}$ and s^2 for the $n = 7$ rats receiving the propranolol. (For this sample $\Sigma \ y = 906$ and $\Sigma \ y^2 = 120{,}766$.)

3.30. Calculate $\bar{y}$ and s for the following measurements:

$$0, \ 4, \ 1, \ 3, \ 3, \ 2, \ 4, \ 0, \ 0, \ 1, \ 3, \ 1, \ 4, \ 5, \ 0, \ 1, \ 2, \ 4, \ 3, \ 1$$

Find the proportion of measurements that fall in the interval $(\bar{y} \pm 2s)$ and compare this result with the Empirical Rule.

3.31. For the 20 measurements 0, 4, 1, 0, 0, 2, 1, 1, 0, 3, 0, 1, 0, 2, 0, 1, 2, 2, 1, 0:
(a) Use the range approximation to guess the value of s.
(b) Calculate s and compare this computed value with the approximation in part (a).

(c) Find the proportion of measurements falling in the interval ($\bar{y} \pm 2s$) and compare this result with the Empirical Rule.

e 3.32. The maximum daily temperature in the vicinity of a major resort area averages 83°F during the summer months. The standard deviation for maximum daily temperatures in this area is 7°F, and assume the measurements are mound-shaped. Describe the variability of the set of maximum daily temperatures by using the Empirical Rule.

3.33. Concerned with the amount of heating oil consumed over the past several years, a homeowner finds that the following amounts (in gallons) per week have been used during the winter months over the past five years: 110, 115, 105, 96, 85. Find the mean and standard deviation for the homeowner's weekly fuel oil consumption over the past five winters.

$ 3.34. An industrial concern uses an employee screening test with average score μ and standard deviation $\sigma = 10$. Assume that the test score distribution is approximately mound-shaped and that a score of 65 qualifies an applicant for further consideration. What is the value of μ such that approximately 2.5% of the applicants qualify for further consideration?

ψ 3.35. The intelligence quotient (IQ) expresses intelligence as a ratio of the mental age to the chronological age multiplied by 100. Thus the average (when mental age equals chronological age) is 100. Construct a relative frequency histogram for the following IQ scores.

100	103	99	101	100	120	109	82
101	112	95	118	118	89	114	113
92	137	130	94	87	93	111	96
93	98	101	96	84	86	89	90

3.36. Refer to exercise 3.35.
(a) Calculate $\bar{y}$ and s.
(b) Use the range approximation for s to check your calculations.
(c) Find the number of scores in the intervals ($\bar{y} \pm s$), ($\bar{y} \pm 2s$), and ($\bar{y} \pm 3s$). Compare the proportions of measurements in these intervals with the proportions specified by the Empirical Rule.

EXPERIENCES WITH REAL DATA

1. Select an area of your undergraduate major that utilizes experimental data. Typical data sources would be chemistry, biology, psychology, geology, or physics laboratories. Or you might seek data contained in the social science or business journals. Either by experimentation or by use of a professional journal, select a sample of at least $n = 25$ observations on some experimental variable.

Experiences with Real Data

    (a)   Define the population from which your sample was drawn.
    (b)   Construct a relative frequency histogram for the data.
    (c)   Calculate $\bar{y}$ and s for the data.
    (d)   Do the data appear to be mound-shaped and thereby satisfy the requirements of the Empirical Rule?
    (e)   What fraction of the observations lie within two standard deviations of $\bar{y}$? Three? Do these results agree with the Empirical Rule?

2. Examine the population of all telephone subscribers in your community (see your local telephone directory). Randomly select a page from the directory and repeat the process until you have selected $n = 30$ pages. Let y be the number of subscribers per page and find y for each of the $n = 30$ pages.
    (a)   Construct a relative frequency histogram for your data.
    (b)   Calculate $\bar{y}$ and s.
    (c)   Find the fraction of observations in the intervals ($\bar{y} \pm s$) and ($\bar{y} \pm 2s$). Does the Empirical Rule provide a satisfactory description of the variability of the data?

    Calculation of $\bar{y}$ and s can be obtained directly on many electronic desk calculators. For large data sets you might wish to use packaged programs and an electronic computer to perform the calculations. Useful packaged programs are available in the Biomed (Biomedical Programs), the SAS (Statistical Analysis Systems), and the SPSS (Statistical Package for the Social Sciences) program libraries (see the References).

4 Probability and Probability Distributions

CHAPTER OUTLINE

4.1 Probability and Inference
4.2 What Is Probability?
4.3 Additivity of Probabilities
4.4 Conditional Probability and Independence
4.5 Random Variables
4.6 The Binomial Probability Distribution
4.7 The Normal Probability Distribution

4.1 PROBABILITY AND INFERENCE

We learned in previous chapters that a population is a set of measurements and that it can be described by a set of numerical measures called parameters. Typical population parameters are its mean μ and standard deviation σ. In most applications of statistics we will not know μ and σ (or other population parameters) but will attempt to make inferences about them based on information contained in a sample. Because a sample will represent only a subset of the measurements needed to calculate the exact value of a parameter, an inference based on a sample will almost always be subject to error.

To illustrate, suppose that a nutritionist wished to estimate the mean weight gain of one-year-old white mice placed on a specific diet for one month. Ten mice, selected from the conceptual set of all one-year-old white mice, are placed on the diet, and the mean weight gain for the sample of 10 is $\bar{y} = 4.3$ grams. Note that it is highly unlikely that the sample mean weight gain, $\bar{y} = 4.3$, will equal μ, the mean weight gain that would be obtained if all one-year-old white mice were placed on the diet. But it is probably close. What is the probability that the sample mean weight gain $\bar{y}$ will lie within .1 gram of the population mean μ? **This example highlights the uncertainty that surrounds every inference, an uncertainty that is measured in terms of probability.**

reliability of an inference but it also plays a fundamental role in all statistical inference-making procedures. What is probability? We will answer this question and provide you with some of the basic concepts of probability in this chapter. And we have not forgotten the nutritionist's problem. We will show you how probability is used to measure the reliability of this estimate in chapter 6.

4.2 WHAT IS PROBABILITY?

Data are obtained either by observation of uncontrolled events in nature or by controlled experimentation. To simplify our terminology, we need a word that will apply to any method of data collection.

DEFINITION 4.1

experiment

An experiment is the process by which an observation (or measurement) is obtained.

Typical examples of experiments are the following:

1. measuring the weight gain for a mouse placed on a specific diet for one month
2. determining a test grade
3. making a measurement of daily rainfall
4. interviewing a voter to obtain the voter's preference prior to an election
5. inspecting a light bulb to determine whether or not it is defective
6. tossing a coin and observing the face that appears
7. counting the number of divorces in a geographical area during a given year

Note from experiments 4, 5, and 6 that an observation need not be numerical.

outcomes
events

Experiments result in outcomes that, except in rare cases, cannot be predicted in advance. These outcomes, called events in statistical terminology, are usually denoted by capital letters. To see how the terminology is used, consider the following experiments.

Experiment 1

Toss two coins and observe the upper face of each coin.
Examples of events associated with this experiment are as follows:

What is Probability?

A: Two heads show.
B: Two tails show.
C: At least one tail shows.
D: A head and a tail show.

Experiment 2

Sample 100 prospective voters and record each voter's preference, for or against, a local referendum.

Examples of events associated with this experiment are as follows:

A: Exactly 50 voters favor the referendum.
B: Fewer than 20 voters favor the referendum.
C: All the voters favor the referendum.

When a particular experiment is conducted only once, some events may be more likely to occur than others. **The uncertainty that we attach to the occurrence of an event is measured by the probability of the event.** For example, if you toss two coins, it is less likely that you will observe event A, "two heads show," than event D, "a head and a tail show."

The determination of the exact meaning of the expression "probability of an event" is a philosophical problem that has not yet been resolved. Although possessing certain limitations, the most commonly accepted definition of probability, and one that is appropriate for a course at this level, utilizes the relative frequency concept. According to this concept, if an experiment is repeated a very large number N of times and, of these, n_A experiments result in event A, then the probability of event A, denoted by the symbol $P(A)$, is defined as follows:

DEFINITION 4.2

probability of an event A

The probability of an event A is

$$P(A) = \frac{n_A}{N}$$

where n_A is the number of experiments resulting in event A and N is the number of times the experiment is repeated.

Because we have defined an event as a relative frequency, the probability of an event, say event A, will always satisfy the property

$$0 \leq P(A) \leq 1$$

4.3 ADDITIVITY OF PROBABILITIES

either A or B occurs

Suppose that A and B represent two experimental events and that you are interested in a new event, the event that either A or B occurs. For example, suppose that we toss a pair of dice and define the following events:

A: A total of 7 shows.
B: A total of 11 shows.

Then the event "either A or B occurs" is the event that you toss a total of either 7 or 11 with the pair of dice.

Notice that for this example the events A and B are mutually exclusive. That is, if you observe event A (a total of 7), you could not at the same time observe event B (a total of 11). Thus if A occurs, B cannot occur (and vice versa).

DEFINITION 4.3

mutually exclusive

Two events A and B are said to be mutually exclusive if (when the experiment is performed a single time) the occurrence of one of the events excludes the possibility of the occurrence of the other event.

The concept of mutually exclusive events is used to specify a second property that the probabilities of events must satisfy. When two or more events are mutually exclusive, then the probability that any one of the events will occur is the sum of the event probabilities. That is, if two events A and B are mutually exclusive, the probability that either event A or event B occurs is

$$P(\text{either } A \text{ or } B) = P(A) + P(B)$$

For example, when we toss a pair of dice, the sum S of the numbers appearing on the dice can assume any one of the values $S = 2, 3, 4, \ldots, 11, 12$. On a single toss of the dice we can observe only one of these values. Therefore, the values $2, 3, \ldots, 12$ represent mutually exclusive events. If we want to find the probability of tossing a sum less than or equal to 4, this probability is

$$P(S \leq 4) = P(2) + P(3) + P(4)$$

For this particular experiment the dice can fall in 36 equally likely different ways. We can observe a 1 on die No. 1 and a 1 on die No. 2, denoted by the symbol (1, 1). We can observe a 1 on die No. 1 and a 2 on die No. 2, denoted by (1, 2). In other words, for this experiment the possible outcomes are

Additivity of Probabilities

$$
\begin{array}{cccccc}
(1, 1) & (2, 1) & (3, 1) & (4, 1) & (5, 1) & (6, 1) \\
(1, 2) & (2, 2) & (3, 2) & (4, 2) & (5, 2) & (6, 2) \\
(1, 3) & (2, 3) & (3, 3) & (4, 3) & (5, 3) & (6, 3) \\
(1, 4) & (2, 4) & (3, 4) & (4, 4) & (5, 4) & (6, 4) \\
(1, 5) & (2, 5) & (3, 5) & (4, 5) & (5, 5) & (6, 5) \\
(1, 6) & (2, 6) & (3, 6) & (4, 6) & (5, 6) & (6, 6)
\end{array}
$$

As you can see, only one of these events, (1, 1), will result in a sum equal to 2. Therefore, we would expect a 2 to occur with a relative frequency of $1/36$ in a long series of repetitions of the experiment, and we let $P(2) = 1/36$. The sum $S = 3$ will occur if we observe either of the outcomes (1, 2) or (2, 1). Therefore, $P(3) = 2/36 = 1/18$. Similarly, we find $P(4) = 3/36 = 1/12$. It follows that,

$$P(S \leq 4) = P(2) + P(3) + P(4) = \frac{1}{36} + \frac{1}{18} + \frac{1}{12} = \frac{1}{6}$$

In most practical data-collecting situations we will make observations on a variable. For example, in recording the diastolic blood pressures of hypertensive patients, the variable of interest is diastolic blood pressure and each patient's diastolic blood pressure represents an observation (measurement) on that variable. The fact that these measurements vary in a seemingly random and unpredictable manner leads us to call the variable a <u>random variable</u>.

random variable

Consider a situation where a single measurement is obtained on a random variable. Since we will observe only one of many possible values of the random variable (e.g., only one diastolic blood pressure), it follows that the values of a random variable represent mutually exclusive events (if one value occurs, the others cannot have occurred). We will make particular use of the additive property of the probabilities of mutually exclusive events when we wish to find the probability that a random variable assumes one of two or more values when it is observed in an experiment.

A third property of event probabilities is stated in terms of an event and its complement.

DEFINITION 4.4

complement

The <u>complement</u> of an event A is the event that A *does not* occur. The complement of A is denoted by the symbol $\overline{A}$.

Thus if we define the complement of an event A as a new event, namely, "A does not occur," it follows that

$$P(A) + P(\overline{A}) = 1$$

For example, refer again to the two-coin-toss experiment. If, in many repetitions of the experiment, the proportion of times you observe event A, "two heads show," is $1/4$, then it follows that the proportion of times you observe the event $\bar{A}$, "two heads do not show," is $3/4$. Thus $P(A)$ and $P(\bar{A})$ will always sum to 1.

The three properties that the probabilities of events must satisfy can be summarized as follows:

Properties of Probabilities

If A and B are any two mutually exclusive events associated with an experiment, then $P(A)$ and $P(B)$ must satisfy the following properties:

1. $0 \leq P(A) \leq 1$ and $0 \leq P(B) \leq 1$.
2. $P(\text{either } A \text{ or } B) = P(A) + P(B)$.
3. $P(A) + P(\bar{A}) = 1$ and $P(B) + P(\bar{B}) = 1$.

4.4 CONDITIONAL PROBABILITY AND INDEPENDENCE

Consider the following situation: The examination of a large number of insurance claims, categorized according to type of insurance and according to whether the claim was fraudulent, produced the results shown in table 4.1. Suppose you are responsible for checking insurance claims—in particular, for detecting fraudulent claims—and you examine the next claim that is processed. What is the probability of the event F, "the claim is fraudulent"? To answer the question, you examine table 4.1 and note that 10% of all claims are fraudulent. Thus assuming that the percentages given in the table are reasonable approximations to the true probabilities of receiving specific types of claims, it follows that $P(F) = .10$. Would you measure the risk that you face a fraudulent claim with the probability .10? We think not, because you have

Table 4.1 Categorization of Insurance Claims

	Type of Policy			
Category	*Fire*	*Auto*	*Other*	*Total*
fraudulent	6%	1%	3%	10%
nonfraudulent	14%	29%	47%	90%
Total	20%	30%	50%	100%

additional information that may affect the assessment of $P(F)$. For example, you would know the type of policy you were examining (i.e., fire, auto, or other).

Suppose that you have the additional information that the claim was associated with a fire policy. Checking table 4.1, we see that 20% (or .20) of all claims are associated with a fire policy and that 6% (or .06) of all claims are fraudulent fire policy claims. Therefore, it follows that the probability that the claim is fraudulent, given that you know the policy is a fire policy, is

$$P(F \mid \text{fire policy}) = \frac{\text{proportion of claims that are fraudulent fire policy claims}}{\text{proportion of claims that are against fire policies}}$$
$$= \frac{.06}{.20} = .30$$

conditional probability

This probability, $P(F|\text{fire policy})$, is called a conditional probability of the event F, that is, the probability of event F given the fact that the event "fire policy" has already occurred. This tells you that 30% of all fire policy claims are fraudulent. The vertical bar in the expression $P(F|\text{fire policy})$ represents the phrase "given that," or simply "given." Thus the expression is read, "the probability of the event F given the event fire policy."

unconditional probability

The probability $P(F) = .10$, called the unconditional probability of the event F, gives the proportion of times a claim is fraudulent, that is, the proportion of times event F occurs in a very large (infinitely large) number of repetitions of the experiment (receiving an insurance claim and determining whether the claim is fraudulent). In contrast, the conditional probability of F, given that the claim is for a fire policy, $P(F|\text{fire policy})$, gives the proportion of fire policy claims that are fraudulent. Clearly, the conditional probabilities of F, given the types of policies, will be of much greater assistance in measuring the risk of fraud than the unconditional probability of F.

Suppose that the probability of event A is the same regardless of whether event B has or has not occurred. That is, suppose

$$P(A|B) = P(A)$$

Then we say that the occurrence of event A is not dependent on the occurrence of event B or, simply, that A and B are independent events.

DEFINITION 4.5

independent events

Two events A and B are independent if

$$P(A|B) = P(A) \quad \text{or if} \quad P(B|A) = P(B)$$

Note: You can show that if $P(A|B) = P(A)$, then $P(B|A) = P(B)$, and vice versa.

Probability and Probability Distributions

The concept of independence is of particular importance in sampling. Subsequently we will draw samples from two (or more) populations in order to compare the population means, variances, or some other population parameters. For most of these applications we will select samples in such a way that the observed values in one sample are independent of the values that appear in another sample. We call these independent samples.

independent samples

4.5 RANDOM VARIABLES

Most experiments in business and research yield outcomes that are numerical measurements on one or more business variables. Since these measurements vary in a seemingly random fashion from one observation to another, the variables themselves are called random variables.

All random variables are one of two types, discrete or continuous. The number y of machine breakdowns in a given week at a manufacturing plant is a discrete random variable, which is a variable that can assume a countable number of values, 0, 1, 2, 3, 4, . . . , ∞. The number y of people in a sample of $n = 200$ that favor a particular piece of labor legislation is a discrete random variable that can assume values 0, 1, 2, 3, . . . , 199, 200.

discrete random variable

Continuous random variables are those that can assume values that correspond to the infinitely large number of points contained in one or more intervals on a line. The height, weight, and length of life of a human are continuous random variables, as are the water level in a reservoir or the percentage of chlorine in a water supply. For example, the water level in a reservoir could be 39.14 feet, 39.1428 feet, 39.14281 feet, and so on, and the number of such possibilities is so large that it is uncountable.

continuous random variables

When we select a sample of n measurements on a random variable y, we describe the sample by using a relative frequency histogram (chapter 2). A relative frequency histogram—one showing the relative frequencies of occurrence of the values of a random variable when a population is sampled over and over again, ad infinitum—gives the probability distribution for the random variable. As noted in chapter 2, the areas over intervals on a relative frequency histogram are proportional to the probabilities of observing values of y in those intervals. Consequently, the relative frequency histogram for a population, adjusted so that the total area under the figure is equal to 1, enables us to find the probability that y will assume values in a particular interval. For example, if y possesses the probability distribution shown in figure 4.1, the total area under the probability distribution is equal to 1. If you randomly sample a single value of y from the population and observe its value, the probability that y will assume a value between 3 and 5 is the shaded area over the interval $3 < y < 5$.

probability distribution

As you will subsequently see, there is a difference between the construction and interpretation of the probability distributions for discrete and continuous

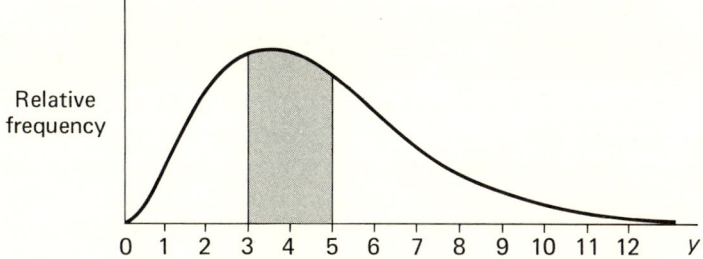

Figure 4.1 The Probability Distribution for a Random Variable y

random variables. These differences will become apparent as we discuss two important probability distributions, one discrete and one continuous, in sections 4.6 and 4.7.

4.6 THE BINOMIAL PROBABILITY DISTRIBUTION

Many populations of interest to business persons and scientists can be viewed as large sets of 0s and 1s. For example, consider the set of responses of all adults in the United States to the question, "Do you favor the development of nuclear energy?" If we disallow "no opinion," the responses will constitute a set of "yes" responses and "no" responses. If we assign a 1 to each "yes" and a 0 to each "no," then the population will consist of a set of 0s and 1s, and the sum of the 1s will equal the total number of persons favoring the development. The sum of the 1s divided by the number of adults in the United States will equal the proportion of people who favor the development.

Gallup and Harris polls are examples of the sampling of 0, 1 populations. People are surveyed and their opinions are recorded. Based on the sample responses, Gallup and Harris estimate the proportions of people in the population who favor some particular issue or possess some particular characteristic.

Similar surveys are conducted in the biological sciences, engineering, and business, but they may be called experiments rather than polls. For example, experiments are conducted to determine the effect of new drugs on small animals, such as rats or mice, before progressing to larger animals and, eventually, to human subjects. Many of these experiments bear a marked resemblance to a poll in that the experimenter records only whether or not the drug was effective. Thus if 300 rats are injected with a drug and 230 show a favorable response, the experimenter has conducted a "poll"—a poll of rat reaction to the drug, 230 "in favor" and 70 "opposed."

Similar "polls" are conducted by most manufacturers to determine the fraction of a product that is of good quality. Samples of industrial products are collected before shipment and each item in the sample is judged "defective" or "acceptable" according to criteria established by the company's quality control department. Based on the number of defectives in the sample, the company can decide whether the product is suitable for shipment. Note that this example, as well as those preceding, has the practical objective of making an inference about a population based on information contained in a sample.

The public opinion poll, the consumer preference poll, the drug-testing experiment, and the industrial sampling for defectives are all examples of a common, frequently conducted sampling situation known as a binomial experiment. The binomial experiment is conducted in all areas of science and business and only differs from one situation to another in the nature of the objects being sampled (people, rats, electric light bulbs, oranges). Thus it is useful to define its characteristics. We can then apply our knowledge of this one kind of experiment to a variety of sampling experiments.

DEFINITION 4.6

binomial experiment

A binomial experiment is one that possesses the following properties:

1. The experiment consists of n identical trials.
2. Each trial results in one of two outcomes. We will label one outcome a success, S, and the other a failure, F.
3. The probability of success on a single trial is equal to p, and p remains constant from trial to trial.
4. The trials are independent—that is, the outcome of one trial does not influence the outcome of any other trial.
5. The experimenter is interested in y, the number of successes observed during the n trials.

Now let us check some experiments to see whether or not they are binomial experiments.

Example 4.1

A survey is conducted to determine the proportion of men in a certain locale who favor the Equal Rights Amendment. A random sample of 300 men is selected from the list of registered voters. They are interviewed and the number of those favoring the ERA is recorded. Is this a binomial experiment?

The Binomial Probability Distribution

Solution

To answer this question, we will check each of the five characteristics of a binomial experiment to determine if they are satisfied.

1. Are there n identical trials? Yes ($n = 300$ interviews are conducted in an identical manner).
2. Does each trial result in one of two outcomes? Yes, each man interviewed either favors or does not favor the ERA.
3. Is the probability of success the same from trial to trial? Yes; if we let success denote a person favoring the ERA, then, assuming the list of registered male voters is large, the probability of selecting a male favoring the ERA will remain (for all practical purposes) constant from trial to trial.
4. Are the trials independent? Yes, the outcome of one interview is unaffected by the results of the other interviews.
5. For this experiment the random variable of interest to the experimenter is y, the number of successes in the sample.

Since all five characteristics are present, the survey represents a binomial experiment.

Example 4.2

A biologist randomly selects 10 portions of water, each equal to .1 cubic centimeter in volume, from the local reservoir and counts the number of bacteria present in each portion. He will then total the number of bacteria for the 10 portions to obtain an estimate of the number of bacteria per cubic centimeter present in the reservoir water. Is this a binomial experiment?

Solution

Check this experiment against the characteristics of a binomial experiment. This experiment consists of $n = 10$ trials, each resulting in the examination of a .1-cubic-centimeter portion of water. However, the second characteristic, that each trial results in only one of two outcomes, is not satisfied. In this experiment we observe the number of bacteria per portion, and this number can be any of the values 0, 1, 2, 3, ... Thus this experiment is not a binomial experiment.

The **binomial random variable** y is a discrete random variable that can assume the values $y = 0, 1, 2, \ldots, n$. The binomial probability distribution, which gives the probabilities associated with each value of y, can be computed by using the following formula.

The Binomial Probability Distribution

$$P(y) = \frac{n!}{y!\,(n-y)!}\, p^y q^{n-y} \quad \text{for} \quad y = 0, 1, 2, \ldots, n$$

where n is the number of trials; p is the probability of a "success" on a single trial; q is the probability of a "failure" on a single trial; and

$$q = 1 - p \qquad q + p = 1$$

The probability of observing y successes in n trials can be calculated by substituting the values for y, n, p, and q into the formula for $P(y)$. The symbol $n!$, read "n factorial," means $n(n-1)(n-2) \cdots (3)(2)(1)$. For example, $5! = (5)(4)(3)(2)(1) = 120$. The quantity $0!$ is defined to be 1.

To see how the formula for the binomial probability distribution can be used to calculate the probability for a specific value of y, consider the following examples.

Example 4.3

Suppose that a sample of households is randomly selected from all the households in the city in order to estimate the percentage in which the head of the household is unemployed. To illustrate the computation of a binomial probability, suppose that the unknown percentage is actually 10% and that a sample of $n = 5$ (we are selecting a small sample to make the calculation manageable) is selected from the population. What is the probability that all 5 of the heads of the households are employed?

Solution

We must carefully define which outcome we wish to call a success. For this example we will define a success as being employed. Then the probability of success when one person is selected from the population is $p = .9$ (because the proportion of unemployed is .1). We wish to find the probability that $y = 5$ (all 5 are employed) in 5 trials. We use the formula

$$P(y) = \frac{n!}{y!(n-y)!} p^y q^{n-y}$$

Thus

$$P(5) = \frac{5!}{5!(5-5)!} (.9)^5 (.1)^0$$

The Binomial Probability Distribution

$$= \frac{5!}{5!0!}(.9)^5(1)$$
$$= (.9)^5 = .590$$

The binomial probability distribution for $n = 5$, $p = .9$ is shown in figure 4.2. The probability of observing 5 employed in a sample of 5 is shaded in the figure.

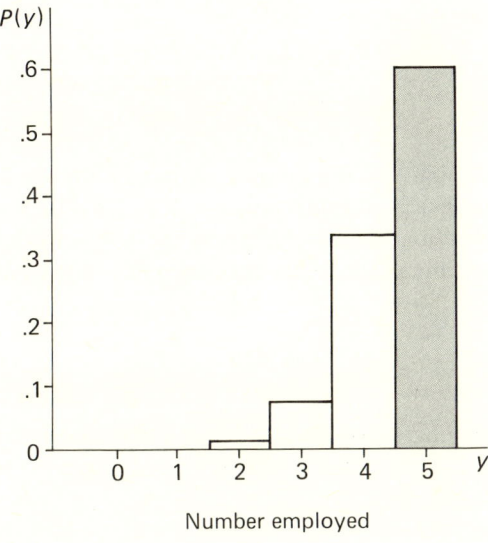

Figure 4.2 The Binomial Probability Distribution for $n = 5$, $p = .9$

Example 4.4

Refer to example 4.3 and calculate the probability that exactly one person in the sample of 5 households is unemployed. What is the probability of one or less being unemployed?

Solution

Since y is the number of employed in the sample of 5, one unemployed person would correspond to 4 employed ($y = 4$). Then

$$P(4) = \frac{5!}{4!(5-4)!}(.9)^4(.1)^1$$
$$= \frac{(5)(4)(3)(2)(1)}{(4)(3)(2)(1)(1)}(.9)^4(.1)$$
$$= 5(.9)^4(.1)$$
$$= .328$$

Thus the probability of selecting 4 employed heads of households in a random sample of 5 is .328, or, roughly, one chance in three.

The outcome "one or less unemployed" is the same as the outcome "4 or 5 employed." Since y represents the number employed, we seek the probability that $y = 4$ or 5. Because the values associated with a random variable represent mutually exclusive events, the probabilities for discrete random variables are additive (this property of probabilities is discussed in section 4.3). Thus we have

$$P(y = 4 \text{ or } 5) = P(4) + P(5)$$
$$= .328 + .590$$
$$= .918$$

That is, the probability that a random sample of 5 households will yield either 4 or 5 employed heads of households is .918. This high probability is consistent with our intuition: we would expect the number of employed in the sample to be large if 90% of all heads of households in the city are employed.

Example 4.5

Suppose that the chance of winning a particular bingo game is 1 in 1000. If the game is played 3 times, what is the probability that the same person wins all 3 times?

Solution

Assuming the games are independent (which seems a realistic assumption), the sampling (with $n = 3$) is a binomial experiment. Defining a success as a "win" in a single game, we wish to find the probability that $y = 3$, where $p = \frac{1}{1000} = .001$. Then we have

$$P(y) = \frac{n!}{y!(n-y)!} p^y q^{n-y}$$
$$= \frac{3!}{3!(3-3)!} (.001)^3 (.999)^0$$
$$= (.001)^3 = .000000001$$

You can see that the chance of one person winning 3 out of 3 games is a bit remote.

Like any relative frequency histogram, a binomial probability distribution possesses a mean μ and a standard deviation σ. Although we omit the derivations, we give the formulas for these parameters.

Mean and Standard Deviation of the Binomial Probability Distribution

$$\mu = np \quad \text{and} \quad \sigma = \sqrt{npq}$$

where p is the probability of success in a given trial, $q = 1 - p$, and n is the number of trials in the binomial experiment.

Knowing p and the sample size n, we can calculate μ and σ to locate the center and describe the variability for a particular binomial probability distribution. Thus we can quickly determine those values of y that are probable and those that are improbable.

Example 4.6

Calculate the mean and standard deviation for a binomial probability distribution with $p = .5$ and $n = 20$. The probability distribution for the number of successes y is shown in figure 4.3.

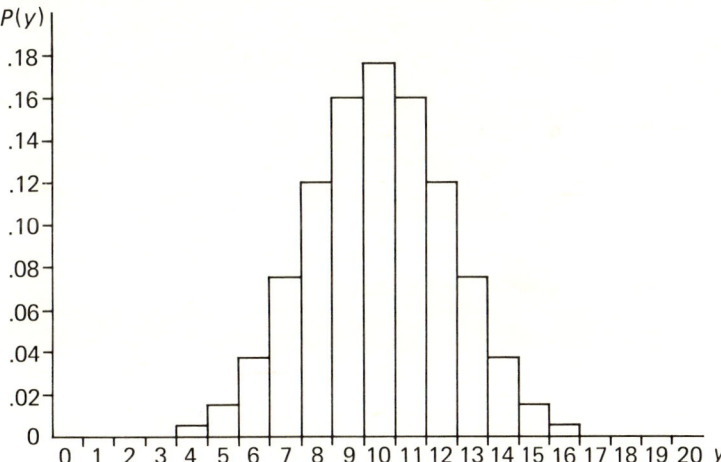

Figure 4.3 Binomial Probability Distribution for y When $n = 20$ and $p = .5$

Solution

Substituting into the formulas, we obtain

$$\mu = np = 20(.5) = 10 \quad (p = .5, q = .5)$$
$$\sigma = \sqrt{npq} = \sqrt{(20)(.5)(.5)} = \sqrt{5} = 2.24$$

Probability and Probability Distributions

Note that $y = 0$ is more than 4σ away from the mean $\mu = 10$. Hence it is highly improbable that in 20 trials we would observe such a small value of y if p really is equal to .5.

Example 4.7

A poll shows that 516 of 1218 voters favor the reelection of a particular political candidate. Do you think that the candidate will win?

Solution

To win the election, the candidate will need at least 50% of the votes. Let us see whether $y = 516$ is too small a value of y to imply a value of p (the proportion of voters favoring the candidate) equal to .5 or larger. If $p = .5$ (and hence $q = .5$),

$$\mu = np = (1218)(.5) = 609$$
$$\sigma = \sqrt{npq} = \sqrt{(1218)(.5)(.5)}$$
$$= \sqrt{304.5} = 17.45$$

and $3\sigma = 52.35$.

You can see from figure 4.4 that $y = 516$ is more than 3σ, or 52.35, away from $\mu = 609$. In fact, if you wish to check, you will see that $y = 516$ is more than 5σ away from $\mu = 609$, the value of μ if p were really equal to .5. Thus it appears that the number of voters in the sample who favor the candidate is much too small if the candidate does, in fact, possess a majority favoring reelection. Consequently, we conclude that he or she will lose. (Note that this conclusion is based on the assumption that the set of voters from which the sample was drawn is the same as the set who will vote. We also must assume that the opinions of the voters will not change between the time of sampling and the date of the election.)

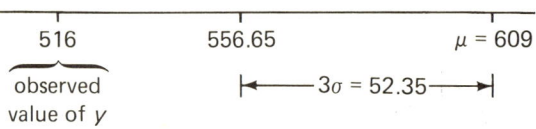

Figure 4.4 Location of the Observed Value of y ($y = 516$) Relative to μ

The purpose of this section is to present the binomial probability distribution so that you can see how binomial probabilities are calculated and so that you can calculate them for small values of n, if you so desire. In practice, n is usually large (in national surveys, sample sizes as large as 1500 are common) and the computation of the binomial probabilities is very tedious.

The Binomial Probability Distribution

Fortunately, these computations can be avoided. Binomial probabilities have been calculated on a computer for various values of n and are readily available in tabulated form (see the References). For our purposes exact values of the probabilities are unnecessary. We will present a simple procedure in chapter 5 for obtaining approximate values to the probabilities we need in making inferences. Or we can use some very rough procedures for evaluating probabilities by using the mean and standard deviation of the binomial random variable y along with the Empirical Rule.

EXERCISES

 4.1. Examine the accompanying newspaper clipping. Does this sampling appear to satisfy the characteristics of a binomial experiment?

Poll Finds Opposition to Phone Taps

New York—People surveyed in a recent poll indicated they are 81 to 13 percent against having their phones tapped without a court order.

The people in the survey, by 68 to 27 percent, were opposed to letting the government use a wiretap on citizens suspected of crimes, except with a court order.

The survey was conducted for 1495 households and also found the following results:

—The people surveyed are 80 to 12 percent against the use of any kind of electronic spying device without a court order.

—Citizens are 77 to 14 percent against allowing the government to open their mail without court orders.

—They oppose, by 80 to 12 percent, letting the telephone company disclose records of long distance phone calls, except by court order.

For each of the questions, a few of those in the survey had no responses.

 4.2. A survey is conducted to estimate the percentage of pine trees in a forest that are infected by the pine shoot moth. A grid is placed over a map of the forest, dividing the area into 25-feet-by-25-feet square sections. One hundred of the squares are randomly selected and the number of infected trees is recorded for each square. Is this a binomial experiment?

 4.3. A survey was conducted to investigate the attitudes of nurses working in Veterans Administration hospitals. A random sample of 1000 nurses was contacted using a mailed questionnaire and the number favoring or opposing a particular issue was recorded. If we confine our attention to the nurses' responses to a single question, would this sampling represent a binomial experiment? As with most mail surveys, some of the nurses will not respond. What effect might nonresponses in the sample have on the estimate of the percentage of all Veterans Administration nurses who favor the particular proposition?

4.4. An experiment is conducted to test the effect of an anticoagulant drug on rats. A random sample of 4 rats is employed in the experiment. If the drug manufacturer claims that 80% of the rats will be favorably affected by the drug, what is the probability that none of the 4 experimental rats are favorably affected? One of the 4? One or less?

 4.5. A criminologist claims that the probability of "reform" for a first-offender embezzler is .9. Suppose that we define "reform" as meaning the person commits no criminal offenses within a five-year period. Three paroled embezzlers were randomly selected from the prison records and their behavioral histories were examined for the five-year period following prison release. What is the probability that all three were reformed? At least two?

4.6. Consider the following experiment: Toss three coins and observe the number of heads y. Repeat the experiment 100 times and construct a relative frequency table for y. Note that these frequencies give approximations to the exact probabilities that $y = 0, 1, 2,$ and 3. (Note: These probabilities can be shown to be $1/8$, $3/8$, $3/8$, and $1/8$, respectively.)

4.7. Refer to exercise 4. Use the formula for the binomial probability distribution to show that $P(0) = 1/8$, $P(1) = 3/8$, $P(2) = 3/8$, and $P(3) = 1/8$.

 4.8. The probability of surviving a rare disease of the nervous system is one in twenty. A new drug has been developed to combat the disease and is tried on three patients. One out of the three survives. To some experimenters, this survival rate (33%) may appear to be a tremendous improvement over the old established rate of 5%. But remember, the 33% survival rate occurred only in a sample of three, not in a large number of cases.

If the survival rate, using the new drug, is still only 5% (that is, the drug is worthless), what is the probability of observing one or more survivals in a sample of three? Examining this probability, what would you be inclined to say (as a matter of intuition) about the effectiveness of the new drug? Explain your reasoning.

4.9. Suppose you match coins with another person a total of 1000 times. What is the mean number of matches? The standard deviation? Calculate the interval ($\mu \pm 3\sigma$). (Hint: The probability of a match in the toss of a single pair of coins is $p = .5$).

 4.10. An immunologist claims that a flu shot is 80% effective against the flu. By this he means that of those having the shot, 80% will be immune to the flu and for 20% the shot will be ineffective. If 8000 people receive the flu shot and all are exposed to the disease, what is the expected number y who would not get the flu? What is the standard deviation of y? Suppose that 6200 survived the winter without contracting the flu. Is this value of y ($y = 6200$) improbable, assuming that the flu shot is 80% effective in immunizing people against the flu?

4.7. THE NORMAL PROBABILITY DISTRIBUTION

normal curve

Many textbooks for introductory statistics note that the distribution of heights for males (or females) can be approximated by a smooth bell-shaped curve that is known as a <u>normal curve</u> or a normal distribution. Searching through the literature we found that R. W. Newman and R. M. White (1951) studied the physical characteristics of army personnel. In figure 4.5 we have presented a histogram from the Newman and White study displaying the relative frequency distribution for the heights (in inches) of a sample of 24,404 United States Army males at the time of their release from the service. We have also superimposed a normal curve over the histogram to show the close approximation obtained from the sample of servicemen.

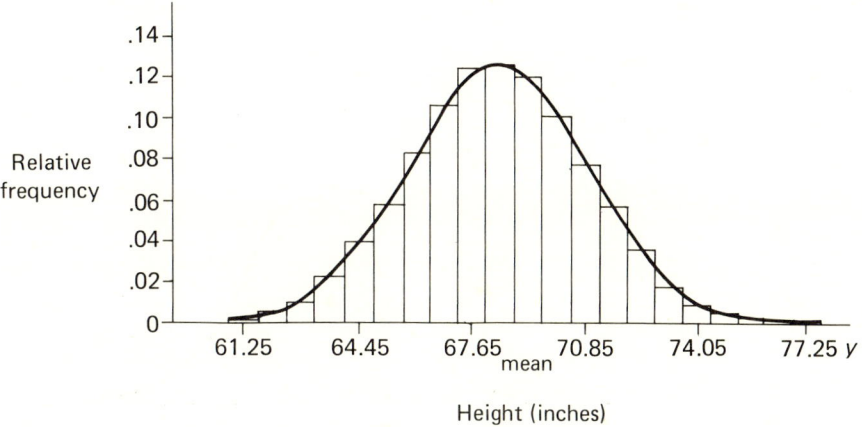

Figure 4.5 Relative Frequency Histogram for the Heights of 24,404 Servicemen, with Normal Curve Superimposed

normal distribution

Many other commonly occurring and important *continuous random variables* possess a normal probability distribution. The <u>normal</u>, or Gaussian, <u>distribution</u> (named for the famous mathematician Karl Friedrich Gauss, 1777–1855) is a continuous bell-shaped curve as shown in figure 4.6. **The total area under the normal curve is equal to 1 and the probability that a normal random variable y assumes a value in a particular interval, say $a < y < b$, is the area under the curve that lies over the interval (see the shaded area in figure 4.6).**

There are infinitely many normal curves, one corresponding to each pair of values that you might assign to μ and σ. But all are symmetrical about the mean and are bell-shaped, and for all such curves the areas within a specified number of standard deviations of the mean are identical. For example, the area within one standard deviation of the mean will always (to two decimal

92 Probability and Probability Distributions

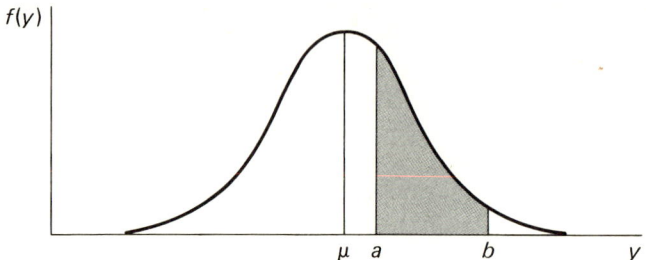

Figure 4.6 A Normal Probability Distribution

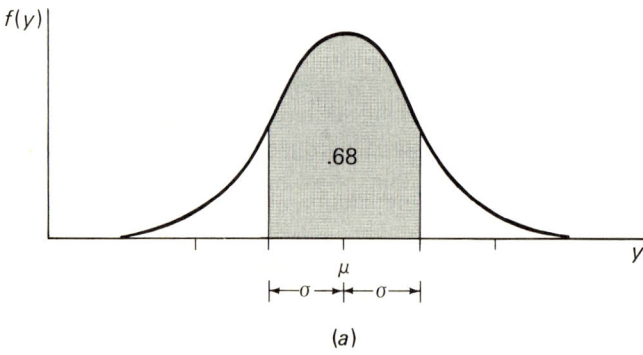

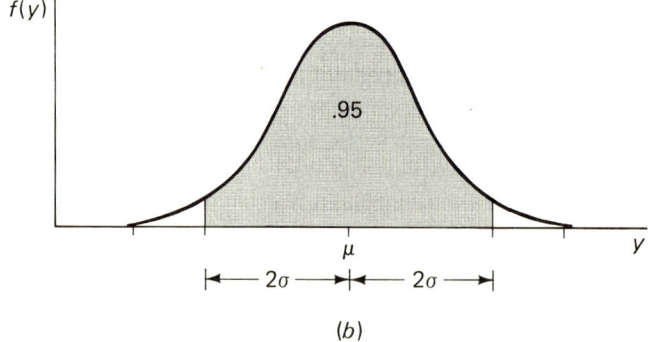

Figure 4.7 Characteristics of a Normal Probability Distribution

The Normal Probability Distribution

places) equal .68. See figure 4.7(a). Likewise, the area within two standard deviations of the mean will always (to two decimal places) equal .95. See figure 4.7(b).

The probability that a normal random variable y assumes values in an interval, say $a < y < b$, can be obtained by using a **table of areas** under the normal curve. Table 1 in the Appendix gives areas under the normal curve between the mean and any number, say z, of standard deviations ($z\sigma$) to the right of μ. A partial reproduction of table 1 of the Appendix is shown in table 4.2. The tabulated area is shown in figure 4.8.

table of areas

Table 4.2 Format of the Table of Normal Curve Areas, Table 1 in the Appendix

z	.00	.01	.02	.03	.04	.05	.06	.07	.08	.09
0.0	.0000	.0040	.0080	.0120	.0160	.0199	.0239	.0279	.0319	.0359
0.1	.0398	.0438	.0478	.0517	.0557	.0596	.0636	.0675	.0714	.0753
0.2	.0793	.0832	.0871	.0910	.0948	.0987	.1026	.1064	.1103	.1141
0.3	.1179	.1217	.1255	.1293	.1331	.1368	.1406	.1443	.1480	.1517
0.4	.1554	.1591	.1628	.1664	.1700	.1736	.1772	.1808	.1844	.1879
.	.	.	.	.	.	.	.	.	.	.
1.0	.3413	.3438	.3461	.3485	.3508	.3531	.3554	.3577	.3599	.3621
1.1	.3643	.3665	.3686	.3708	.3729	.3749	.3770	.3790	.3810	.3830
1.2	.3849	.3869	.3888	.3907	.3925	.3944	.3962	.3980	.3997	.4015
.	.	.	.	.	.	.	.	.	.	.
1.6	.4452	.4463	.4474	.4484	.4495	.4505	.4515	.4525	.4535	.4545
.	.	.	.	.	.	.	.	.	.	.
2.0	.4772	.4778	.4783	.4788	.4793	.4798	.4803	.4808	.4812	.4817

In the table, areas to the left of the mean need not be tabulated because the normal curve is symmetric about the mean. Thus the area between the mean and a point 2σ to the right of the mean is the same as the area between the mean and a similar point 2σ to the left.

The number z of standard deviations is given to the nearest tenth in the left-hand column of table 1. Adjustments to take z to the nearest hundredth are given in the top row of the table. Entries in the table are the areas corresponding to particular values of z. For example, the area between the mean and a point $z = 2$ standard deviations to the right of the mean is shown in the second

94 Probability and Probability Distributions

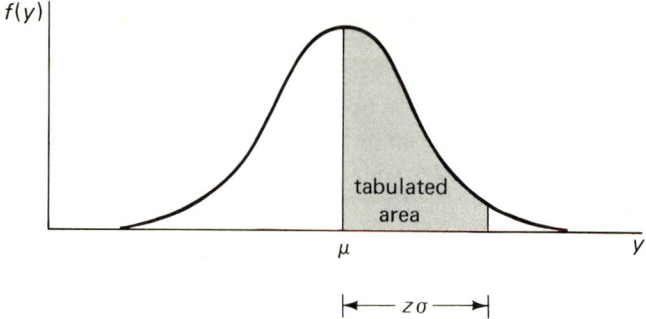

Figure 4.8 Tabulated Area Under the Normal Curve

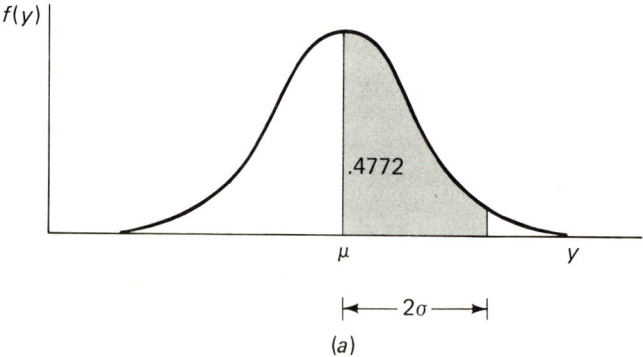

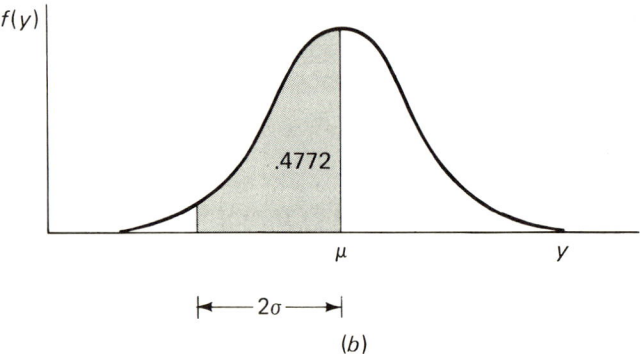

Figure 4.9 Tabulated Area Corresponding to $z = 2$

column of the table opposite $z = 2.0$. This area, shaded in figure 4.9(a), is .4772. Likewise, the area between the mean and a point two standard deviations to the left of the mean, shown in figure 4.9(b), is also .4772. Then the area within two standard deviations of the mean is $2(.4772) = .9544$. This explains the origin of the "approximately 95%" in the Empirical Rule.

Similarly, the area between the mean and a point one standard deviation to the right of the mean (that is, $z = 1$) is .3413. The area within one standard deviation of the mean is .6826, or approximately 68%, as stated in the Empirical Rule. This area is shown in figure 4.10.

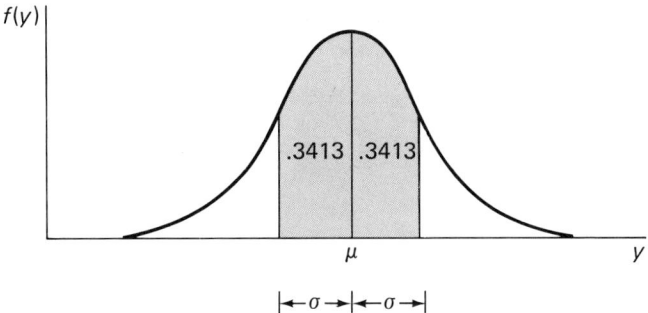

Figure 4.10 Area Within One Standard Deviation of the Mean

Suppose we wish to find the area corresponding to $z = 1.64$. Proceed down the left column to the row $z = 1.6$ and across the top of the page to the .04 column. The intersection of the $z = 1.6$ row with the .04 column gives the desired area, .4495. This area is shown in figure 4.11.

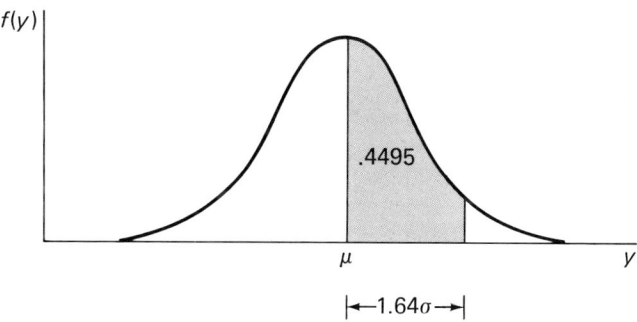

Figure 4.11 Area Corresponding to $z = 1.64$

Probability and Probability Distributions

To determine how many standard deviations a measurement y is from the mean μ, we first determine the distance between y and μ. Recall that this distance can be represented as

$$\text{distance} = y - \mu$$

The distance between y and μ can then be converted into a number of standard deviations by dividing by σ, the standard deviation of y. This standardized distance is often called a **z score**.

$$z = \frac{\text{distance}}{\text{standard deviation}} = \frac{y - \mu}{\sigma}$$

The probability distribution for z, which has $\mu = 0$ and $\sigma = 1$, is called the **standard normal distribution** (see figure 4.12), and the variable z is called a **standard normal random variable**. The area under the curve between $z = 0$ and a specified value of z, say z_0, has been tabulated in table 1 in the Appendix and is shown in figure 4.12. These tabulated areas can be used to find the area under any normal curve if you know the mean μ and the standard deviation σ.

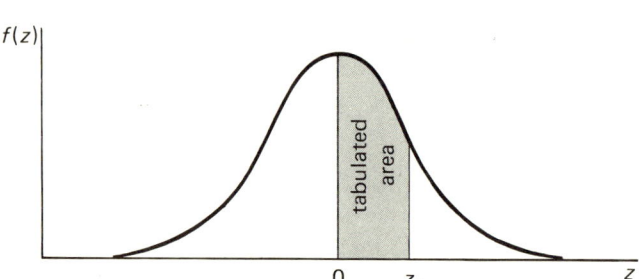

Figure 4.12 Standard Normal Distribution

To calculate the area under a normal curve between the mean μ and a specified value y to the right of the mean, we first determine the number z of standard deviations y is from μ, using

$$z = \frac{y - \mu}{\sigma}$$

We then refer to table 1 in the Appendix and obtain the entry corresponding to the calculated value of z. This entry is the desired area (probability) under the curve between μ and the specified value of y.

We illustrate the use of the table of normal curve areas with a simple example and then proceed to more practical applications.

The Normal Probability Distribution

Example 4.8

Suppose that y is a normally distributed random variable with mean $\mu = 8$ and standard deviation $\sigma = 2$. Find the probability that y is in the interval from 8 to 11. That is, what fraction of the total area under the curve is between 8 and 11 (see the shaded portion of figure 4.13)?

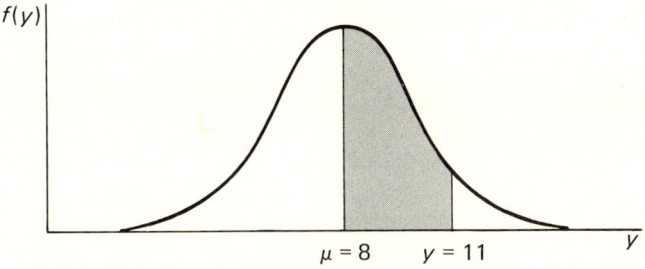

Figure 4.13 Area Under the Curve over the Interval from $\mu = 8$ to $y = 11$

Solution

To determine the desired area, we compute the number of standard deviations that separate $y = 11$ from the mean $\mu = 8$:

$$z = \frac{y - \mu}{\sigma} = \frac{11 - 8}{2} = 1.5$$

The corresponding area can then be determined from the entry in table 1 in the Appendix opposite $z = 1.5$. The desired area is .4332 (see figure 4.14). Therefore, if a single value of y is selected at random from its population, the probability that y lies between 8 and 11 is equal to .4332.

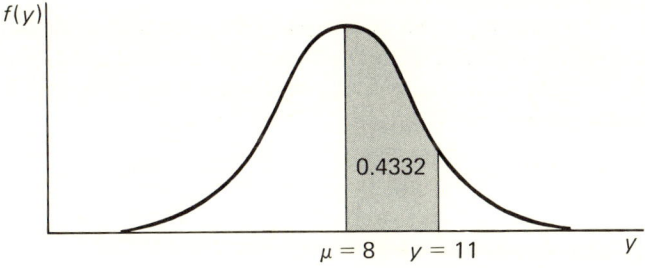

Figure 4.14 Area Between $\mu = 8$ and $y = 11$

Example 4.9

The quantitative portion of a nationally administered achievement test is scaled so that the mean score is 500 and the standard deviation is 100.

(a) If we assume the distribution of scores is normal (bell-shaped), what percentage of the students throughout the country should score between 500 and 682?

(b) What percentage should score between 340 and 682?

Solution

Consider figure 4.15.

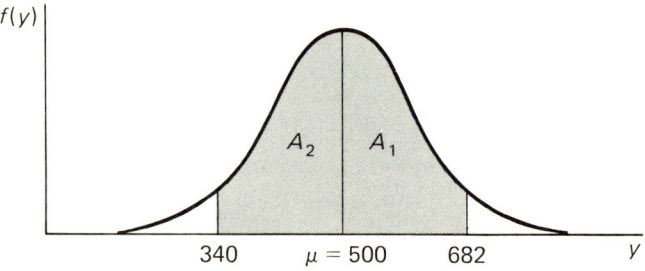

Figure 4.15 Area Between $y = 340$ and $y = 682$

(a) To determine the percentage of the students that should score between 500 and 682, we must compute the area A_1 between $\mu = 500$ and $y = 682$:

$$z = \frac{y - \mu}{\sigma} = \frac{682 - 500}{100} = 1.82$$

The tabulated area for this value of z (table 1, Appendix) is $A_1 = .4656$. Thus we expect 46.56% of the students to score between 500 and 682.

(b) To determine the percentage of students that should score between 340 and 682, note that the area between 340 and 682 is equal to the sum of A_1 and A_2 in figure 4.15. To find A_2, we compute the number of standard deviations that separate $y = 340$ from $\mu = 500$. Hence

$$z = \frac{y - \mu}{\sigma} = \frac{340 - 500}{100} = -1.6$$

(Negative values of z indicate a point to the left of the mean.) The appropriate area, found by ignoring the negative sign, is then .4452. Thus we would expect $A_1 + A_2 = .4656 + .4452 = .9108$ or 91.08% of the students to score between 340 and 682 on the examination.

The Normal Probability Distribution

Example 4.10

Records maintained by the office of the budget in a particular state indicate that the amount of time elapsed between the submission of travel vouchers and the reimbursement of funds has an approximately normal distribution with a mean equal to 45 days and a standard deviation equal to 5 days.

(a) What is the probability that the elapsed time between submission and reimbursement will exceed 58 days for a travel expense report selected at random?

(b) If you had submitted a travel expense report and had still not been reimbursed after 58 days, what would you conclude?

Solution

A sketch of the desired area, A_2, is shown in figure 4.16.

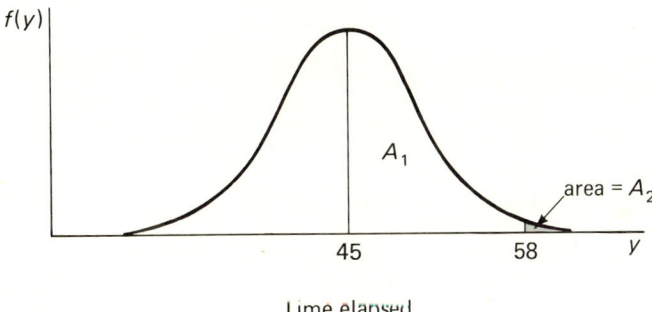

Time elapsed

Figure 4.16 Probability That the Elapsed Time Will Exceed 58 Days

(a) To determine the area A_2, we must first compute a z score:

$$z = \frac{y - \mu}{\sigma} = \frac{58 - 45}{5} = 2.6$$

Thus the value 58 is 2.6 standard deviations above the mean $\mu = 45$. The area under a normal curve from the mean out to a point 2.6 standard deviations above the mean is, from table 1, .4953. This area is indicated by A_1 in figure 4.16. Since the total area under a normal curve to the right of the mean is .5, we can compute A_2 by subtracting $A_1 = .4953$ from .5. Thus

$$A_2 = .5 - .4953 = .0047$$

(b) Since the probability of having to wait more than 58 days is so small, .0047, we would conclude that something has happened to the travel expense

Probability and Probability Distributions

report. Perhaps it was lost in the mail or misplaced by a person in the office of the budget.

In this section we have computed areas under a normal curve. To do this, we converted the distance between a measurement y and the mean μ into a number of standard deviations and then referred to table 1 in the Appendix. The resulting areas are equal to the probabilities that measurements will fall in particular intervals.

EXERCISES

4.11. Use table 1 of the Appendix to find the area under the normal curve between these values:
(a) $z = 0$ and $z = 1.3$
(b) $z = 0$ and $z = -1.9$

4.12. Repeat exercise 4.11 for these values:
(a) $z = 0$ and $z = .7$
(b) $z = 0$ and $z = -1.2$

4.13. Repeat exercise 4.11 for these values:
(a) $z = 0$ and $z = 1.29$
(b) $z = 0$ and $z = -.77$

4.14. Repeat exercise 4.11 for these values:
(a) $z = -.21$ and $z = 1.35$
(b) $z = .37$ and $z = 1.20$

4.15. Repeat exercise 4.11 for these values:
(a) $z = 1.43$ and $z = 2.01$
(b) $z = -1.74$ and $z = -.75$

4.16. Find the probability that z is greater than 1.75.

4.17. Find the probability that z is less than 1.14.

4.18. Find a value for z, say z_0, such that $P(z > z_0) = .5$.

4.19. Find a value for z, say z_0, such that $P(z > z_0) = .025$.

4.20. Find a value for z, say z_0, such that $P(z > z_0) = .0089$.

4.21. Find a value for z, say z_0, such that $P(z > z_0) = .05$.

4.22. Find a value for z, say z_0, such that $P(-z_0 < z < z_0) = .95$.

4.23. Let y be a normal random variable with mean equal to 100 and standard deviation equal to 8. Find these probabilities:
(a) $P(y > 100)$

(b) $P(y > 110)$
(c) $P(y < 115)$
(d) $P(88 < y < 112)$
(e) $P(100 < y < 108)$

 4.24. Suppose that a population of measurements on the potency of a drug is normally distributed with a mean equal to 5% and a standard deviation of .6%.
(a) Find the probability that a measurement of drug potency will be less than 4%.
(b) Find the probability that a measurement of drug potency will be larger than 7%.
(c) To maintain consistency in the preparation of the drug, the manufacturer wants the potency to be between 4.5% and 5.5%. What percentage of the time will the manufacturer meet this standard?

4.25. To some viewers, the lengths of television commercials seem to average 3 minutes, with a standard deviation of 1.2. If this is true, what is the probability that a commercial break will exceed 5 minutes? (Assume that the lengths of breaks are normally distributed and occur near the end of a feature-length evening film.)

4.26. Let y be a normally distributed random variable with mean $\mu = 26$ and standard deviation $\sigma = 4$. If a value of y is chosen at random from the population, find the probability that y falls in the interval $y = 20$ to $y = 32$.

 4.27. The mean and standard deviation for a national achievement test are 410 and 60, respectively. Find the fraction of the total number of students scoring less than 300. (Assume that the test scores are normally distributed.)

SUMMARY

Chapter 4 presents some of the basic concepts of probability and the concept of a random variable. The probability distributions for two particularly important random variables, the binomial and the normal, are presented along with examples showing how they can be applied.

Probability, which measures our belief in the occurrence of a particular outcome of an experiment, is viewed conceptually as the relative frequency of occurrence of the outcome (event) when the experiment is repeated over and over again. Probability is important to statistics because the observation of a sample selected from a population is an experiment. Based on the probability of an observed sample, statisticians (and you) make inferences about the population from which the sample was selected. For example, if we theorize that only 5% of all telephone subscribers are delinquent in paying their telephone bills and then sample 10 bills and find that all are overdue, we would reject our theory and infer that the proportion of overdue bills in the population is larger than 5%. We reach this conclusion not because it is impossible to draw 10 unpaid bills in a sample of 10, assuming our 5% theory is correct, but because it is highly improbable.

Three probabilistic concepts play important roles in sampling and statistical inferences. These are the additive property of the probabilities for mutually exclusive events, conditional probability, and independent events. As noted in this chapter, most experiments result in numerical events obtained by the observation of random variables. These events, the values that a random variable may assume, are mutually exclusive. Therefore, to find the probability that a random variable when observed will assume one of a set of specific values, we sum the probabilities corresponding to these values.

The concept of independence plays an important role in sampling, because the manner in which a sample is selected affects the probability of its occurrence. We will very often speak of independent random sampling when we select a sample from each of two (or more) populations to compare the population means or some other population parameters. By independent random sampling we mean that the probability of observing a particular set of measurements in one sample does not depend on the values observed in the second sample. Particularly, two events (sample outcomes) A and B are independent if the conditional probability that event A will occur, given that event B has occurred, is equal to the unconditional probability that event A will occur.

Random variables observed in real life experiments or surveys can be discrete (usually data representing counts, such as the number of bacteria per cubic centimeter of water) or continuous (such as the length of time required to obtain service at a medical clinic). Two of the most important are the discrete binomial random variable and the continuous normal random variable. In addition to noting the characteristics of these two random variables and the situations to which they might apply, we gave their probability distributions. The probability distribution for the binomial random variable gives the probability associated with each of the values $0, 1, 2, \ldots, n$ that the random variable can assume. To find the probability that a discrete random variable y will assume one of a set of these mutually exclusive numerical events (one event corresponding to each value), we simply sum the probabilities corresponding to the values. Thus, for example,

$$P(3 \leq y \leq 5) = P(3) + P(4) + P(5)$$

The probability distribution for the normal random variable is a smooth bell-shaped curve which is essentially the theoretical relative frequency distribution for a normally distributed population of measurements. For a continuous random variable y, the probability that y will assume a value in the interval $a \leq y \leq b$ is defined to be the area under the probability distribution curve over the interval $a \leq y \leq b$. Probability distributions for continuous random variables possess one peculiar property that may disagree with our intuition but does not affect their utility when employed in statistical inference. Since there is no area lying above a single point, say $y = a$, it follows that $P(y = a) = 0$. However, this situation causes no difficulty. If we are looking for the probability that y will assume a value equal to or near the value $y = a$, we will select a small interval that includes $y = a$ and find the area lying above the interval. In actual practice, we will be interested in probabilities of the form $y \leq a$, $y \geq a$, or $a \leq y \leq b$ rather than in the probability that y assumes some specific value, say $y = a$.

In the following chapters we will employ sample statistics to make inferences about population parameters. The probability distribution of a sample statistic, called its *sampling distribution,* plays a key role in selecting good statistics and in evaluating their reliability. Sampling distributions are the topic of chapter 5.

REFERENCES

Freund, J. E. *Statistics: A First Course.* 2d ed. Englewood Cliffs, N.J.: Prentice-Hall, 1976.

Handbook of Tables for Probability and Statistics. 2d ed. Cleveland, Ohio: The Chemical Rubber Co., 1968.

Huntsberger, D. V., and Billingsley, P. *Elements of Statistical Inference.* 4th ed. Boston: Allyn and Bacon, 1977.

Mendenhall, W. *Introduction to Probability and Statistics.* 5th ed. N. Scituate, Mass.: Duxbury Press, 1979. Chapters 4, 7.

National Bureau of Standards. *Tables of the Binomial Probability Distribution.* Washington, D.C.: Government Printing Office, 1949.

Newman, R. W., and White, R. M. *Reference Anthropometry of Army Men.* Report no. 180, Environmental Climatic Research Laboratory, Lawrence, Mass.

Walpole, R. E. *Introduction to Statistics.* 2d ed. New York: Macmillan, 1974. Chapters 2, 7.

SUPPLEMENTARY EXERCISES

4.28. Class exercise: Have each student perform the following experiment 20 times.

Experiment: Toss 5 coins and observe the number of heads.

The possible values of y, the number of heads in 5 coin tosses, are 0, 1, 2, 3, 4, and 5. Each student should keep track of the number of times each outcome is observed and combine his or her results with the rest of the class to construct a table of the form shown on page 104.

y	Frequency	Relative Frequency (Approximate Probability)	P(y)
0			.031
1			.156
2			.313
3			.313
4			.156
5			.031

You will note that the exact value of the probability, $P(y)$, associated with each value of y is listed in the last column of the table. Your relative frequencies computed from this experiment should be approximately the same as the exact probabilities. As mentioned previously, the degree of accuracy increases as the number of repetitions of the experiment increases.

4.29. Refer to exercise 4.28 and the observed relative frequencies. Approximate the probability of observing $y = 0$ or 1. What is the actual probability of observing 0 or 1 head in a toss of 5 coins?

 4.30. A community has an equal number of black families and white families. A random sample of 5 families is selected from the community; the families will be interviewed concerning the advisability of rezoning a school district. What is the probability that all 5 families chosen are either all white or all black? If all families chosen are either all white or all black, would you suspect a racial bias in the selection of the sample? (Note: Your decision should be made on the basis of probability.)

4.31. Identify the following variables as being either discrete or continuous.
(a) the number of patient arrivals per hour at a medical clinic
(b) the number of accidents at a given intersection for each year
(c) the average amount of electricity (measured in kilowatthour units) consumed per household per month in New York City
(d) the number of deaths per year attributed to lung cancer
(e) the age of freshman U.S. senators when they take the oath of office
(f) the gross national product for the United States per year

4.32. The weather forecaster has been the object of many jokes due to errors in forecasting. Of course, these errors are not completely avoidable. After obtaining information (measurements) on many different variables such as wind direction, wind velocity, and barometric pressure from local sources and satellite communications, the forecaster must interpret these data and supply an inference (weather forecast). How effectively weather forecasters have employed statistics to prepare their forecasts is open to question, but it is clear that probability has become an integral part of a weather forecast. We've all heard weather forecasts that state, "There is a 50 percent chance of rain this morning, decreasing to a 30 percent chance this afternoon and evening." Give your interpretation of this statement.

Area Forecast

Partly cloudy to occasionally cloudy through tomorrow with a chance of thundershowers mainly during the afternoons and evenings. Low today and tomorrow will be near 80. High will be near 90. Winds will be southwest to west 5 to 15 m.p.h. stronger and gusty near showers. Rain probability is 50 percent today and 30 percent tonight.

4.33. Read the accompanying news clipping. Explain why this might or might not be a binomial experiment. What information, missing in the article, is needed to conclude firmly that the survey is a binomial experiment?

Study of Divided Families Shows Positive Attitudes

Chicago—A study of divorced mothers and their children has revealed some positive attitudes among members of divided families. Perhaps a broken home is not the psychological disaster for family members that society has suspected.

The study, involving 20 mothers with one or more children between the ages of 6 and 18, was conducted to determine the basic concerns of divorced mothers and their children. There were 20 mothers and 35 children involved in the study.

All the women were working full time. Most of them had made plans toward bettering their earning power. The women had been divorced from 3 months to 15 years. The educational level of the women in the study was high, compared to the national average: 12 years to 18 years of education.

A key aim of the study was to determine the feelings of the women and their children about their acceptance in society.

Eighty-six percent of the children felt that at school they were treated the same as children whose parents were married. Children aged 10 through 12 especially preferred that teachers and friends be told about the home situation. They wanted news of the divorce not to come as a surprise to others or to be a source of embarrassment for them.

In general, the children were doing well in school and even excelled in some areas.

Although the trend among most of the women was to socialize mainly with single persons, eighty percent of them felt accepted in their neighborhoods. Half of them said they felt accepted at church.

Among the children, ninety-one percent indicated they were treated no differently at Sunday school. Ninety percent of the sample were active church members.

Most of the women, eighty-five percent, said that after their divorces their attitudes toward divorce had shifted from negative to positive. The same proportion saw advantages for their children, in terms of understanding life and people, as a result of the divorce.

4.34. Suppose you are the personnel manager for a manufacturing concern and are responsible for safety procedures in your plant. Records are maintained on the number of accidents on a daily basis and these are totaled by the month. Explain why these data are or are not measurements on a binomial random variable.

 4.35. A recent survey suggests that Americans anticipate a reduction in living standards and that a steadily increasing consumption no longer may be as important as it was in the past. Suppose that a poll of 2000 people indicated 1373 in favor of forcing a reduction in the size of American automobiles by legislative means. Would you expect to observe as many as 1373 in favor of this proposition if, in fact, the general public was split 50–50 on the issue? Why?

 4.36. An experiment was conducted to test for the presence or absence of a fungus on tobacco plants. Four hundred plants were observed in a field and 242 were observed to have been infected by the fungus.
(a) Does this appear to be a binomial experiment? Explain why it might or might not satisfy the characteristics of a binomial experiment.
(b) Suppose the characteristics of a binomial experiment are satisfied. What interpretation can you give to p?
(c) Previous experience suggests the fungus affects 50% of a planting of tobacco seedlings. What is the mean value of y? The standard deviation of y? if p really equals .5, is it probable that the observed number of infected plants could be as large (or larger) than $y = 242$? Explain.

 4.37. For the survey discussed in the accompanying news article, answer the following questions.
(a) Does this appear to be a binomial experiment?
(b) Explain why it might or might not satisfy the five characteristics of a binomial experiment.

Alcoholism Reported Up in Army

Large numbers of young American soldiers are becoming alcoholics. This parallels the increase of alcohol abuse among young civilians, researchers report.

In a study of 1873 Army men randomly selected from bases of the United States, nearly 2 out of every 5 soldiers were found to be either actual alcoholics, borderline alcoholics, or potential alcoholics.

The study showed that the largest percentage of problem drinkers were under age 20 and had ranks below sergeant.

 4.38. Experience has shown that a lie detector will show a positive reading (indicate a lie) 10% of the time when a person is telling the truth and 95% of the time when a person is lying. Suppose a sample of five suspects is subjected to a lie detector test regarding a recent one-person crime.
(a) What is the probability of observing no positive reading if all suspects plead innocent and are telling the truth?

(b) What is the probability of observing one or more positive readings if all the suspects plead innocent and are telling the truth?
(c) Do the results for parts (a) and (b) disturb you? Why?

4.39. A large stadium utilizes 1500 floodlights to illuminate the field. The supplier of the lights claims that 99% or more of all the manufactured floodlights will last at least 45 hours. We assume the floodlights in the stadium represent a random sample from the manufacturer and that exactly 99% of all the floodlights manufactured will last at least 45 hours.
(a) What is the probability that 1450 or fewer of the 1500 floodlights in the stadium will last at least 45 hours? (Hint: Use the Empirical Rule.)
(b) If the stadium finds that only 1450 of the 1500 floodlights last at least 45 hours, what might you conclude about the supplier's claim? [Hint: Base your answer on the result of part (a).]

4.40. Define z, the standard normal random variable, in terms of y, a normally distributed random variable with mean μ and standard deviation σ.

4.41. Using table 1 in the Appendix, calculate the area under the normal curve between the following:
(a) $z = 0$ and $z = 1.5$
(b) $z = 0$ and $z = 1.8$

4.42. Repeat exercise 4.41 for the following:
(a) $z = 0$ and $z = 2.5$
(b) $z = -1.5$ and $z = 0$

4.43. Repeat exercise 4.41 for the following:
(a) $z = -0.8$ and $z = 0$
(b) $z = -0.8$ and $z = 0.8$

4.44. Repeat exercise 4.41 for the following:
(a) $z = -1.96$ and $z = 1.96$
(b) $z = -2.58$ and $z = 2.58$

4.45. Repeat exercise 4.41 for the following:
(a) $z = -0.12$ and $z = 1.8$
(b) $z = 1.65$ and $z = 2.0$

4.46. Find the value of z such that 30% of the area lies to its right. (Note: This is the 70th percentile of the standard normal distribution.)

4.47. Find the value of z such that 5% of the area lies to its right.

4.48. Find the value of z such that 2.5% of the area lies to its right.

4.49. A normally distributed variable y possesses a mean and standard deviation equal to 7 and 2, respectively. Find the z value corresponding to $y = 6$.

4.50. Refer to exercise 4.49. Find the value of z corresponding to $y = 8.5$.

4.51. Refer to exercise 4.49. Find the probability that y lies in the interval 6 to 8.5.

4.52. The local union hourly wage for a particular type of construction worker possesses a national average of $5.30 and a standard deviation of .63. Find the fraction of local unions for which the compensation is more than $6.00 per hour. (You can assume, with good reason, that the relative frequency distribution of union wages throughout the nation may be approximated by a normal distribution.)

4.53. The dollar sales per salesman for a large company average $60,000 per year, with a standard deviation of $7,000. What fraction of the salesmen might be expected to sell less than $50,000 per year?

4.54. The length of time y to complete a standard achievement test possesses a mean of 58 minutes, with a standard deviation of 9.5 minutes. If the professor wants to time the exam so that it will be completed by 90% of the students, how long an examination period must be scheduled? (Hint: Using table 1 of the Appendix, we can find the z value corresponding to an area of .4. Then from the formula

$$z = \frac{y - \mu}{\sigma}$$

we can solve for the required value of y. See figure 4.17.

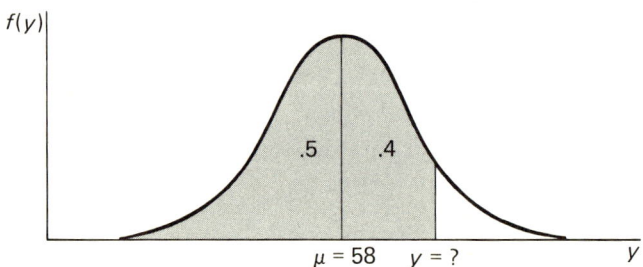

Figure 4.17 Probability Distribution of the Length of Time to Complete Achievement Test (Exercise 4.54)

4.55. Refer to exercise 4.54. Find the 80th percentile for the length of time to complete the test. (Note: The 80th percentile here means that 80% of the students would have completion times less than or equal to this value.) Find the lower quartile.

4.56. Give the mean and standard deviation of a binomial random variable with $n = 30$ and $p = .2$.

EXPERIENCES WITH REAL DATA

Suppose that 50% of all apartment dwellings in a city possess one or more violations of the fire code. If 10 apartments are selected at random, what is the probability that 7 or more will be found to be in violation of the fire code?

Although we can solve this problem using the formula of section 4.6, we can easily obtain an approximate answer (and thereby verify our formula) by a coin-tossing experiment. Tossing a single coin once is analogous to selecting a single apartment in the population described above. A head could correspond to an apartment in violation of the fire code, a tail to one that is in conformance with the code. Flipping the coin 10 times (or tossing 10 coins once) would be equivalent to sampling 10 apartments from the population.

Select 10 coins, mix thoroughly, toss, and record the number y of heads (apartments in violation of the code) in the sample. Record whether y is 7 or larger. Repeat this process $N = 100$ times and count the number of times y is 7 or larger. This will give you the number of times n_A your event of interest was observed. Then calculate an approximate value for the probability of observing 7 or more apartments in violation of the code in a sample of 10, using

$$P(A) = P(7 \text{ or more}) \approx \frac{n_A}{N} = \frac{n_A}{100}$$

This approximation may be poor because $N = 100$ repetitions of the experiment is not large enough to acquire an accurate approximation to $P(A)$.

To obtain a more accurate approximation to $P(A)$, combine the data collected by all the members of your class and use these data to approximate $P(A)$. This value of P should be close to the exact value of $P(A)$, which is .172. [You can compute the exact value by using the binomial formula $P(y)$ for $y = 7, 8, 9,$ and 10.]

To see how the experimentally obtained approximations for P vary, collect the approximations from each member of your class and describe them by using a relative frequency histogram. Notice how they cluster about the exact value $P(A) = .172$. Although we only have one approximation based on the larger value of N for the combined data, you can see that this value will tend to fall closer to .172 than the approximations based on $N = 100$ repetitions of the experiment.

5 Sampling Distributions

CHAPTER OUTLINE

5.1 Introduction

5.2 Random Sampling

5.3 The Central Limit Theorem and the Sampling Distribution for a Sample Mean

5.4 The Sampling Distribution of a Sample Proportion

5.1 INTRODUCTION

As noted in earlier chapters, a *statistic* is a numerical descriptive measure of the sample, whereas a *parameter* is a numerical descriptive measure of the population. We will use statistics to make inferences about population parameters—particularly, to <u>estimate</u> the value of a population parameter or to make a <u>decision</u> about its value. As you will subsequently learn, we will need to know the probability distribution, called a <u>sampling distribution</u>, for the statistic in order to evaluate the reliability of inferences. For example, we will want to know the probability that a statistic will give an estimate that will fall close to (say within some specified distance of) the actual value of the population parameter. Or, if we base a decision about a population parameter on the observed value of a sample statistic, we will want to know the probability that the sample statistic has led us to an incorrect decision.

    This chapter is about sampling and sampling distributions. We will define the simplest and most common method of sampling and will give the sampling distributions for some important statistics computed from this type of sample. As we proceed through this chapter, you will see how probability and probability distributions (chapter 4) play key roles in statistical inference.

estimate
decision
sampling distribution

5.2 RANDOM SAMPLING

The Environmental Protection Agency (EPA) recently announced that Alachua County, along with nine other Florida areas, had been tentatively designated as an Air Quality Maintenance Area. This designation means that local officials in the affected areas must develop a ten-year master plan to improve the air quality.

Upon receiving the EPA announcement, the Alachua County Engineer stated that he planned to fight the EPA designation because the sampling stations used were ill placed and did not provide accurate readings of the air quality. Further investigation showed that at least one of the stations was located near a dusty clay road. This alone could have accounted for the poor air quality of samples at that location. After acquiring additional information to substantiate the fact that the sampling was biased and tended to show a much higher rate of pollution than actually existed, the engineer sent an appeal through proper channels, and the air quality maintenance designation for Alachua County was lifted by the EPA.

This example illustrates the important role that sampling plays in statistical inferences. Particularly, it shows that statistical inferences depend on how the sample is selected.

The most common type of sampling is called random sampling. It is defined as follows:

DEFINITION 5.1

random sample, one measurement

A random sample of one measurement from a population of N measurements is one in which each of the N measurements has an equal probability of being selected.

The notion of random sampling can be extended to the selection of more than one measurement, say n measurements. For purposes of illustration we will take a population that contains a very small number of measurements,

Table 5.1 Six Possible Samples from a Population of Four Measurements

Possible Samples	Measurements in Sample
1	1, 2
2	1, 3
3	1, 4
4	2, 3
5	2, 4
6	3, 4

Random Sampling

say $N = 4$, the 4 measurements being 1, 4, 3, and 2. We will assume further that we wish to select a random sample of $n = 2$ from the 4. How many different and distinct samples could we select? The six possible samples are listed in table 5.1. A random sample of $n = 2$ measurements taken from the population of $N = 4$ measurements is one in which each of the six different samples listed in table 5.1 has an equal chance of being selected.

DEFINITION 5.2

random sample, n measurements

A random sample of n measurements from a population is one in which every different subset of size n from the population has an equal probability of being selected.

table of random numbers

It is unlikely that we would draw a sample that satisfies exactly our definition of a random sample, but we can achieve a very close approximation to random sampling in many situations. When a population is finite, we can use a table of random numbers, such as table 7 in the Appendix, to select (approximately) a random sample. A reproduction of table 7 is shown in table 5.2.

When you look at table 7, concentrate on the digits 0, 1, 2, 3, $\cdots$, 9 rather than on the groups of digits that form larger numbers. Proceed across a row of the table, down a column, or in any other path that you desire and accumulate a large number of digits. If you were to calculate the proportions of 0s, 1s, 2s, $\cdots$, 9s in this set, you would find that the proportions are near .1. In other words, if you let your finger fall on any digit in the table, the probability that it will hit a particular digit, say a 7, is (with a high degree of accuracy) near .1. For any *pair* of digits selected from table 7—the numbers 0, 1, 2, $\cdots$, 99—the probability of each is approximately equal to .01. Similarly, for any group of three digits—the numbers 0, 1, 2, $\cdots$, 999—the probability of each is approximately equal to .001.

We now illustrate how to use table 7 to select a random sample of n from a population containing N measurements.

Example 5.1

Select a random sample of $n = 5$ measurements from a population that contains $N = 2000$ measurements.

Solution

Number the 2000 population measurements in sequence, starting with 0 and ending with 1999. To identify which $n = 5$ of these $N = 2000$ measurements

Table 5.2 Format of the Random Number Table, Table 7 in the Appendix

Line	1	2	3	4	Column 5	6	7	8	9	10
1	75029	50152	25648	02523	84300	83093	39852	91276	88988	12439
2	73741	30492	19280	41255	74008	72750	70420	67769	72837	27098
3	07049	98408	27011	76385	15212	03806	85928	81312	14514	55277
4	01033	08705	42934	79257	89138	21506	26797	67223	62165	67981
5	48399	78564	35787	07647	23794	73938	29477	11420	03228	16586
6	70459	73480	06740	79124	14078	72352	07410	93292	93057	18715
7	74770	80185	08181	27417	90866	98444	72870	51219	51481	47916
8	24167	13753	65011	66288	12633	79199	61497	56186	83643	96184
9	24316	80240	62592	53393	57028	61626	56508	84407	97873	27571
10	84565	59254	94435	33322	50014	00180	50954	04099	66005	59141
11	60794	32497	47830	94509	36576	68874	84062	84503	50454	42199
12	99104	14833	97062	48867	19645	78069	91602	46991	57523	22219
13	15604	93654	21487	86036	22827	62637	70378	58539	17827	80108
14	20204	00253	19678	15789	17628	63667	23348	67083	92361	50413
15	71233	73676	00958	42662	47344	00104	74530	46238	06655	23791
16	82846	82954	52107	66054	27358	69664	71760	03577	75622	21536
17	48613	97858	49627	17036	55574	80116	80533	62146	48083	29177
18	42313	91287	66900	79817	76803	42462	63542	99089	22655	44130
19	60879	68102	60700	51281	61386	06782	88214	68246	15552	79093
20	34593	95713	62942	16236	30933	39470	58423	95304	46017	18364
21	96033	10917	01205	08978	43021	77321	76736	64527	96534	98457
22	21932	45476	75464	43497	81807	99369	59945	65349	52588	27386
23	91019	99635	78638	75114	42943	81629	03283	85036	80666	18675
24	86053	48238	14952	55565	98821	92843	67663	70387	13356	46650
25	59700	38346	92770	11506	34101	01051	99390	86884	26788	78768

should appear in the random sample, we randomly select a starting point in table 7 and select $n = 5$ four-digit numbers (we only need four-digit numbers because the largest number used to identify the population measurements is 1999). For example, suppose that we decide to start with the first four digits of the random number appearing in the first row and third column of table 5.2 and to proceed down the column, *discarding numbers larger than 1999*. We obtain 1928, 0674, 0818, 1967, and 0095. This tells us that the five measurements that are numbered 1928, 674, 818, 1967, and 95 should appear in our random sample.

It is relatively easy to select a random sample from a population when all the population measurements are available and can be numbered for purposes of identification. But when the sample measurements are generated by experimentation, such as measuring the acidity (pH) of a solution in a chemistry laboratory, the population (pH readings for the solution) is conceptual. That is, for this situation the population consists of the very large number of pH measurements that we could make on the solution if we were able to repeat the process ad infinitum. A sample of $n = 5$ pH readings would be obtained by thoroughly mixing the solution, drawing five specimens from the container and measuring the pH of each. But considering the infinitely large number of different solution specimens that could be selected from the container (with one pH reading corresponding to each specimen), there is no guarantee that our method of drawing the specimens will give every different sample of five pH readings an equal probability of selection. But with thorough mixing of the solution, we would expect the sampling procedure to approximate, very closely, random sampling.

Other types of laboratory measurements, selected from conceptual populations—populations that we can visualize but are not within our grasp—yield samples that are not so easily visualized as random samples. For example, if we measure the blood pressure of five experimental rats treated with a new drug, we obtain five measurements from the conceptual population of measurements for all rats that might be treated with the drug. Whether the sample of $n = 5$ measurements is a random sample from the conceptual population is a matter for conjecture.

So what do we do in practice when we wish to draw a random sample? We do the best that we can. If the population is finite and is available, we use a table of random numbers to select the sample. When the data arise as the result of experimentation, we take every possible precaution to avoid biasing the sample data. Particularly, we try to select the data so that the observation of any one measurement will not influence the observation of any others. We hope that independent selection of the sample measurements will give every subset of n measurements an equal probability of selection. Most important, if we have doubts that the sampling procedure satisfies the requirements of random sampling, we make note of this fact when we report particular inferences derived from the sample. Then persons reading the conclusions of a statistical analysis of experimental data or a sample survey will be aware of the conditional nature of the conclusions.

EXERCISES

5.1. Suppose that you wish to randomly sample the opinions of $n = 10$ persons from a population of 800. Use table 7 of the Appendix to identify the persons to appear in your sample.

 5.2. Suppose that you wish to sample the opinions of the homeowners in a community regarding the desirability of increasing local expenditures to improve the quality of the public schools. You randomly select the households by using a table of random numbers, and you discard any households in your sample for which the homeowner is not at home when visited by the interviewer. Do you think this process is likely to approximate random sampling? Explain.

5.3. Use table 7 of the Appendix, to identify the measurements to be included in a random sample of $n = 10$ from a population containing $N = 1000$ measurements.

5.4. Randomly select a point in table 7 of the Appendix and proceed across the row (or down the column) to collect 1000 random digits. Calculate the proportion of times you observe each of the digits 0, 1, 2, $\cdots$, 9. Are these proportions near .1 (as would be expected)?

5.3 THE CENTRAL LIMIT THEOREM AND THE SAMPLING DISTRIBUTION FOR A SAMPLE MEAN

In the previous section we defined the term "random sample" and showed how we could select a random sample of n measurements from a population. Having obtained a sample, we can use the methods of chapters 2 and 3 to describe the data graphically and numerically. For example, if the relative frequency histogram for the sample measurements is mound-shaped, the intervals $(\bar{y} \pm s)$, $(\bar{y} \pm 2s)$, and $(\bar{y} \pm 3s)$ obtained from the Empirical Rule can be used to describe the variability of the sample measurements.

Suppose now that in addition to describing the sample data, we want to estimate the mean of the population. How close will the sample mean $\bar{y}$ be to the actual population mean μ? While the intervals mentioned above [$(\bar{y} \pm s)$, $(\bar{y} \pm 2s)$, etc.] are useful in describing the variability of individual sample measurements, they cannot be used to describe the variability of the sample mean $\bar{y}$. To do this we need to know how $\bar{y}$ would vary if computed from many different random samples. The relative frequency histogram for the $\bar{y}$'s, computed from random samples of size n, is called the **sampling distribution of $\bar{y}$**. Knowing how the values of $\bar{y}$ from these many samples are distributed about the population mean μ will enable us to describe how close the value of $\bar{y}$ is to μ when only a single sample is collected.

The Central Limit Theorem, one of the most important theorems in statistics, provides information on the sampling distribution of $\bar{y}$. It states that the sampling distribution of sums, or means, based on repeated random samples of measurements from a population will be approximately bell-shaped. This idea can best be illustrated with an example: Recall that a die (one of a pair of dice) is a cube with the faces of the cube showing from 1 to 6 dots. Thus there are six possible values that could appear face up when a die is rolled (tossed). We can simulate a population of outcomes by throwing a die a large number of times. The resulting relative frequency histogram for y, the number of dots appearing face up, would be as indicated in figure 5.1. (If the die were per-

Central Limit Theorem and Sampling Distribution for a Sample Mean

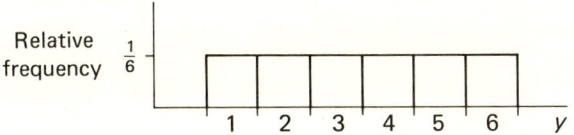

Figure 5.1 Relative Frequency Distribution for a Population of Die Tosses

fectly balanced, the probability distribution for y, the number of dots appearing on the upper face, also would appear as shown in figure 5.1.)

Suppose we now draw samples of five measurements ($n = 5$) from this distribution by tossing the die five times and recording the value of y observed each time. (See table 5.3; note that the values of y observed for the first sample of five measurements are 3, 5, 1, 3, 2, respectively.) We repeat this process 100 times to obtain 100 samples of five measurements. The sum $\Sigma\, y$ and sample mean $\bar{y}$ are shown for each sample in columns three and seven, and four and eight, respectively.

We can construct a relative frequency histogram for $\bar{y}$, the mean of the five sample measurements, by first preparing a frequency table from the data in table 5.3. The tabulation of relative frequencies is shown in table 5.4, and the resulting relative frequency histogram is shown in figure 5.2(b). According to the Central Limit Theorem, this relative frequency histogram of $\bar{y}$ should be approximately normal.

It is interesting to note that although the relative frequencies for the individual values of y are equal and hence the relative frequency distribution is flat [see figure 5.2(a)], the distribution of the sample means is mound-shaped and even somewhat bell-shaped [see figure 5.2(b)]. You can visually compare these two relative frequency distributions by observing figure 5.2. The irregularities in the curve of figure 5.2(b) are due to the small number of samples used to illustrate this concept. These irregularities would be less obvious if the sampling were conducted a large number of times. (Such extensive sampling is a time-consuming task to perform manually, but it can be done easily with the aid of an electronic computer.) The result would verify the Central Limit Theorem, which we now state as it applies to means.

Theorem 5.1: The Central Limit Theorem, Applied to Means

If random samples containing a fixed number n of measurements are repeatedly drawn from a population with finite mean μ and standard deviation σ, then if n is large, the sample means will have a distribution that is approximately normal (bell-shaped), with mean $\mu_{\bar{y}} = \mu$ and standard deviation $\sigma_{\bar{y}} = \sigma/\sqrt{n}$.

Table 5.3 The Sums and Means for 100 Samples of 5 Die Tosses

Sample Number	Sample Measurements	Σy	$\bar{y}$	Sample Number	Sample Measurements	Σy	$\bar{y}$
1	3, 5, 1, 3, 2	14	2.8	45	4, 4, 5, 4, 4	21	4.2
2	3, 1, 1, 4, 6	15	3.0	46	5, 4, 5, 5, 4	23	4.6
3	1, 3, 1, 6, 1	12	2.4	47	6, 6, 6, 2, 1	21	4.2
4	4, 5, 3, 3, 2	17	3.4	48	2, 1, 5, 5, 4	17	3.4
5	3, 1, 3, 5, 2	14	2.8	49	6, 4, 3, 1, 5	19	3.8
6	2, 4, 4, 2, 4	16	3.2	50	4, 4, 4, 4, 4	20	4.0
7	4, 2, 5, 5, 3	19	3.8	51	2, 3, 5, 3, 2	15	3.0
8	3, 5, 5, 5, 5	23	4.6	52	1, 1, 1, 2, 4	9	1.8
9	6, 5, 5, 1, 6	23	4.6	53	2, 6, 3, 4, 5	20	4.0
10	5, 1, 6, 1, 6	19	3.8	54	1, 2, 2, 1, 1	7	1.4
11	1, 1, 1, 5, 3	11	2.2	55	2, 4, 4, 6, 2	18	3.6
12	3, 4, 2, 4, 4	17	3.4	56	3, 2, 5, 4, 5	19	3.8
13	2, 6, 1, 5, 4	18	3.6	57	2, 4, 2, 4, 5	17	3.4
14	6, 3, 4, 2, 5	20	4.0	58	5, 5, 4, 3, 2	19	3.8
15	2, 6, 2, 1, 5	16	3.2	59	5, 4, 4, 6, 3	22	4.4
16	1, 5, 1, 2, 5	14	2.8	60	3, 2, 5, 3, 1	14	2.8
17	3, 5, 1, 1, 2	12	2.4	61	2, 1, 4, 1, 3	11	2.2
18	3, 2, 4, 3, 5	17	3.4	62	4, 1, 1, 5, 2	13	2.6
19	5, 1, 6, 3, 1	16	3.2	63	2, 3, 1, 2, 3	11	2.2
20	1, 6, 4, 4, 1	16	3.2	64	2, 3, 3, 2, 6	16	3.2
21	6, 4, 2, 3, 5	20	4.0	65	4, 3, 5, 2, 6	20	4.0
22	1, 3, 5, 4, 1	14	2.8	66	3, 1, 3, 3, 4	14	2.8
23	2, 6, 5, 2, 6	21	4.2	67	4, 6, 1, 3, 6	20	4.0
24	3, 5, 1, 3, 5	17	3.4	68	2, 4, 6, 6, 3	21	4.2
25	5, 2, 4, 4, 3	18	3.6	69	4, 1, 6, 5, 5	21	4.2
26	6, 1, 1, 1, 6	15	3.0	70	6, 6, 6, 4, 5	27	5.4
27	1, 4, 1, 2, 6	14	2.8	71	2, 2, 5, 6, 3	18	3.6
28	3, 1, 2, 1, 5	12	2.4	72	6, 6, 6, 1, 6	25	5.0
29	1, 5, 5, 4, 5	20	4.0	73	4, 4, 4, 3, 1	16	3.2
30	4, 5, 3, 5, 2	19	3.8	74	4, 4, 5, 4, 2	19	3.8
31	4, 1, 6, 1, 1	13	2.6	75	4, 5, 4, 1, 4	18	3.6
32	3, 6, 4, 1, 2	16	3.2	76	5, 3, 2, 3, 4	17	3.4
33	3, 5, 5, 2, 2	17	3.4	77	1, 3, 3, 1, 5	13	2.6
34	1, 1, 5, 6, 3	16	3.2	78	4, 1, 5, 5, 3	18	3.6
35	2, 6, 1, 6, 2	17	3.4	79	4, 5, 6, 5, 4	24	4.8
36	2, 4, 3, 1, 3	13	2.6	80	1, 5, 3, 4, 2	15	3.0
37	1, 5, 1, 5, 2	14	2.8	81	4, 3, 4, 6, 3	20	4.0
38	6, 6, 5, 3, 3	23	4.6	82	5, 4, 2, 1, 6	18	3.6
39	3, 3, 5, 2, 1	14	2.8	83	1, 3, 2, 2, 5	13	2.6
40	2, 6, 6, 6, 5	25	5.0	84	5, 4, 1, 4, 6	20	4.0
41	5, 5, 2, 3, 4	19	3.8	85	2, 4, 2, 5, 5	18	3.6
42	6, 4, 1, 6, 2	19	3.8	86	1, 6, 3, 1, 6	17	3.4
43	2, 5, 3, 1, 4	15	3.0	87	2, 2, 4, 3, 2	13	2.6
44	4, 2, 3, 2, 1	12	2.4	88	4, 4, 5, 4, 4	21	4.2

Central Limit Theorem and Sampling Distribution for a Sample Mean

Table 5.3 The Sums and Means for 100 Samples of 5 Die Tosses (continued)

Sample Number	Sample Measurements	Σy	$\bar{y}$	Sample Number	Sample Measurements	Σy	$\bar{y}$
89	2, 5, 4, 3, 4	18	3.6	95	6, 1, 4, 2, 2	15	3.0
90	5, 1, 6, 4, 3	19	3.8	96	1, 1, 2, 3, 1	8	1.6
91	5, 2, 5, 6, 3	21	4.2	97	6, 2, 5, 1, 6	20	4.0
92	6, 4, 1, 2, 1	14	2.8	98	3, 1, 1, 4, 1	10	2.0
93	6, 3, 1, 5, 2	17	3.4	99	5, 2, 1, 6, 1	15	3.0
94	1, 3, 6, 4, 2	16	3.2	100	2, 4, 3, 4, 6	19	3.8

We hope the die-tossing data of table 5.3 clarify the meaning of the Central Limit Theorem. Although proof is omitted, it can be shown that the mean and the standard deviation for y in the die-tossing distribution [figure 5.2(a)] are $\mu = 3.50$ and $\sigma = 1.71$. For these values the Central Limit Theorem states that the sampling distribution for sample means based on $n = 5$ measurements

Table 5.4 Relative Frequency Table for 100 Values of $\bar{y}$

Class	Class Boundaries	Frequency	Relative Frequency
1	1.3–1.5	1	1/100
2	1.5–1.7	1	1/100
3	1.7–1.9	1	1/100
4	1.9–2.1	1	1/100
5	2.1–2.3	3	3/100
6	2.3–2.5	4	4/100
7	2.5–2.7	6	6/100
8	2.7–2.9	10	10/100
9	2.9–3.1	7	7/100
10	3.1–3.3	9	9/100
11	3.3–3.5	11	11/100
12	3.5–3.7	9	9/100
13	3.7–3.9	11	11/100
14	3.9–4.1	10	10/100
15	4.1–4.3	7	7/100
16	4.3–4.5	1	1/100
17	4.5–4.7	4	4/100
18	4.7–4.9	1	1/100
19	4.9–5.1	2	2/100
20	5.1–5.3	0	0
21	5.3–5.5	1	1/100

120 **Sampling Distributions**

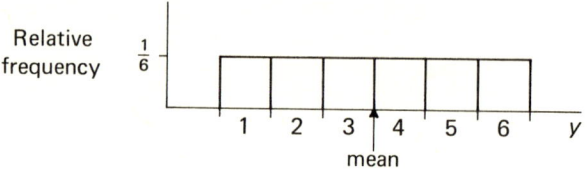

(a) Relative frequency distribution for y

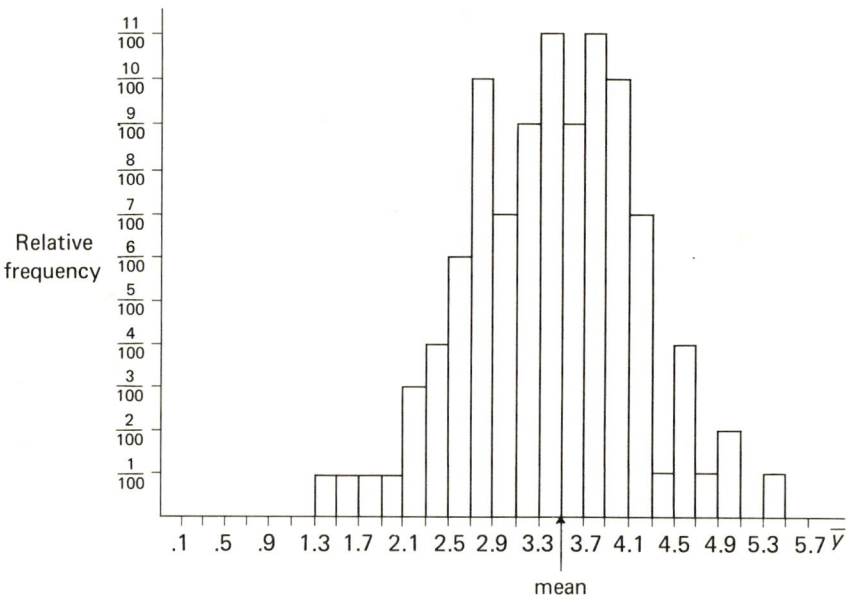

(b) Relative frequency distribution for the 100 sample means given in table 5.3

Figure 5.2 Illustration of the Central Limit Theorem

possesses the same mean as the original population: $\mu_{\bar{y}} = \mu = 3.5$. The standard deviation is

$$\sigma_{\bar{y}} = \frac{\sigma}{\sqrt{n}} = \frac{1.71}{\sqrt{5}} = .76$$

(Note: Because the standard deviation of $\bar{y}$ differs from the population standard deviation σ, we denote its value by the symbol $\sigma_{\bar{y}}$.)

Our experimental die-tossing sampling verifies these results. You can see that the mean of the distribution of $\bar{y}$'s, figure 5.2(b), is approximately 3.5. In addition, the range of the $\bar{y}$'s is $(5.4 - 1.4) = 4.0$, so the standard deviation is approximately $(4.0)/4$, or 1.0. Hence the standard deviation of the distribution of 100 sample means [figure 5.2(b)] is near the value stated by the Central Limit Theorem, $\sigma/\sqrt{n} = .76$. As noted earlier, the relative frequency histogram of figure 5.2(b) is approximately bell-shaped. Thus the die-tossing sampling experiment provides us with practical verification of the Central Limit Theorem.

The Central Limit Theorem also applies to the distribution of sums of sample observations, and the die-tossing data can also be used to provide empirical evidence to support this version of the Central Limit Theorem. We leave it to you to construct a histogram of the sample sums Σy if you seek graphic evidence of the validity of the Central Limit Theorem as it applies to sums.

Theorem 5.2: The Central Limit Theorem, Applied to Sums

If random samples containing a fixed number n of measurements are repeatedly drawn from a population with finite mean μ and standard deviation σ, then if n is large, the sums of the sample measurements will have a distribution that is approximately normal (bell-shaped), with mean equal to $n\mu$ and standard deviation equal to $\sigma\sqrt{n}$.

Careful reading of the Central Limit Theorem suggests its broad applicability—it applies to the distribution of sample means drawn from any population with a finite mean μ and standard deviation σ. The resulting distribution of means will be approximately normal, with mean and standard deviation related not only to the mean and standard deviation of the population from which the samples are drawn but also to the sample size n. **Note that the normal approximation to the distribution of sample means or sums becomes more and more accurate as the sample size n increases.** Recall, however, that the approximation was quite good for an n as small as 5 in the die-tossing experiment.

The significance of the Central Limit Theorem is twofold:

1. **It explains why many measurements have bell-shaped frequency distributions.** For example, we might imagine that the test score for an individual on a national aptitude test is influenced by random factors or variables such as the amount of sleep the individual had the night before, the length of time he prepared for the exam, his IQ, and so forth. If each of these factors in some way affects the final score, that score is the sum of random variables. The Central Limit Theorem may then help to explain why such scores are approximately normally distributed.

2. **It is useful in statistical inference.** Many estimators of population parameters used for purposes of statistical inference are sums or averages of the sample measurements. (For example, we will use the sample mean $\bar{y}$ to estimate a population mean μ.) Where sums or averages are involved and the

Sampling Distributions

sample measurements. (For example, we will use the sample mean $\bar{y}$ to estimate a population mean μ.) Where sums or averages are involved and the sample size n is large, the many estimates generated in repeated sampling can be expected to possess a bell-shaped normal distribution. We can then use the properties of the normal distribution to describe the behavior of $\bar{y}$. (We will illustrate this application of the Central Limit Theorem in section 6.2.)

To summarize, the Central Limit Theorem tells us the nature of the *sampling* distribution of $\bar{y}$ when the sample size is large. This information is as follows:

Properties of the Sampling Distribution of a Sample Mean $\bar{y}$

1. The sampling distribution of $\bar{y}$ is approximately normal for large sample sizes (n large).

2. The mean of the sampling distribution is equal to the population mean: $\mu_{\bar{y}} = \mu$.

3. The standard deviation of the sampling distribution is $\sigma_{\bar{y}} = \sigma/\sqrt{n}$, where σ is the standard deviation of the sampled population.

How large must the sample size be before the Central Limit Theorem will hold? The answer depends on the population from which we are sampling and on our use of the approximation. We can, however, take comfort in the fact that the theorem worked very well for a sample as small as $n = 5$ in the die-tossing experiment. Therefore, it should work even better for a larger sample size. Recall also that the relative frequency histogram for $\bar{y}$ was approximately bell-shaped, demonstrating that the Central Limit Theorem can work for some small sample sizes. We will give sample size requirements for specific applications of the Central Limit Theorem as we encounter them later in the text.

Example 5.2

Reclaimed phosphate land in Polk County, Florida, has been found to emit a higher mean radiation level than other nonmining land in the county. Suppose that the true mean radiation level for the reclaimed land is $\mu = 5.0$ working levels (WL) with a standard deviation of .5 WL. Suppose that 20 houses built on reclaimed land are randomly selected and the radiation level measured in each. What is the probability (approximately) that the sample mean for the 20 houses exceeds 5.2 WL?

Solution

According to the Central Limit Theorem the sample mean of $n = 20$ randomly selected radiation level measurements will be approximately normally distributed with

Central Limit Theorem and Sampling Distribution for a Sample Mean

$$\mu_{\bar{y}} = \mu = 5.0 \text{ WL}$$

$$\sigma_{\bar{y}} = \frac{\sigma}{\sqrt{n}} = \frac{.5}{\sqrt{20}} = .11$$

The distribution will be as shown in figure 5.3.

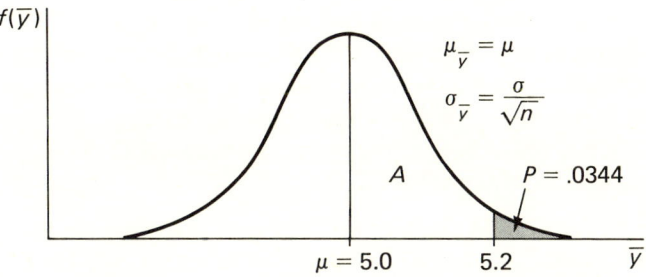

Figure 5.3 Sampling Distribution of $\bar{y}$ for Samples of $n = 20$ Randomly Selected from a Population with $\mu = 5.0$ and $\sigma = .5$

The probability that $\bar{y}$ exceeds 5.2 WL is the shaded area shown in figure 5.3. To find this area, we must determine how many standard deviations ($\sigma_{\bar{y}} = .11$) the point 5.2 lies to the right of $\mu = 5.0$. This number is

$$z = \frac{\bar{y} - \mu}{\sigma_{\bar{y}}} = \frac{5.2 - 5.0}{.11} = 1.82$$

Turning to table 1 of the Appendix, we find the area A corresponding to $z = 1.82$ is

$$A = .4656$$

Then since the total area to the right of the mean is equal to .5, the probability that $\bar{y}$ will exceed 5.2 WL is P, where

$$P = .5 - A = .5 - .4656 = .0344$$

EXERCISES

5.5. A random sample of $n = 100$ measurements is obtained from a population with $\mu = 55$ and $\sigma = 20$. Describe the sampling distribution for $\bar{y}$, giving $\mu_{\bar{y}}$ and $\sigma_{\bar{y}}$.

5.6. A random sample of $n = 60$ measurements is obtained from a population with $\mu = 192$ and $\sigma = 43$. Describe the sampling distribution for $\bar{y}$.

Sampling Distributions

5.7. Refer to table 5.3. Using the sample data from the 100 samples of $n = 5$ die tosses, construct a frequency table similar to table 5.4 for Σy, the sum of the sample measurements. Use an interval width of 1 with a starting point of 6.5.

5.8. Using the frequency table of exercise 5.7, construct a relative frequency histogram for Σy. The mean of the distribution of y, the number of dots appearing on the upper face, can be shown to be $\mu = 3.5$, and the size of each sample is $n = 5$. Thus by the Central Limit Theorem the relative frequency histogram should be mound-shaped, with a mean approximately equal to $n(\mu) = 5(3.5) = 17.5$. Because the relative frequency histogram of Σy is based on only 100 samples, the approximation could be improved by using more repetitions of the experiment.

5.9. A random sample of $n = 25$ measurements is selected from a population with mean equal to 80 and standard deviation equal to 7.
(a) What is the probability that the sample mean will exceed 81?
(b) What is the probability that the sample mean will fall in the interval $79 \leq \bar{y} \leq 81$?
(c) What is the probability that the sample mean will be less than 78?

e **5.10.** The oxygen content in water must exceed some minimum value in order to support aquatic life. Suppose that this value is approximately 6.0 parts per million (ppm). In one experiment $n = 5$ jars of water are randomly selected from a stream with a mean oxygen content equal to $\mu = 6.0$ ppm and a standard deviation equal to $\sigma = .7$ ppm.
(a) What is the probability that the sample mean exceeds 6.5 ppm?
(b) Suppose that the sample mean equals 7.0 ppm. Intuitively, what would you conclude about the oxygen content of the stream? (Note: In chapter 7 we will explain how to answer this question by using a statistical decision procedure.)

5.4 THE SAMPLING DISTRIBUTION OF A SAMPLE PROPORTION

We stated in section 4.6 that a binomial population could be viewed as a very large collection of 0s and 1s, a 0 corresponding to each failure and a 1 corresponding to each success. Then if we select a random sample of n measurements from the population, it follows that the *total* number y of observed successes will equal the *sum* of the 0s and 1s in the sample. This enables us to conclude that for samples of adequate size the distribution for y will be approximately normally distributed (because of the Central Limit Theorem). As noted in section 4.6, the mean and standard deviation of the distribution of y will be as follows:

Mean and Standard Deviation of the Binomial Random Variable y

$$\mu = np \quad \text{and} \quad \sigma = \sqrt{npq}$$

where p is the probability of success in a given trial, $q = 1 - p$, and n is the sample size.

To illustrate how well the distribution of y can be approximated by a normal distribution, we show the binomial probability distribution for $n = 20$ and $p = .5$ in figure 5.4, with the approximating normal curve superimposed. For this situation the mean and standard deviation are

$$\mu = np = (20)(.5) = 10$$
$$\sigma = \sqrt{npq} = \sqrt{(20)(.5)(.5)} = 2.24$$

As you can see, this approximation is surprisingly good, considering the small value of n. However, when p is close to 0 or 1, we would need a much larger sample in order for the normal distribution to provide a good approximation to the binomial probability distribution.

Example 5.3

Use the normal approximation to the binomial probability distribution, figure 5.4, to find the probability that $y \geq 13$.

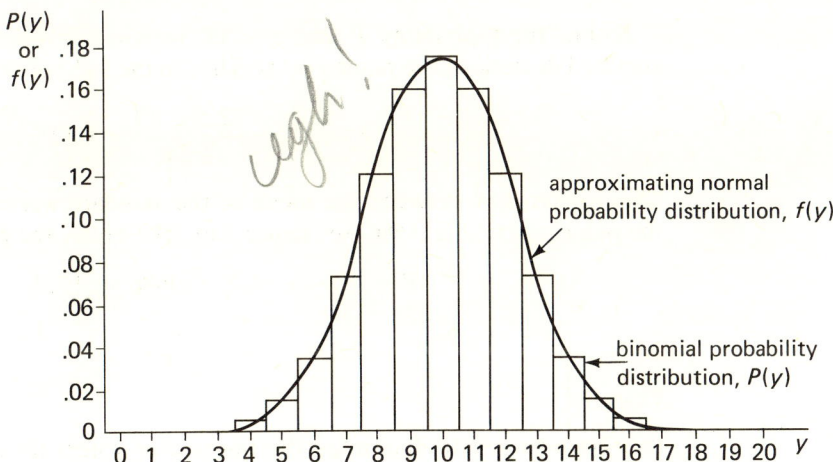

Figure 5.4 Comparison of a Binomial Probability Distribution, for $n = 20$ and $p = .5$, and the Approximating Normal Curve

Solution

The probability that we wish to approximate is the sum of the probability rectangles, figure 5.4, corresponding to $y = 13, 14, \ldots, 20$. To approximate the area corresponding to these rectangles, we need to find the area under the normal curve to the right of $y = 12.5$; see figure 5.5. (Notice that using only the area to the right of $y = 13$ would be incorrect, because then we would be omitting the area corresponding to the left half of the rectangle for $y = 13$.)

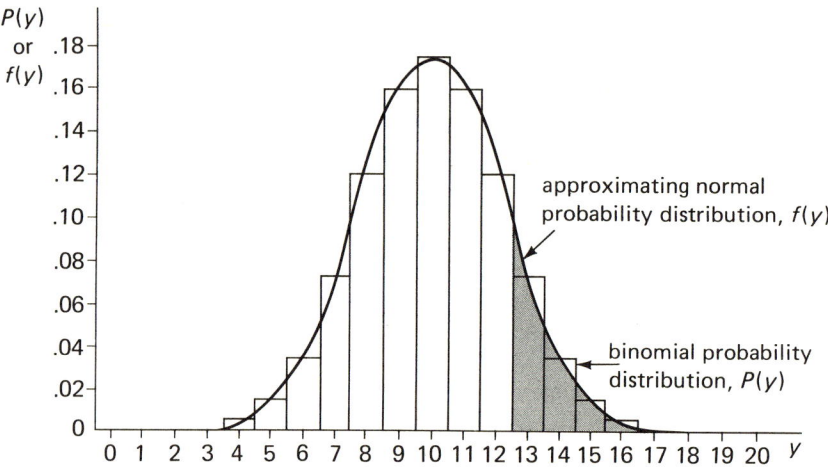

Figure 5.5 Required Area for Example 5.3

To find the probability P that $y \geq 13$, we must first determine how many standard deviations the point $y = 12.5$ lies to the right of $\mu = np = 10$. This is

$$z = \frac{y - \mu}{\sigma} = \frac{12.5 - 10}{2.24} = 1.12$$

Then the area A between the mean of the standard normal distribution and the point $z = 1.12$ is .3686 (see figure 5.6). Therefore, the probability is

$$P = .5 - A = .5 - .3686 = .1314$$

Example 5.4

In a state survey to determine the fraction of voters favoring a Republican congressional candidate, a sample of $n = 1000$ voters was polled to determine their preferences. Let y denote the number of persons favoring the Republican

The Sampling Distribution of a Sample Proportion

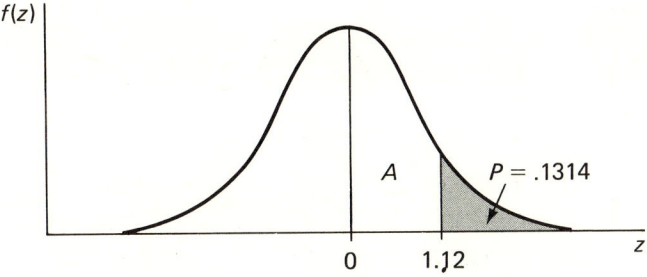

Figure 5.6 Standardized Normal Curve for Example 5.3

in the 1000 sampled. Assuming 50% of the registered voters favor him, find the mean and standard deviation of the random variable y and use this information to describe the variability of y in repeated sampling.

Solution

This election survey satisfies the properties of a binomial experiment ($n = 1000$, $p = .5$). The mean and the standard deviation of the binomial random variable y are

$$\mu = np = 1000(.5) = 500$$
$$\sigma = \sqrt{npq} = \sqrt{1000(.5)(.5)} = \sqrt{250} = 15.81$$

By the Empirical Rule approximately 95% of the values of y in repeated sampling will fall in the interval ($\mu - 2\sigma$) to ($\mu + 2\sigma$); that is,

$$[500 - 2(15.81)] \quad \text{to} \quad [500 + 2(15.81)]$$

or
$$468.38 \quad \text{to} \quad 531.62$$

You will readily see how inference could enter this example. Suppose that y assumed an improbable value; that is, suppose we drew a sample of $n = 1000$ and observed fewer than $y = 468$ favoring the Republican candidate. From figure 5.7 we see that the probability of observing a value of 468 or less is approximately .025, assuming that 50% (or more) of the voters favor the Republican candidate. Since this probability is so small, an observed value of y less than or equal to 468 is contradictory to the assumption that 50% (or more) of the voters favor him. Because of this contradiction, we conclude that he will receive less than 50% of the votes.

Binomial populations are sampled in order to make inferences about p, the proportion of successes in the population. The most obvious statistic to select to make these inferences is the proportion of successes in the sample, denoted

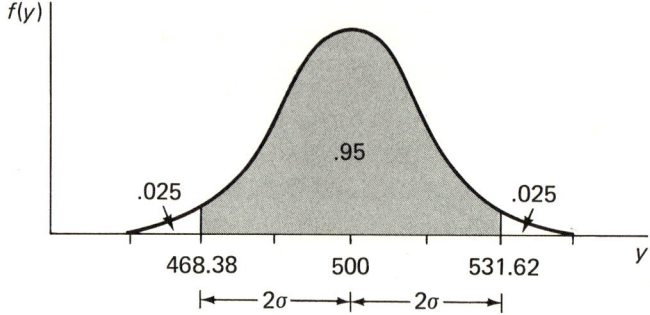

Figure 5.7 Area Under the Normal Curve Within 2σ of $\mu = 500$

by the symbol $\hat{p}$ (read, "p hat"; the "hat" over the symbol p is used to denote "estimator of p"). What can we say about the sampling distribution of the sample proportion $\hat{p}$?

We have stated that the number y of successes in the sample can be viewed as the sum of the (0 and 1) sample measurements. Therefore, it follows that the sample proportion,

$$\hat{p} = \frac{y}{n} = \frac{\text{sum of the sample measurements}}{n}$$

is the sample mean and that the sampling distribution of $\hat{p}$ will be approximately normally distributed when n is large (because of the Central Limit Theorem). It can be shown (proof omitted) that the mean and the standard deviation of this approximating normal distribution are

$$\mu_{\hat{p}} = p$$
$$\sigma_{\hat{p}} = \sqrt{\frac{pq}{n}}$$

The sampling distribution for the sample proportion $\hat{p}$ is as shown in figure 5.8.

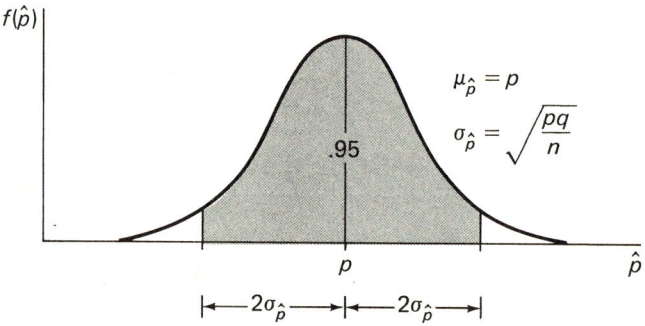

Figure 5.8 The Sampling Distribution for $\hat{p}$

The Sampling Distribution of a Sample Proportion

Properties of the Sampling Distribution of a Binomial Sample Proportion $\hat{p}$

1. The sampling distribution for $\hat{p}$ is approximately normal for large sample sizes (n large).
2. The mean $\mu_{\hat{p}}$ of the sampling distribution is equal to p, the population proportion of successes.
3. The standard deviation of the sampling distribution for $\hat{p}$ is equal to

$$\sigma_{\hat{p}} = \sqrt{\frac{pq}{n}}$$

In the next example we show how we can use this information in a practical application.

Example 5.5

Two thousand new automobile steering mechanisms were tested in order to estimate the proportion p of all the steering mechanisms that might be faulty. If the true proportion is $p = .03$, what is the probability that the sample proportion will be within .01 of p?

Solution

The 2000 steering mechanisms can be viewed as a random sample. Thus the sampling distribution for $\hat{p}$ will appear as shown in figure 5.8, with mean

$$\mu_{\hat{p}} = p = .03$$

and standard deviation

$$\sigma_{\hat{p}} = \sqrt{\frac{pq}{n}} = \sqrt{\frac{(.03)(.97)}{2000}} = .0038$$

The probability that the sample proportion $\hat{p}$ falls within .01 of the population proportion $p = .03$ is the shaded area shown under the sampling distribution of $\hat{p}$ in figure 5.9. If A is the area over the interval .03 to .04, then the probability that $\hat{p}$ falls in the interval .02 to .04 is $2A$.

To find the number of standard deviations ($\sigma_{\hat{p}}$) between $p = .03$ and a value of $\hat{p}$ equal to .04, we calculate the z score:

$$z = \frac{\hat{p} - p}{\sigma_{\hat{p}}} = \frac{.04 - .03}{.0038} = 2.63$$

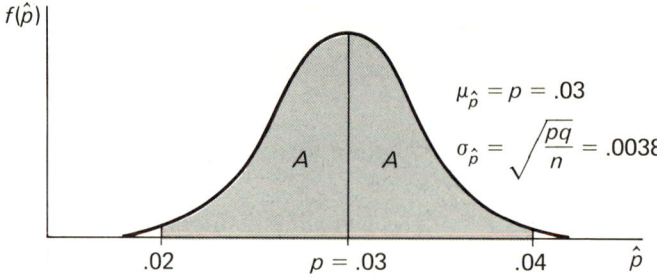

Figure 5.9 The Sampling Distribution for $\hat{p}$; $n = 2000$ and $p = .03$

Since the area A to the right of the mean, corresponding to $z = 2.63$, is .4957, it follows that

$$P(.02 \leq \hat{p} \leq .04) = 2A = 2(.4957) = .9914$$

In other words, if the population proportion of defective steering mechanisms is $p = .03$, the proportion of malfunctioning steering mechanisms in a sample of $n = 2000$ should be within .01 of p with a high probability—namely, .9914. Therefore, you can see that in this case the sample proportion will provide an accurate estimate of p.

EXERCISES

5.11. A random sample of $n = 1000$ measurements is obtained from a binomial population with $p = .7$. Describe the sampling distribution for the sample proportion, giving $\hat{p}$, $\mu_{\hat{p}}$, and $\sigma_{\hat{p}}$.

5.12. Refer to exercise 5.11. Find the probability that $\hat{p}$ lies in the interval from .6 to .8. (Hint: Use the normal approximation.)

5.13. Refer to figure 5.4 and use the normal approximation to the binomial probability distribution to find the probability that $6 \leq y \leq 12$. (Hint: Be sure to shade the desired area in figure 5.4, as we did in solving example 5.3.)

5.14. Suppose that you select a sample of $n = 10$ from a binomial population with $p = .5$.
(a) Use the binomial probability distribution to calculate the probability that y falls in the interval $3 \leq y \leq 5$.
(b) Use the normal approximation to the binomial probability distribution to find the approximate probability that $3 \leq y \leq 5$. Compare this result with your answer to part (a).

$ 5.15. An insurance company states that 10% of all fire insurance claims are fraudulent. Suppose that the company is correct and that it receives 100 claims.

(a) What is the probability that more than 12 will be fraudulent?
(b) What is the probability that 15 or more will be fraudulent?

5.16. Refer to exercise 5.15.
(a) What assumptions must be made in order that your answers in part (a) be valid?
(b) Suppose that only one claim in the sample was fraudulent. Would you have any doubts about the company's statement? Explain.

 5.17. Suppose that you wish to conduct a poll to estimate the proportion of adult Americans who favor less governmental control of private business. Further, suppose that this unknown proportion is really $p = .5$. If you draw a random sample of $n = 1000$ adults from the U.S. population, what is the probability that the sample proportion will differ from the population proportion ($p = .5$) by more than .02?

SUMMARY

This chapter presents important results on the sampling distributions (the probability distributions) of two statistics: the sample mean $\bar{y}$ and the sample proportion $\hat{p}$. The sampling distribution for a statistic tells us how the statistic will behave in repeated sampling. This information is useful because when a statistic is used to estimate a population parameter, we will want to know the probability that the statistic will be within some specified distance of the parameter. This probability, which can be obtained from the sampling distribution of the statistic, gives us a measure of the *reliability* of the estimate.

The Central Limit Theorem establishes, for large samples, the approximate normality of the sampling distributions of the three statistics discussed in this chapter: the sample mean, the sample sum, and a sample proportion.

REFERENCES

Mendenhall, W., and Scheaffer, R. *Mathematical Statistics with Applications*. N. Scituate, Mass.: Duxbury Press, 1973.

SUPPLEMENTARY EXERCISES

5.18. A random sample of $n = 25$ measurements is selected from a population with mean $\mu = 3$ and standard deviation $\sigma = 1$.
(a) Find the approximate probability that $\bar{y} \geq 3.1$.
(b) Find the approximate probability that $2.8 \leq \bar{y} \leq 3.2$.
(c) Find the approximate probability that $\Sigma y \geq 80$.

Sampling Distributions

5.19. The daily shrinkage due to theft in the inventory of a men's department store possesses a probability distribution with mean equal to $320 and standard deviation equal to $80. The store reports the shrinkage as a total T for a four-week period. Describe the probability distribution for the shrinkage for the four-week period and justify your conclusions.

5.20. Refer to exercise 5.19. Suppose that the mean daily shrinkage μ is unknown to you and that you randomly sample $n = 50$ days to estimate its value. What is the probability that the sample mean $\bar{y}$ will deviate from μ by more than $20?

5.21. Due to temperature variations, the expansion or contraction of a gas pipeline per 1000 feet is normally distributed with mean $\mu = 1$ inch and standard deviation $\sigma = .5$ inch. If the pipeline is 50,000 feet long, what is the probability that the pipe might expand more than 55 inches?

5.22. From returns in previous years it has been found that approximately 70% of the tax returns in a given income category are incorrectly filed. A spot check of 5000 returns drawn at random shows that only $y = 2600$ have been filed incorrectly. Assuming that 70% of the returns will be incorrect this year also, find the mean and standard deviation of the random variable y and use this information to describe its variability in repeated sampling.

5.23. Judging from the results obtained in exercise 5.22, would you anticipate approximately 70% of the returns this year to be incorrectly filed? Explain.

5.24. A manufacturer claims that 95% of the components that it is supplying for a new jet transport meet a specified rigid standard of performance. Suppose 400 of the components are put on test. Find the probability of observing 30 or more that do not meet the standard of performance. (Use the normal approximation to the binomial probability distribution.)

5.25. Refer to exercise 5.24. Suppose that the test of 400 components was performed and that 30 of the components fell below the rigid standard of performance. What might you conclude about the manufacturer's claim?

5.26. An airline has found over the past several years that 10% of the persons making reservations on a particular flight will not show up at flight time. On a given day records indicate that the flight is fully booked at 300, with 35 more people waiting on standby. What is the probability that all 35 people on standby will have a seat available at flight time?

5.27. Data collected over a long period of time indicate that a particular birth defect occurs in one of every 1,000 live births. Data collected from a medical center in a particular section of the country indicate 10 children with the birth defect from the total of 20,000 birth records examined.
(a) If we assume the 20,000 records examined represent a random sample of

birth records, what is the probability of observing 10 or fewer children with birth defects in the sample?

(b) How might you explain the apparent lower rate of birth defects from this area of the country?

 5.28. Research on sales displays in grocery stores has shown that 70% of all people who pick up a particular item will purchase it. As part of a class project in a small community, a random sample of 300 shoppers is observed to see how many handle and then purchase the sale item on display at the front of the grocery store. Assuming 70% of all shoppers who handle the item will eventually purchase it, what is the probability of observing fewer than 200 who handle and then buy the item in the sample of 300 shoppers?

5.29. If you randomly sample $n = 400$ observations from a binomial population with parameter $p = .3$, what is the probability that the sample proportion will differ from $p = .3$ by more than .02?

 5.30. A supplier of a particular semiconductor claims that no more than .5% are defective. If you randomly sample and test 10,000 of these semiconductors and find .9% defective, what would you conclude about the manufacturer's claim? Explain your reasoning. (Note: In chapter 8 we will give a statistical procedure for making this decision.)

EXPERIENCES WITH REAL DATA

Each three-digit number shown in a random number table appears with a probability near .001. That is, they possess a probability distribution as shown in figure 5.10.

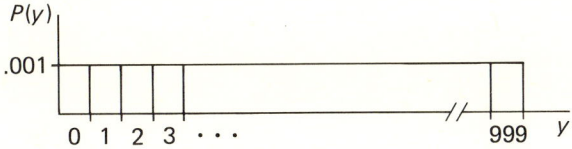

Figure 5.10 A Probability Distribution for Three-Digit Random Numbers Selected from a Random Number Table

Simulate random sampling from a population possessing the probability distribution of figure 5.10 by selecting $n = 3$ three-digit random numbers from table 7 of the Appendix and calculate the sample range,

range = difference between the largest and the smallest measurements in the sample

Repeat this process 200 times to obtain 200 sample ranges, and construct the relative frequency histogram. This histogram will be similar to the sampling distribution for the sample range based on samples of $n = 3$ measurements selected from a population possessing a relative frequency histogram similar to the one in figure 5.10. You could obtain a better approximation to this sampling distribution by increasing the number of samples used to construct your histogram—say, 1000 samples of $n = 3$ measurements each rather than 200.

As noted in chapter 3, the range is a good measure of data variation for small samples. In fact, for small samples it is easier to compute and it is almost as good an estimator of the population standard deviation σ as is the sample standard deviation s. For this reason the range is often used to measure data variation in industrial quality control, where a manufacturing process is monitored by taking frequent small samples of one or more measures of product quality over time (for example, one sample every hour). The sampling distribution that you have derived would be useful in evaluating the properties of the sample range for quality control measurements on a random variable that possesses the probability distribution shown in figure 5.10.

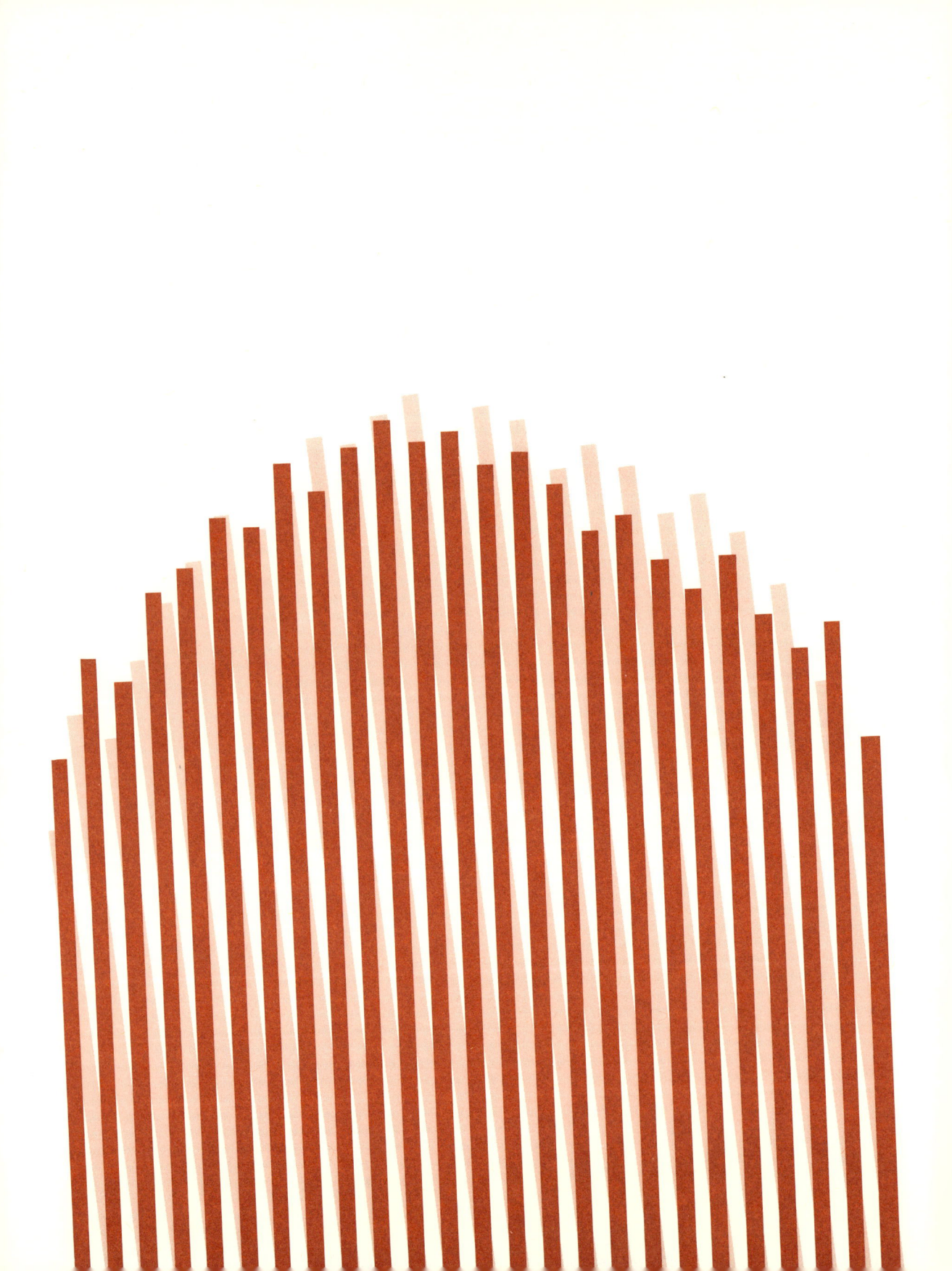

6

Making Inferences: Estimation

CHAPTER OUTLINE

6.1 Introduction
6.2 Point Estimation of a Population Mean
6.3 Interval Estimation of a Population Mean
6.4 Point Estimation of the Binomial Parameter p
6.5 Interval Estimation of the Binomial Parameter p

6.1 INTRODUCTION

Inference, specifically decision making and prediction, is centuries old and plays a very important role in our lives. Each of us is faced daily with personal decisions and situations that require predictions concerning the future. The government is concerned with predicting the flow of gold to Europe. A stockbroker wants to know how the stock market will behave. A metallurgist seeks to use the results of an experiment to determine whether a new type of steel is more resistant to temperature changes than another. A veterinarian investigates the effectiveness of a new product for treating worms in cattle. The inferences that these individuals make should be based on relevant facts, which we call observations, or data.

In many practical situations the relevant facts are abundant, seemingly inconsistent, and, in many respects, overwhelming. As a result, a careful decision or prediction is often little better than an outright guess. You need only refer to the "Market Views" section of the *Wall Street Journal* to observe the diversity of expert opinion concerning future stock market behavior. Similarly, a visual analysis of data by scientists and engineers will often yield conflicting opinions regarding conclusions to be drawn from an experiment.

Many individuals tend to feel that their own built-in inference-making equipment is quite good. But experience suggests that most people are incapable of utilizing large amounts of data, mentally weighing each bit of relevant infor-

Making Inferences: Estimation

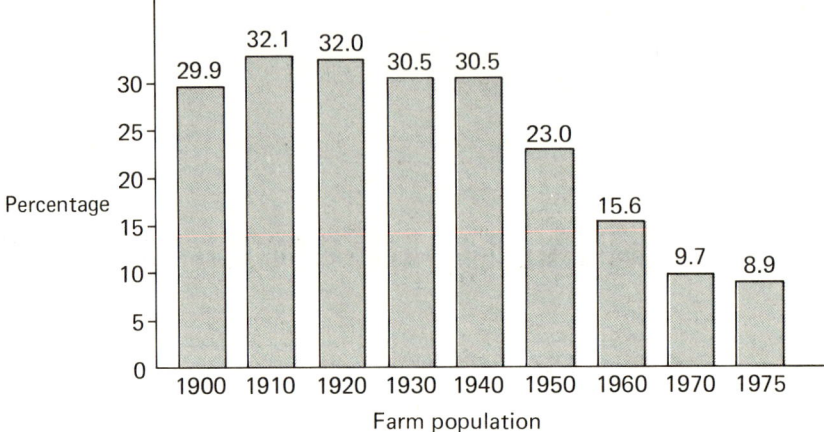

Given the Data Shown Here for the Years 1900 Through 1940, Do You Think That Sociologists, Statisticians, and Agricultural Experts Would Have Predicted the Decreases in the Farm Population Shown for 1950, 1960, 1970, and 1975?

mation, and arriving at a good inference. (You may test your own inference-making ability by using the exercises in chapters 6 and 7. Scan the data and make an inference before you use the appropriate statistical procedure. Then compare the results.) The statistician, rather than relying upon his or her own intuition, uses statistical results to aid in making inferences. Although we have touched upon some of the notions involved in statistical inference in preceding chapters, we will now collect our ideas in a presentation of some of the basic ideas involved in statistical inference.

The objective of statistics is to make inferences about a population based on information contained in a sample. Populations are characterized by numerical descriptive measures called parameters. Typical population parameters are the mean μ, the standard deviation σ, the area under the probability distribution to the right (or left) of some value of the random variable, or the area between two values of the variable. Most practical inferential problems you will encounter can be phrased to imply an inference about one or more parameters of a population. For example, in an experiment in which we wish to predict the average amount of money paid to welfare recipients in a given year, the population of interest is the set of all yearly welfare payments, and we are interested in estimating the value of the population mean μ.

Methods for making inferences about parameters fall into one of two categories. Either we will **estimate** (predict) the value of the population parameter of interest or we will **test an hypothesis** about the value of the parameter. These two methods of statistical inference—estimation and testing an hypothesis—involve different procedures, and, more important, they answer two different questions about the parameter. In estimating a population parameter

Introduction

we are answering the question, "What is the value of the population parameter?" In testing an hypothesis we are answering the question, "Is the parameter value equal to this specific value?"

Consider a study where an investigator is interested in examining the effectiveness of a drug product in reducing anxiety levels of anxious patients. A screening procedure is employed to identify a group of anxious patients. After the patients are admitted into the study, each one's anxiety level is measured on a rating scale immediately before he or she receives the first dose of the drug and then at the end of one week of drug therapy. These sample data can be used to make inferences about the population from which the sample was drawn by either estimation or a statistical test:

Estimation: Information from the sample can be used to estimate (or predict) the mean decrease in anxiety ratings for the set of all anxious patients who may conceivably be treated with the drug.

Statistical test: Information from the sample can be used to determine whether the population mean decrease in anxiety ratings is greater than zero.

Notice that the inference related to estimation is aimed at answering the question, "What is the mean decrease in anxiety ratings for the population?" In contrast, the statistical test attempts to answer the question, "Is the mean drop in anxiety ratings greater than zero?"

Estimation procedures can be classified into two categories: point estimation and interval estimation. For example, we might use the sample data to specify a single number—say 15—as an estimate of the mean decrease in anxiety ratings. Alternatively, we might estimate that the mean decrease in anxiety ratings is in some interval—say from 12 to 18. See figure 6.1.

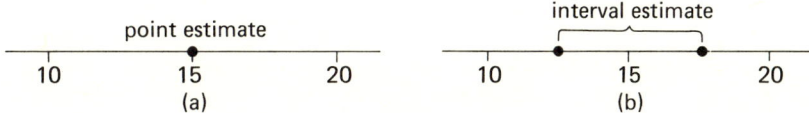

Figure 6.1 Point Estimate and Interval Estimate

We define two types of estimates as follows:

DEFINITION 6.1

point estimate

A single number (or point) used to estimate a population parameter is called a point estimate.

140 Making Inferences: Estimation

DEFINITION 6.2

interval estimate

An estimate of a population parameter formed by two numbers that determine an interval on a line is called an <u>interval estimate</u>.

Any statistical inference, whether related to estimation or testing, would be incomplete without a measure of how good the inferential procedure is. For example, not only would we like to estimate the population mean decrease in anxiety ratings, but also we would like to know how accurate our estimate is. Is the estimate within a specified distance (say 5 mm Hg) of the actual mean decrease? Don't be concerned now about how we will determine the accuracy of a statistical inference; we will discuss that later. Just remember that statistical inference-making procedures involve two elements: the inference and a measure of its goodness (accuracy).

Before concluding this introductory discussion of inference, we should consider a question that frequently disturbs the beginner: Which method of inference should be used? That is, should the value of the parameter be estimated or should we test an hypothesis concerning its value? The answer to this question is dictated primarily by the research question that has been posed or in some cases, is a personal preference. Many substantive problems involve testing hypotheses about parameters; others involve making estimates. We will employ estimation procedures in this chapter and tests of hypotheses in chapter 7. In both chapters we will confine our attention to inferences based on a single large sample selected from a population. Inferences based on small sample sizes will be presented in chapter 9.

EXERCISES

 6.1. A researcher is interested in estimating the percentage of registered voters in her state who have voted in at least one election over the past two years.
(a) Identify the population of interest to the researcher.
(b) How might you select a sample of voters to gather this information?

6.2. Refer to exercise 6.1. Is the researcher faced with a problem related to estimation or testing an hypothesis? What is the parameter of interest?

 6.3. A manufacturer claims that the average lifetime of a particular fuse is 1500 hours. Information from a sample of 35 fuses shows that the average lifetime is 1380 hours. What can be said about the manufacturer's claim?

(a) Identify the population of interest to us.
(b) Would an answer to the question posed involve estimation or testing an hypothesis?

6.4. Refer to exercise 6.3. How might you select a sample of fuses from the manufacturer to test the claim?

6.2 POINT ESTIMATION OF A POPULATION MEAN

Many practical problems lead to the estimation of a population mean μ. For example, a psychologist may want to estimate the average reaction time of patients given a particular stimulus, an economist may want to estimate the mean increase in wages for various unions, and a sociologist may wish to estimate the mean family income in a ghetto area. We will discuss point estimation of a population mean μ in this section and interval estimation of μ in section 6.3.

As we stated previously, statistical inferences differ from most other inferences in that we not only make an inference but we also provide a measure of how good our inference is. For example, the value of the sample mean $\bar{y}$ computed from the sample measurements will be a point estimate of μ, the population mean. This is our guess for the value of μ. However, since there are many possible values of $\bar{y}$, and since it is unlikely that the observed value of $\bar{y}$ will exactly equal μ, we must say how close $\bar{y}$ should be to μ, the unknown parameter we're trying to estimate. To do this, we must refer to the sampling distribution of $\bar{y}$ (discussed in section 5.3). The properties of the sampling distribution of $\bar{y}$ are summarized below.

Properties of the Sampling Distribution of $\bar{y}$

The sampling distribution of $\bar{y}$, obtained by drawing many samples of size n from the same population and calculating $\bar{y}$ for each sample, has the following properties. These properties hold regardless of the form of the population from which the sample measurements are drawn.

1. The average (mean) of the distribution of sample means is $\mu_{\bar{y}} = \mu$; that is, it is the same as the mean for the population from which the sample measurements are drawn (see figure 6.2).

2. The standard deviation of the sampling distribution of the $\bar{y}$'s is

$$\sigma_{\bar{y}} = \frac{\sigma}{\sqrt{n}}$$

standard error of the mean

where σ is the standard deviation of the original population and n is the sample size. Note that the symbol $\sigma_{\bar{y}}$ is used to denote the standard deviation of the sampling distribution of $\bar{y}$; $\sigma_{\bar{y}}$ is called the standard error of the mean.

3. When n is large (30 or more), the sampling distribution of $\bar{y}$ is approximately normally distributed (see figure 6.2).

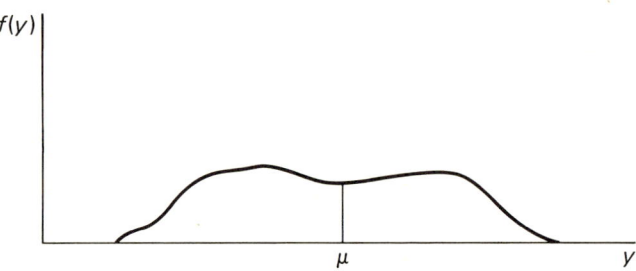

(a) Probability distribution of y, with mean μ and standard deviation σ

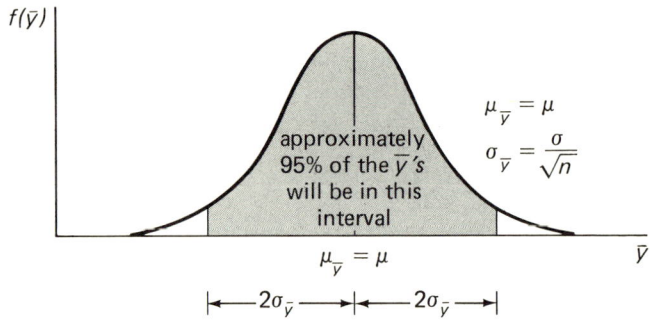

(b) Sampling distribution of $\bar{y}$, with mean $\mu_{\bar{y}} = \mu$ and standard deviation $\sigma_{\bar{y}} = \sigma/\sqrt{n}$

Figure 6.2 The Probability Distribution of y Compared with the Sampling Distribution of $\bar{y}$

From knowledge of the sampling distribution for $\bar{y}$ and of areas under the normal curve, we know that approximately 95% of the sample means calculated from repeated samples will be within $2\sigma_{\bar{y}}$ of μ (see figure 6.2).

error of estimation

The error of estimation for a single point estimate $\bar{y}$ is defined as the difference $(\bar{y} - \mu)$, ignoring the sign of the result. Stated differently, the error of estimation is the absolute value of the difference between what we think the parameter value is $(\bar{y})$ and what it actually is (μ). We designate this quantity by $|\bar{y} - \mu|$.

$$\text{error of estimation} = |\bar{y} - \mu|$$

Point Estimation of a Population Mean

bound on error

Of course, since μ is unknown, we will never know the exact value of the error of estimation. But by knowing that the sampling distribution for $\bar{y}$ is normal, we can say that $\bar{y}$ will be within a distance of $2\sigma_{\bar{y}}$ from μ approximately 95% of the time. Or, in other words, the error of estimation will be less than $2\sigma_{\bar{y}}$ approximately 95% of the time. We call $2\sigma_{\bar{y}}$ a bound on the error of estimation. This quantity is a measure of the accuracy of our point estimate. The smaller the bound on error, the better the inference is.

For any point estimation problem we will give the point estimate *and* the bound on error. Because we don't know the value of the population parameter we're trying to estimate, the bound on error gives us an indication of how close the point estimate is to the parameter. The procedure for point estimation of μ is shown below.

Point Estimation of μ

$$\text{point estimate:} \quad \bar{y}$$
$$\text{bound on error:} \quad 2\sigma_{\bar{y}} = \frac{2\sigma}{\sqrt{n}}$$

Note: When σ is unknown and when n is 30 or more, you may substitute the sample standard deviation s for σ in the bound on error.

Example 6.1

A farmer is interested in estimating the average weight gain over a two-week period for 10,000 chickens fed on a new ration. To estimate this mean gain, he draws a random sample of $n = 64$ chickens and measures the gain in weight for each. Over the two weeks he finds that the sample mean and standard deviation (in grams) are $\bar{y} = 40$ and $s = 6$. Obtain a point estimate of the average weight gain μ and place a bound on the error of estimation.

Solution

The point estimate of the average weight gain is $\bar{y} = 40$ grams. The bound on the error of estimation is

$$2\sigma_{\bar{y}} = \frac{2\sigma}{\sqrt{n}} = \frac{2\sigma}{\sqrt{64}}$$

Although we do not know σ, we may approximate its value by using s, which is calculated from the sample. (Recall that this approximation to σ will be reasonably good for $n \geq 30$.) Thus the bound on the error of estimation is approximately

$$\frac{2s}{\sqrt{n}} = \frac{2(6)}{\sqrt{64}} = \frac{12}{8} = 1.5$$

We take this to imply that the probability is approximately .95 that the error of estimation will be less than two standard errors of the mean ($2\sigma_{\bar{y}}$). Thus we feel quite confident that our estimate, $\bar{y} = 40$ grams, is within 1.5 grams of the true population mean weight gain μ.

Three points deserve further comment

1. Beginners sometimes use 2σ as a bound on the error of estimation rather than $2\sigma_{\bar{y}}$. Remember that if we want to describe the variability of the sampling distribution of $\bar{y}$, we should use its standard error $\sigma_{\bar{y}}$. The quantity σ is the standard deviation of the population from which the sample was drawn.

2. We sometimes have to approximate σ in the bound on the error of estimation. In example 6.1 we used the sample standard deviation s to estimate σ. This approximation will be reasonably good when n is large (30 or more).

3. You will note that the standard error of the sampling distribution for $\bar{y}$ depends on the standard deviation σ of the population and the sample size n, through the formula

$$\sigma_{\bar{y}} = \frac{\sigma}{\sqrt{n}}$$

Although it is not obvious that the formula should take this form, it is clear that the greater the variation of the population as measured by σ, the greater

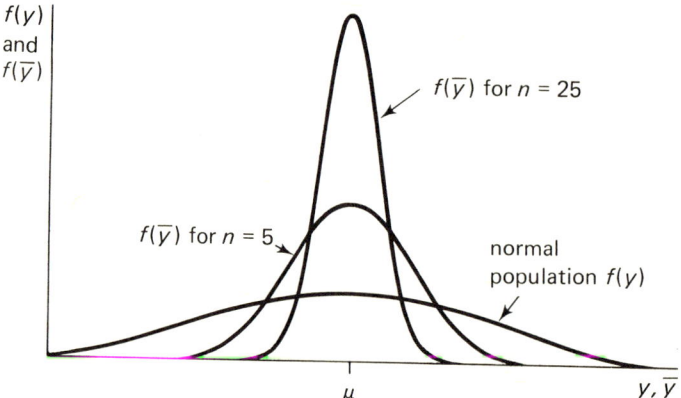

Figure 6.3 Sampling Distribution for $\bar{y}$ Based on Samples of $n = 5$ and 25 from a Normal Population

Point Estimation of a Population Mean

the variation in the sampling distribution for the sample mean $\bar{y}$. We would also expect to obtain more information as the sample size n is increased, and hence we would expect the variability of the sampling distribution of the $\bar{y}$'s to decrease as n increases. This is shown in figure 6.3.

The sampling distribution for $\bar{y}$ based on samples of 5 and 25 from a normal population are superimposed on the probability distribution for a normal population in figure 6.3. Notice how the spread of the distribution of the sample mean $\bar{y}$ decreases as the sample size n increases.

EXERCISES

6.5. Define the terms "point estimate" and "error of estimation."

6.6. A random sample of 64 measurements was obtained from a population. If the sample mean and sample standard deviation are, respectively, 12.2 and 3.0, give a point estimate of the population mean μ and the bound on error.

6.7. Repeat exercise 6.6 with $n = 40$, $\bar{y} = 63.1$, and $s = 12.5$.

6.8. Class exercise: This experiment will illustrate the notion of an estimation error and the concept of a sampling distribution for $\bar{y}$.

Experiment: Estimate some descriptive measures of the class, for example, the average height μ of all students in the class.

Unlike most practical sampling situations, the exact value of μ can be determined by recording the heights of each student and calculating the class average.
(a) To illustrate the error of estimation, draw a random sample of 5 students from the class, record their heights, and calculate the estimate $\bar{y}$. Calculate the error of estimation.
(b) To illustrate the notion of repeated sampling, draw more random samples of 5 students and compute the sample average $\bar{y}$ for each. Show how these estimates vary about μ by using a dot diagram or a frequency histogram. If this sampling process were repeated a large number of times (infinitely large), you would obtain the sampling distribution for the sample mean $\bar{y}$.

6.9. In a particular society, marriage counselors are interested in the length of time from a first marriage to a separation for divorced couples. A random sampling of 80 divorced couples showed an average time to separation of 6.3 years with a standard deviation of 2.9 years. Estimate the mean length of time to separation and place a bound on the error of estimation.

6.10. Geologists are interested in shifts and movements of the earth's surface indicated by fractures (cracks) in the earth's crust. One of the most famous large fractures is the San Andreas fault (moving fracture) in California.

A geologist attempting to study the movement of the relative shifts in the earth's crust at a particular location found many fractures in the local rock structure. Attempting to determine the mean angle of the breaks, he sampled 50 fractures and found the sample mean and standard deviation to be 39.8° and 17.2°, respectively. Estimate the mean angular direction of the fractures and place a bound on the error of estimation.

  6.11. Over a period of several years a large bank of data was accumulated for blood chemistry variables on rats that had been used as controls for a series of experiments aimed at examining the toxicity of certain substances in drinking water. The mean and standard deviation (milliunits per milliliter) for the variable SGOT (serum glutamic oxaloacetic transaminase, which is a measure of liver function), based on 110 rats of a given strain of rats, are, respectively, 160 and 50. Use these data to estimate the population mean SGOT level and place a bound on the error of estimation.

6.12. If the experimenter of exercise 6.11 desired a bound on the error of estimation no larger than 5, determine the number of control rats needed. How many additional values are needed in the data bank?

6.3 INTERVAL ESTIMATION OF A POPULATION MEAN

We can use the point estimate $\bar{y}$ to form an interval estimate for the population mean μ. From previous work we know that when n is 30 or more, the sampling distribution for $\bar{y}$ will be approximately normal with mean μ and standard deviation $\sigma_{\bar{y}}$. Thus the interval $(\mu \pm 2\sigma_{\bar{y}})$, or, more precisely, $(\mu \pm 1.96\sigma_{\bar{y}})$, includes 95% of the $\bar{y}$'s in repeated sampling. See figure 6.4.

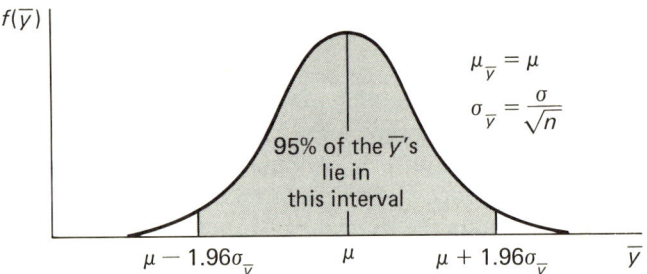

Figure 6.4 Sampling Distribution for $\bar{y}$

Consider the interval $(\bar{y} \pm 1.96\sigma_y)$. Any time $\bar{y}$ lies in the interval $(\mu \pm 1.96\sigma_{\bar{y}})$, the interval $(\bar{y} \pm 1.96\sigma_{\bar{y}})$ will contain (or capture) the parameter μ (see figure 6.5), and this will occur with probability .95. The interval $(\bar{y} \pm 1.96\sigma_{\bar{y}})$ represents an interval estimate of μ.

Interval Estimation of a Population Mean

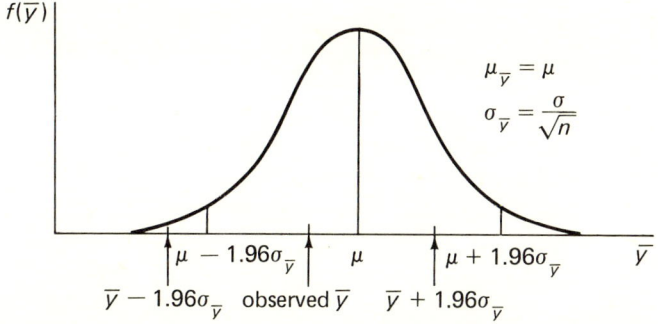

Figure 6.5 When the Observed Value of $\bar{y}$ is in the Interval $(\mu \pm 1.96\sigma_{\bar{y}})$, the Interval $(\bar{y} \pm 1.96\sigma_{\bar{y}})$ Contains the Parameter μ

We evaluate the goodness of an interval estimation procedure by examining the fraction of times in repeated sampling that the intervals contain the parameter being estimated. This fraction, called the <u>confidence coefficient</u>, is .95 when using the formula $(\bar{y} \pm 1.96\sigma_{\bar{y}})$. That is, 95% of the time in repeated sampling, intervals calculated by using the formula $(\bar{y} \pm 1.96\sigma_{\bar{y}})$ will contain the population mean μ.

confidence coefficient

This situation is illustrated in figure 6.6. Twenty different samples were drawn from a population with mean μ and variance σ^2. For each sample an interval estimate was computed by using the formula $(\bar{y} \pm 1.96\sigma_{\bar{y}})$. Note that although the intervals bob about, most of them intersect the vertical line and hence contain μ. In fact, if we repeated this process over and over again, approximately 95% of the intervals formed would contain μ.

In a given experimental situation we will calculate only one such interval. This interval, called a <u>95% confidence interval</u>, represents an interval estimate of μ.

95% confidence interval

Example 6.2

A consulting firm was contracted to study the hourly wages in a particular labor union. A random sample of $n = 70$ employee records are examined, and the sample mean and standard deviation are found to be $\bar{y} = \$13.00$ and $s = \$2.20$. Estimate μ, the average hourly wage for the entire labor union, using a 95% confidence interval.

Solution

The appropriate 95% confidence interval is computed by using the formula $(\bar{y} \pm 1.96\sigma_{\bar{y}})$. Since $n > 30$, we can substitute s for σ in the formula for $\sigma_{\bar{y}}$).

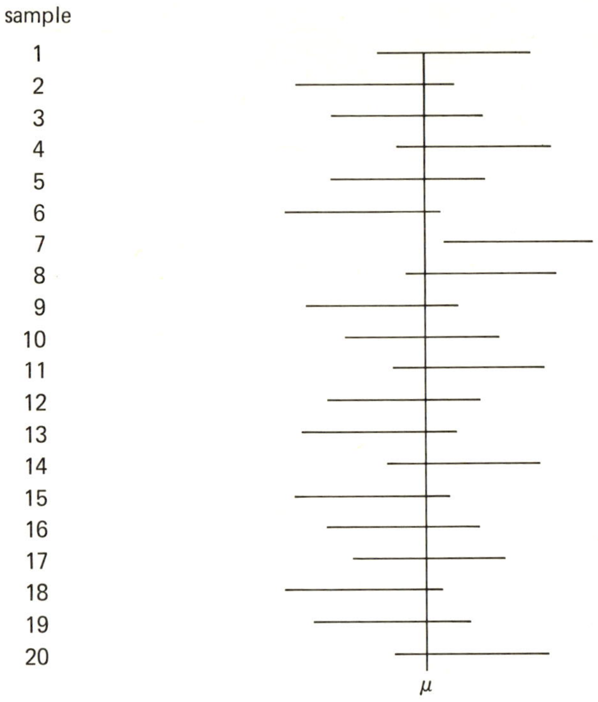

Making Inferences: Estimation

[Figure showing 20 horizontal confidence interval lines numbered 1-20, centered around vertical line labeled μ]

Figure 6.6 Twenty Interval Estimates Computed by Using $(\bar{y} \pm 1.96\sigma_{\bar{y}})$

lower confidence limit

The lower point for the confidence interval, called the lower confidence limit, is

$$\bar{y} - 1.96\sigma_{\bar{y}}$$

where $\sigma_{\bar{y}} = \sigma/\sqrt{n}$. Substituting s for σ, we find that the lower confidence limit is

$$13.00 - 1.96\left(\frac{2.20}{\sqrt{70}}\right) = 13.00 - .515 = 12.485$$

upper confidence limit

Similarly, the upper confidence limit is $(\bar{y} + 1.96\sigma_{\bar{y}})$, and in this example its value is

$$13.00 + 1.96\left(\frac{2.20}{\sqrt{70}}\right) = 13.515$$

Then the 95% confidence interval for the mean hourly wage is $12.49 to $13.52.

The general formula for a confidence interval for μ is given by $(\bar{y} \pm z\sigma_{\bar{y}})$. Different values of z are used depending on the desired degree of confidence.

Large-Sample Confidence Interval for μ ($n \geq 30$)

$$\bar{y} \pm z\sigma_{\bar{y}}$$

where $\sigma_{\bar{y}} = \sigma/\sqrt{n}$. Note: The values of z for a 90%, a 95%, or a 99% confidence interval for μ are 1.645, 1.96, or 2.58, respectively. When $n \geq 30$, you may substitute s for σ in the formula for $\sigma_{\bar{y}}$. Refer to chapter 9 when $n < 30$.

EXERCISES

6.13. Refer to the data for example 6.2. Construct a 90% confidence interval for μ, the average hourly wage for the entire labor union.

6.14. Refer to exercise 6.13. Use the same data to compute a 99% confidence interval for μ. Compare the 90%, 95%, and 99% confidence intervals for these data.

6.15. In a psychological study of depth perception 42 students were asked to judge the distance between two stationary objects. For each student's response the difference between the actual distance and the perceived distance (ignoring sign) was calculated. The mean and the standard deviation of the sample differences (in feet) were $\bar{y} = 6.5$ and $s = 2.6$. Use the sample data to estimate the population mean difference, using a 95% confidence interval.

6.16. Refer to exercise 6.15. Use the same sample data to compute a 99% confidence interval for μ. Compare your result to that for exercise 6.15.

6.17. In a study of the effects of a diuretic 30 healthy adult males were given a single dose of the drug and closely monitored to determine their urinary output over the next 24 hours. Construct a 95% confidence interval for μ if the mean urinary output (in milliliters) for the sample was 3300, with $s = 500$.

6.18. Refer to exercise 6.17. Assuming s remains approximately the same, determine the width of a 95% confidence interval based on samples of size 30, 60, 90, and 120. Discuss the effect of sample size on the width of a confidence interval.

6.19. The 1974 World Food Conference, convened in Rome, Italy, focused attention on the critical shortage of food and the ever-increasing danger of world famine. An associated statistical problem of great interest to the United States government is finding a method to estimate the total amount of grain crops that will be produced throughout the world in a particular year.

One method of predicting total crop yields is based on satellite photographs of the earth's surface. Because a scanning device will read the total acreage of a particular type of grain with error, it will be necessary to have the device read many equal-sized plots of a particular planting in order to calibrate the reading on the scanner with the actual acreage. Satellite photographs of fifty 100-acre plots of wheat were read by the scanner and gave a sample average and standard deviation equal to

$$\bar{y} = 3.27 \quad s = .23$$

Find a 95% confidence interval for the mean scanner reading for 100 acres of wheat. Explain the meaning of this interval.

6.20. Another food problem concerns the production of protein, an important component of human and animal diets. While it is common knowledge that grains and legumes contain high amounts of protein, it is not so well known that certain grasses provide a good source of protein. For example, Bermuda grass contains approximately 20% protein by weight. In a study to verify these results 100 1-pound samples were analyzed for protein content. The mean and standard deviation of the sample were

$$\bar{y} = .18 \text{ pounds} \quad s = .08 \text{ pounds}$$

Estimate the mean protein content per pound for the Bermuda grass from which this sample was selected. Use a 95% confidence interval. Explain the meaning of this interval.

6.4 POINT ESTIMATION OF THE BINOMIAL PARAMETER p

The point estimate we will use for the binomial parameter p is one that you would choose intuitively. Let p be the proportion of elements in the population that are classified as successes. Then the best estimate of p would appear to be the proportion of observed successes in the sample. If y represents the number of successes in a sample of n trials, then the sample proportion of successes, denoted by the symbol $\hat{p}$, is

$$\hat{p} = \frac{\text{number of successes}}{\text{number of trials}} = \frac{y}{n}$$

The value of $\hat{p}$ computed from a sample of n measurements will be our estimate of p. How close will this value be to the unknown parameter p? To answer this question, we must refer to the sampling distribution of $\hat{p}$.

In chapter 5 we noted that as n becomes large, the binomial random variable y possesses a mound-shaped probability distribution that approaches the normal curve. We also noted that the sampling distribution for $\hat{p}$ will possess the same shape as the probability distribution for y but that the mean and the standard deviation will be different. The mean and the standard deviation for the sampling distribution of $\hat{p}$ are as follows:

Point Estimation of the Binomial Parameter p

Mean and Standard Deviation of the Sampling Distribution of $\hat{p}$

mean: $\mu_{\hat{p}} = p$

standard deviation: $\sigma_{\hat{p}} = \sqrt{\dfrac{pq}{n}}$

where $q = 1 - p$.

---

error of estimation

The reasoning applied in evaluating the goodness of $\hat{p}$ as the point estimate of p is identical to the logic employed for evaluating how well $\bar{y}$ estimates μ. From knowledge of the sampling distribution of $\hat{p}$ we know that approximately 95% of the sample proportions calculated from repeated samples will be within $2\sigma_{\hat{p}}$ of p (see figure 6.7). The error for a single point estimate is the absolute difference between what we think the population proportion is and what it actually is:

$$\text{error of estimation} = |\hat{p} - p|$$

bound on error

Because, by the Empirical Rule, the error of estimation will be less than $2\sigma_{\hat{p}}$ approximately 95% of the time, we call $2\sigma_{\hat{p}}$ a bound on the error of estimation. Since the value of the population proportion is never known, the bound on the error of estimation indicates to us how close we think $\hat{p}$ is to p. The smaller the bound on the error of estimation, the better the inference is.

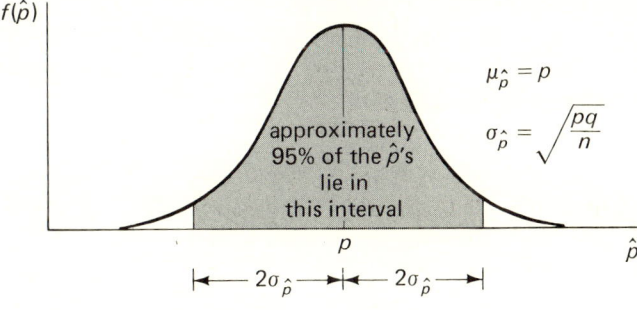

Figure 6.7 Sampling Distribution of $\hat{p}$

Point Estimation of a Population Proportion p

point estimate of p: $\hat{p} = \dfrac{y}{n}$

bound on error: $2\sigma_{\hat{p}} = 2\sqrt{\dfrac{pq}{n}}$

Making Inferences: Estimation

Note: Since p is unknown, we can obtain an approximate bound on error by substituting $\hat{p}$ and $\hat{q}$ for p and q in the formula for $\sigma_{\hat{p}}$. Provided both $n\hat{p}$ and $n\hat{q}$ are 10 or more, little error will be introduced by this substitution.

Example 6.3

A random sample of 100 seniors from a large university was selected to estimate the fraction of graduating seniors going on to graduate school. The sample produced 15 students who planned to attend a graduate school. Estimate the proportion p of the entire class planning to attend graduate school, and place a bound on the error of estimation.

Solution

The point estimate of p is

$$\hat{p} = \frac{y}{n} = \frac{15}{100} = .15$$

The corresponding bound on the error of estimation is

$$2\sqrt{\frac{pq}{n}}$$

Note that we do not know p. However, we can obtain an approximate bound on the error of estimation by substituting $\hat{p}$ for p. Provided both $n\hat{p}$ and $n\hat{q}$ are 10 or more, little error will be introduced by this substitution. Hence we obtain

$$2\sigma_{\hat{p}} = 2\sqrt{\frac{pq}{n}} \approx 2\sqrt{\frac{(.15)(.85)}{100}} = 2(.0357) = .0714$$

We estimate that the proportion of seniors who plan to attend a graduate school is .15. The bound on the error of estimating p is .0714.

EXERCISES

 6.21. A sample of 180 voters go to the polls and 50 vote for the Republican candidate. Give a point estimate of the population proportion p casting votes for the Republican candidate.

6.22. Use the information in exercise 6.21 to place a bound on the error of estimation.

Point Estimation of the Binomial Parameter *p*

6.23. Use the data in example 6.3 and assume that a second sample of 100 students was drawn with 20 students planning to attend a graduate school. Combine the data to estimate *p* for the entire class, and place a bound on the error of estimation.

6.24. Pharmacologists sometimes use a series of tests in the screening of compounds for potential use in the mental health area. Those compounds that pass the screening test are then studied in more detail. One such test in a screening process is concerned with the proportion of rats that exhibit increased self-cleaning activity (called pernicious preening) after being given a single dose of a compound. In one study 27 of 40 rats exhibited signs of pernicious preening. Use these data to construct a point estimate of *p* with a corresponding bound on the error of estimation.

6.25. An accurate estimate of the percentage unemployed (unemployment rate) in a given area is an essential measure of social well-being. A random sample of 1000 employees in Alachua County, Florida, showed that 47 were unemployed. Estimate the proportion unemployed in the county and place a bound on your error of estimation.

6.26. The accompanying news article appeared recently in the press. You will note that a survey found that 80% of those interviewed (2500 adults) had adopted the new social values.

Self Culture

Washington—A recent news survey involving approximately 2500 adults reveals some fascinating changes in the mood of the country. Questions related to one's values, attitudes, and lifestyles were posed. The results show that clearly the Puritan ethic, characterized by self-denial, conformity, hard work, and upward mobility, has given way to a new set of values: self-expression, self-gratification, and self-fulfillment. In the survey 80% of those interviewed indicated that they have adopted the new value system to some extent. This is in contrast to results from a similar survey in the mid 1970s, when only 60% said they had adopted the new social values in some way.

(a) If this survey is to be of value to an unbiased observer, the sample proportion, .80, should be an estimate of the proportion of people all over the country who have adopted the new value system. If this is the case, what assumptions must be made concerning the sample of 2500 adults?

(b) How good is the point estimate .80? (Place a bound on the error of estimation.)

6.27. The accompanying news report describes the results of a recent national survey.

Only 56.8% Feel Job Security

New York—A recently conducted telephone poll of 1688 people showed that only 56.8 percent of U.S. adults feel that they—or their spouses—are safe from job layoffs over the next six months. This figure represents the highest degree of job apprehension since the 1930s.

Other attitudes emerged in the study. Here are some of the principal results.

• Data for the months November through January show that for the first time since World War II, less than 75 percent of Americans feel they have job security.

• In the polled sample, 56.8 percent said that the main family supporter had no chance of being laid off. However, 36.9 percent felt that there was some chance—even a good chance—that they would lose their jobs.

• During the last year, the percentage of adults expressing job security has fallen from 85.9 percent, the sharpest drop since the 1930s.

(a) How accurate is the estimate of the proportion of U.S. adults who feel that they or their spouses are safe from job layoffs over the next six months? (Hint: Find a bound on the error of estimation.)

(b) Does this article provide all the information you need to know to be certain that the survey is valid? Explain.

6.28. Class exercise: If a balanced coin is flipped a very large number of times, the proportion of outcomes resulting in heads will (for all practical purposes) equal $1/2$.

(a) To illustrate the notion of sampling error, flip a coin 25 times and estimate the probability p of tossing a head. Calculate a bound on your error of estimation. Since you know $p = 1/2$, then

$$\sigma_{\hat{p}} = \sqrt{\frac{pq}{n}} = \sqrt{\frac{(1/2)(1/2)}{25}} = .1$$

(b) Have 20 different students repeat the process to obtain several estimates. Note how their estimates vary about $1/2$. Most of the errors of estimation should be less than $2\sigma_{\hat{p}}$.

(c) Combine all your sample results into one sample to obtain a more accurate estimate of p. Calculate a bound on the error of estimation. This bound will be smaller than for the samples of 25 because it is based on a larger number of tosses.

 6.29. A survey of cancer patients treated in ten regional hospitals shows that of 874 patients treated for cancer, 257 have survived for a period of five years or longer. Estimate the proportion of cancer patients treated at one of these regional hospitals who will survive for a period of at least five years and place a bound on your error of estimation. In order for this inference to be valid, what assumptions must be made regarding the sample data?

6.5 INTERVAL ESTIMATION OF THE BINOMIAL PARAMETER p

A 95% confidence interval for estimating the population proportion p has the same form as a 95% confidence interval for a population mean μ (see section 6.3). We know from the sampling distribution of $\hat{p}$ that $\mu_{\hat{p}} = p$ and that the interval $(\mu_{\hat{p}} \pm 2\sigma_{\hat{p}})$, or, more precisely, $(\mu_{\hat{p}} \pm 1.96\sigma_{\hat{p}})$, includes 95% of the sample proportions $\hat{p}$ found in repeated sampling. Any time $\hat{p}$ lies in the interval $(\mu_{\hat{p}} \pm 1.96\sigma_{\hat{p}})$, the interval $(\hat{p} \pm 1.96\sigma_{\hat{p}})$ will contain $\mu_{\hat{p}} = p$. This occurs with a probability equal to .95. This gives us confidence (measured by the confidence coefficient .95) that in a practical situation the interval calculated from a single sample when using the formula $(\hat{p} \pm 1.96\sigma_{\hat{p}})$ will enclose the parameter p. The interval estimate is called a 95% confidence interval.

Example 6.4

A sample of 1000 working-class people in Great Britain was interviewed to determine each person's political party affiliation. If 680 identified with the major left-of-center party, use a 95% confidence interval to estimate the true proportion p of Great Britain's working class that identified with the left-of-center party.

Solution

The point estimate of p is

$$\hat{p} = \frac{y}{n} = \frac{680}{1000} = .68$$

Substituting $\hat{p} = .68$ and $\hat{q} = 1 - \hat{p} = .32$ for p and q, respectively, we find the 95% confidence interval has a lower limit of

$$\hat{p} - 1.96\sqrt{\frac{\hat{p}\hat{q}}{n}} = .68 - 1.96\sqrt{\frac{(.68)(.32)}{1000}}$$
$$= .68 - 1.96\sqrt{.000218} = .68 - .03 = .65$$

and an upper limit of

$$\hat{p} + 1.96\sqrt{\frac{\hat{p}\hat{q}}{n}} = .68 + .03 = .71$$

The 95% confidence interval for p is then .65 to .71. We do not know whether this interval contains the parameter p. However, since 95% of the intervals $(\hat{p} \pm 1.96\sigma_{\hat{p}})$ contain p in repeated sampling, we are confident that the interval .65 to .71 includes the true proportion of Great Britain's working class that identify with the left-of-center political party.

Making Inferences: Estimation

In section 6.4 we indicated that it is possible to construct a general large-sample confidence interval for μ by using the formula $(\bar{y} \pm z\sigma_{\bar{y}})$. The corresponding general formula for estimating p is as follows:

Large-Sample Confidence Interval for p

$$\hat{p} \pm z\sigma_{\hat{p}}$$

where

$$\sigma_{\hat{p}} = \sqrt{\frac{pq}{n}} \approx \sqrt{\frac{\hat{p}\hat{q}}{n}}$$

Note: The z values corresponding to a 90%, a 95%, or a 99% confidence interval are, respectively, 1.645, 1.96, or 2.58. This confidence interval is valid only if $n\hat{p}$ and $n\hat{q}$ are both 10 or greater.

EXERCISES

6.30. Refer to example 6.4. Construct a 90% confidence interval for p and compare this result to the corresponding 95% confidence interval.

6.31. A firm wishes to determine the proportion of new washing machines that require servicing sometime during their first six months of operation. A random sample of $n = 50$ sales records is examined. The records disclose that $y = 13$ of the machines have been serviced within six months of the date of purchase. Use this information to construct a 95% confidence interval for the proportion of all machines requiring servicing within six months.

6.32. Use the data of exercise 6.31 to construct a 99% confidence interval for p. Compare your results.

6.33. It is not uncommon for patients to complain of certain side effects that accompany the use of a drug product. For example, certain muscle relaxants can cause drowsiness in some individuals and not in others. In a sample of 150 users of a given product, 38 complained of a particular side effect. Estimate the population proportion who would experience this side effect. How might your estimate differ if information about the side effect was solicited rather than volunteered?

6.34. A federal highway study was undertaken to evaluate the effectiveness of a new commuter bus system for relieving highway congestion and reducing gasoline consumption. One measure of the effectiveness of the system is the proportion of people working in the destination areas that make use of the bus

Summary

system. A random sample of $n = 1200$ people working in these areas was selected, using census information. Of the 1200 people interviewed 80 used the system. Construct a 95% confidence interval for the proportion of workers using the system.

6.35. A survey was conducted to investigate the proportion of registered nurses in a particular state that are actively employed. A random sample of 400 nurses selected from the state registry showed 274 actively employed. Find a 95% confidence interval for the proportion of registered nurses actively employed.

6.36. A survey conducted to determine the proportion of college students favoring "more than equal" job-rights opportunities for women (to offset past injustices) showed 258 of a random sample of 1000 favoring the proposal. Estimate the proportion of all college students favoring the proposal, using a 95% confidence interval.

6.37. The results of a recent national survey are presented in the news report shown here.
(a) Does the article provide all the information you need to be certain that the survey is valid? Explain.
(b) Can you place a bound on the error of estimation for the proportion of people who feel that the administration is not doing enough about energy? Explain.

55% of People Disenchanted with Energy Policy in Washington

Washington—According to a recent national survey, 55% of the adults who were questioned indicated that the administration is not doing enough to lessen our dependence on foreign oil supplies.

Following the meeting of OPEC ministers last month, a survey of adults in many regions of the United States was conducted. More than half of those questioned reported a lack of confidence in the depth and the extensiveness of the administration's energy program. This finding was not confined to one socioeconomic demographic group but rather was uniform throughout all subgroups examined.

The survey also showed that 63% of the people think we have the ability to develop sources for artificial fuels to help alleviate the shortages of natural gas and crude oil. One interesting finding of the survey was that only 20% of the people feel that the energy crisis is a hoax.

SUMMARY

Point estimation and interval estimation of a population mean μ and binomial proportion p based on large sample sizes are discussed in this chapter. Given the appropriate sample data, anyone can construct estimates of the desired population parameters by using statistics (an objective procedure) or by using a subjective procedure. Regardless of how an estimate is obtained, the question is, how much reliance can we place on it? Is the estimate reasonably close to the true value of the population parameter?

Subjective procedures based on experience or intuition may or may not yield good estimates. They may be given by people who truly possess sufficient experience to yield satisfactory estimates (but how will you know when you have a person with this ability and how will you know how accurate the estimate is?) or they may be presented by those people who are inadequately prepared to do the job.

One of the major contributions of statistics is the insight that all inferences must be accompanied by a measure of their goodness. We talk of point estimates with a bound on the error of estimation or of confidence intervals with specified confidence coefficients. This is the great advantage of statistics. By using statistical estimation procedures, we make an estimate and we know how much we can rely on it.

In this chapter we emphasized the basic concepts of statistical estimation and illustrated these concepts by considering inferences based on single samples: estimation of population means and estimation of proportions. Having given a brief introduction to estimation, we turn next to a second method of making inferences—testing hypotheses.

REFERENCES

Barr, A. J.; Goodnight, J. H.; Sall, J. P.; Helwig, J. T. *A User's Guide to SAS 76.* Raleigh, N.C.: SAS Institute, Inc., 1976.

Dixon, W. J. *BMDP: Biomedical Computer Programs.* Berkeley: University of California Press, 1975.

Freund, J. E. *Statistics: A First Course.* 2d ed. Englewood Cliffs, N.J.: Prentice-Hall, 1976.

Mendenhall, W. *Introduction to Probability and Statistics.* 5th ed. N. Scituate, Mass.: Duxbury Press, 1979. Chapter 8.

Nie, N.; Hull, C. H.; Jenkins, J. G.; Steinbrenner, K.; and Bent, D. H. *Statistical Package for the Social Sciences.* 2d ed. New York: McGraw-Hill, 1975.

Ryan, T. A.; Joiner, B. L.; and Ryan, B. F. *Minitab Student Handbook.* N. Scituate, Mass.: Duxbury Press, 1976.

Service, J. *A User's Guide to the Statistical Analysis System.* Raleigh, N.C.: Student Supply Stores, North Carolina State University, 1972.

Walpole, R. E. *Introduction to Statistics.* 2d ed. New York: Macmillan, 1974. Chapter 9.

SUPPLEMENTARY EXERCISES

6.38. What are parameters and how do they differ from statistics?

 6.39. Is there a need for estimation procedures in situations where the sam-

Supplementary Exercises

ple constitutes the population of interest? For example, would there be a need to determine a bound on error for an opinion survey involving all the U.S. senators? Why or why not?

6.40. Distinguish between a point and an interval estimate. What are the two basic procedures for making inferences? How do we measure the goodness of a point estimation procedure? An interval estimation procedure?

6.41. A random sample of insurance records of $n = 500$ physicians, selected from the files of an insurance company, shows that 10% of the physicians have been involved in one or more lawsuits. Estimate the proportion of all physicians covered by the insurance company who have been involved in lawsuits. Place a bound on the error of estimation.

6.42. Projected due dates for expectant mothers have, over the years, been notoriously bad. In a recent survey of 100 mothers it was found that the average number of days to birth beyond the projected due date was 9.2, with a standard deviation of 12.4. Use these data to find a 95% confidence interval for the mean number of days to birth beyond the due date.

6.43. A pharmacist decides to reevaluate the prices he charges. To do this, he attempts to determine the average cost of filling a prescription over the past year. A random sample of 110 prescription records is examined to determine the cost of filling each order. The sample mean and standard deviation are $3.50 and $1.40, respectively. Estimate the mean cost μ for all prescriptions written during the past year. Place a bound on the error of estimation.

6.44. Refer to exercise 6.43 and find a 95% confidence interval for μ. What do we mean by saying we are 95% confident that the interval will enclose the true mean μ?

6.45. A sample of 100 department store accounts receivable gave a mean and standard deviation of $\bar{y} = \$25.83$ and $s = \$16.12$. Estimate the mean value for accounts receivable, and place a bound on the error of estimation.

6.46. Refer to exercise 6.45 and note that s is rather large in comparison to $\bar{y}$. This difference in size is due to the fact that the distribution of accounts receivable is nonsymmetric and skewed to the right. Assuming this to be true, why can you assume that the sampling distribution of $\bar{y}$ will be normal? (Hint: See section 6.2.)

6.47. Give a 95% confidence interval for μ in exercise 6.45.

6.48. A corporation maintains a large fleet of company cars for its salespeople. In order to determine the average number of miles driven per month by all salespeople, a random sample of 70 records was obtained. The mean and the standard deviation for the number of miles driven were 3250 and 420, respectively. Estimate μ, the average number of miles driven per month for all the salespeople within the corporation, using a 99% confidence interval.

160 Making Inferences: Estimation

 6.49. The length of time to assemble an electronic fuse was measured for 50 assemblers. The mean and the standard deviation were 3.2 and .3 minutes, respectively. Give a 90% confidence interval for the mean length of time to assemble a fuse.

 6.50. The accompanying news report gives the results of a recent national survey, which indicate that a frighteningly large number of teenagers either have driven while drunk or have been passengers in cars with drivers who had been drinking heavily.
(a) Are you given enough information in the article to be certain that the survey is valid? Explain.
(b) Can you place a bound on the error of estimation for the proportion of teenagers who say they have driven once or twice when they knew themselves to be too drunk to drive? Explain.
(c) Might there be a tendency for some teenagers to understate the number of times they had been driving while under the influence of alcohol? Explain.
(d) How might the report of this survey be improved to allow the reader to draw his or her own conclusions?

Study of Teenage Drinkers Completed

Washington—A recent study conducted by the National Highway Traffic Safety Administration revealed that an "alarming . . . frightening" number of teenagers have either driven while drunk or have been passengers in cars with heavy-drinking drivers.

The study found that neither scare tactics nor legal threats discourage teenage drinking and driving. The legal consequences of being stopped by police are not considered as serious by teenagers. They also do not consider death or a crippling injury a likely consequence, according to the study.

The study indicated other aspects of teenage drinking:

• Many teenagers admit they have frequently driven when they are "pretty drunk."

• One-fourth say they have driven once or twice when they knew themselves to be too drunk to drive.

• Another one-fourth have driven three or more times when drunk.

• Of the teenagers interviewed, 32% said they were passengers at least once a month in a car operated by a heavily drinking driver.

The agency's spokesman said that its future activities would be directed toward encouraging social pressures against drinking and driving.

 6.51. Two hundred voters were randomly selected and 110 were found to favor candidate *A*. Estimate the fraction of voters favoring candidate *A*, using a 95% confidence interval.

6.52. Response to an advertising display was measured by counting the number of prospects who purchased out of the total exposed to the display. If 330 purchased out of a total of 870 exposed, estimate the probability of purchase, using a 95% confidence interval.

6.53. In a random sample of 150 students on a college campus, 72 students were in favor of an increase in their activities fee to help fund a proposed coliseum. Estimate p, the fraction of the entire student body in favor of the fee increase. Place a bound on the error of estimation.

6.54. The state highway patrol wishes to estimate the fraction of automobiles with badly worn tires that use a particular turnpike. In a spot check (random sample) of 300 cars, 58 cars were observed to have bad tires. Estimate p, the fraction of the automobiles on the turnpike with badly worn tires, using a 90% confidence interval.

6.55. A random sample of 400 fuses was selected from a day's production and tested; 40 were found to be defective. Estimate the fraction defective in the day's production, using a 99% confidence interval.

6.56. A recent Gallup poll was taken to determine the proportion of Americans that are aware of the separatist movement in Canada, which is in favor of independence for the province of Quebec. The sample data, based on personal interviews with 1508 adults 18 years of age or older, show that 784 were aware of the separatist movement. Use these data to construct a 99% confidence interval for the proportion of all adult Americans who are aware of the Canadian separatist movement.

6.57. Refer to exercise 6.56. The same adults were asked whether they favored independence for Quebec and 181 said yes. Construct a 95% confidence interval for the proportion of American adults favoring independence for the province of Quebec.

6.58. Americans (as well as the people of most highly industrialized nations) have had to live with the realities of inflation for several years, and people are still concerned with the problem. In a recent survey of 1540 American adults 18 years of age or older, 830 indicated that they felt inflation was the most important problem facing the nation today. Use those data to construct a 95% confidence interval for p, the proportion of adults who rate inflation as the major problem facing our nation.

EXPERIENCES WITH REAL DATA

Conduct a sample survey to determine the attitude of your student body on a major political, social, or campus issue. Use the methods of section 5.2 to select a large random sample of n students (say $n = 400$ or more) from the student directory. Since each of the n students in the sample must be contacted by telephone to determine his or her attitude concerning the question, this chore should be divided so that each student (or a small group of students) is responsible for contacting a fixed number, say 25, of the total.

When the data have been collected, pool the n student responses and obtain an estimate of the proportion of students in the student body who favor the issue. Place a bound on your error of estimation.

To observe sampling variation, let each student (or team) use the 25 students whom he or she was to contact to estimate the proportion in favor of the issue. Collect the estimates from each team and construct a histogram of the estimates. Locate the estimate based on all n students on the graph. Notice how the estimates based on the samples containing 25 students tend to vary and that the large-sample estimate (based on all n students) falls near the center of the histogram.

The calculations for this experiment can most easily be accomplished on an electronic desk calculator. However, for other analyses considered in this chapter, you might wish to use packaged programs and an electronic computer to perform the calculations. Useful packaged programs are available in the Biomed (Biomedical Programs), the SAS (Statistical Analysis Systems), and the SPSS (Statistical Package for the Social Sciences) program libraries (see the References).

7

Making Inferences: Testing Hypotheses

CHAPTER OUTLINE

7.1 Introduction
7.2 Testing an Hypothesis About a Population Mean
7.3 Testing an Hypothesis About a Binomial Parameter
7.4 The Level of Significance of a Statistical Test

7.1 INTRODUCTION

In chapter 6 we discussed large-sample estimation procedures for a population mean μ and a population proportion p. These procedures were aimed at answering the question, "What is the value of μ (or p)?" A second kind of inferential procedure is the statistical test, in which we attempt to answer a specific question about the value of the unknown population parameter. In this chapter we will discuss the elements of a statistical test and illustrate the procedure for a large-sample test about μ or p.

A statistical test of an hypothesis can be likened to a court trial. We begin with a **research hypothesis**, something that we wish to verify. For example, in the court trial the research hypothesis is that the defendant is guilty. The prosecuting attorney attempts to verify this hypothesis by showing that its antithesis—that the defendant is innocent—is false. This latter hypothesis, called a **null hypothesis**, is the crux of a statistical test. If we can collect evidence to show that the null hypothesis is false, we can conclude that the research hypothesis (also called **alternative hypothesis**) is true.

How does the statistician, or the court, decide which hypothesis is true: the null hypothesis or the alternative hypothesis? In both cases evidence is collected; the evidence for the statistician is the information contained in a sample selected from the population. Then this evidence is weighed (or considered) so that a decision can be made. In the court trial a jury functions as a decision maker, weighing the evidence to reach a decision. In a statistical test

test statistic of an hypothesis we utilize a test statistic, some quantity computed from the sample measurements, to assist us in reaching a decision. We use this quantity in the following way: **If the test statistic takes a value that is contradictory to the null hypothesis, we reject the null hypothesis and conclude that the alternative is true.** Similarly, in a court trial, if the evidence presented to a jury is highly contradictory to the hypothesis of innocence (null hypothesis), the jury rejects the null hypothesis and concludes that the defendant is guilty (that is, that the alternative hypothesis is true).

A statistical test is made up of four parts: a null hypothesis, an alternative (research) hypothesis, a test statistic, and a rejection region. These four elements are defined as follows:

The Four Elements of a Statistical Test

null hypothesis
1. A null hypothesis: an hypothesis about a population parameter. We sometimes designate a null hypothesis by the symbol H_0.

alternative hypothesis
2. An alternative (research) hypothesis: an hypothesis we will accept if the null hypothesis is rejected. We sometimes designate an alternative hypothesis by the symbol H_a.

test statistic
3. A test statistic: a quantity computed from the sample data.

rejection region
4. A rejection region: a set of values for the test statistic that are contradictory to the null hypothesis and imply its rejection.

To summarize, we test an hypothesis in much the same manner as a court tries a defendant. We begin by specifying the null hypothesis and the alternative hypothesis. Then a random sample is drawn from the population of interest and the value of a test statistic is computed from the sample values. The decision to accept or reject the null hypothesis depends upon the computed value of the test statistic.

How do we decide which values of the test statistic imply rejection of the null hypothesis and which do not? The answer is that we consider the set of all values that the test statistic could possibly assume. Then we divide the set **acceptance region** into two regions: one corresponding to a rejection region and one to an acceptance region. (How this division is made will be explained subsequently.) This situation is shown symbolically in figure 7.1.

A point on the horizontal line (figure 7.1) corresponds to a possible value of the test statistic. We symbolically divide these values with a vertical line to obtain two sets, one corresponding to rejection and the other to acceptance of the null hypothesis. **If the computed value of the test statistic falls in the rejec-**

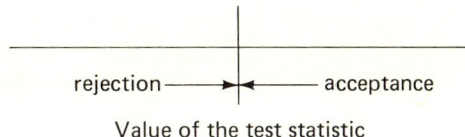

Figure 7.1 All Possible Values of a Test Statistic

tion region, the null hypothesis is rejected (and the alternative hypothesis verified. Otherwise, the null hypothesis is accepted. (See p. 169 for further discussion.) Notice that figure 7.1 shows how the rejection region might be located for one type of test. As you will subsequently see, for some tests the rejection region might lie entirely to the right of the acceptance region; for other tests it might be divided on either side of the acceptance region.

7.2 TESTING AN HYPOTHESIS ABOUT A POPULATION MEAN

In this section we will present a statistical test that will lead to an answer to the question, "Is the population mean μ equal to μ_0 (a specified value)?" For example, in studying the properties of a new antihypertensive drug product, a pharmaceutical company might want to examine the drop in diastolic blood pressure following a single fixed dose of the antihypertensive product. In particular, it may be important to show that the average drop in standing diastolic blood pressure two hours after administration of a single fixed dose is greater than 10 millimeters Hg, the average drop for a previously studied, weak antihypertensive product. Thus for this example the research hypothesis of interest to the pharmaceutical company is $\mu > 10$. To verify the alternative (research) hypothesis, we try to contradict the null hypothesis, $\mu = 10$.

Once the alternative and null hypotheses have been stated, a random sample of hypertensive patients (those patients having a standing diastolic blood pressure between 105 millimeters Hg and 130 millimeters Hg) would be screened and entered into the trial. The standing diastolic blood pressure would be recorded for each patient immediately prior to administration of the single fixed dose of medication and again two hours afterward. The response of interest is the drop in standing diastolic blood pressure at two hours postmedication.

The decision to accept or reject the null hypothesis in favor of the alternative hypothesis would be based on the value of a test statistic computed from the sample data. A logical test statistic for a test of hypothesis related to μ is $\bar{y}$, the sample mean drop in standing diastolic blood pressure.

Making Inferences: Testing Hypotheses

If we choose $\bar{y}$ as the test statistic, and if we assume the null hypothesis is true, we know that the sampling distribution of $\bar{y}$ will be approximately normal, with mean $\mu = 10$. Values of $\bar{y}$ that are contradictory to the null hypothesis (and favor the alternative hypothesis) will be those values that lie in the upper tail of the sampling distribution. See figure 7.2.

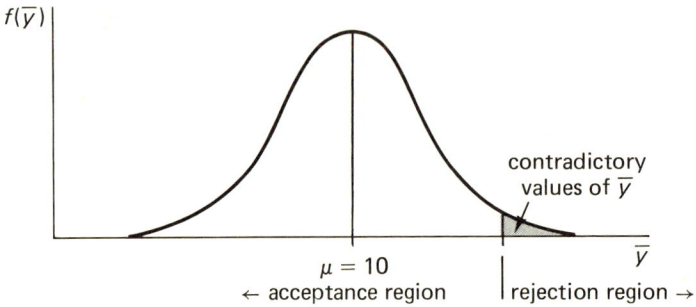

Figure 7.2 Assuming H_0 Is True, Contradictory Values of $\bar{y}$ Are in the Upper Tail

The contradictory values of $\bar{y}$ form a rejection region for our statistical test. If the observed value of $\bar{y}$ falls in the rejection region of figure 7.2, we will reject the null hypothesis ($\mu = 10$) in favor of the alternative hypothesis ($\mu > 10$). Note that we are attempting to verify the alternative hypothesis by contradicting the null hypothesis. If the observed value of $\bar{y}$ falls in the acceptance region, we will accept the null hypothesis.

Before we give a precise location for the acceptance and rejection regions of figure 7.2, we should consider the two types of errors that can be made while performing a statistical test of an hypothesis. As with any two-way decision, such as a jury decision in a court trial, we can make an error by falsely rejecting the null hypothesis (convicting an innocent defendant) or by falsely accepting the null hypothesis (acquitting a guilty defendant). These errors are called type I and type II errors, respectively (see table 7.1).

Table 7.1 Decisions and Corresponding Errors

	Null Hypothesis	
Decision	False	True
reject the null hypothesis	correct	type I error
accept the null hypothesis	type II error	correct

**type I error
probability α**

DEFINITION 7.1

A *type I error* is committed if the null hypothesis is rejected when it is true. The *probability* of a type I error is denoted by the Greek letter *α (alpha)*.

**type II error
probability β**

DEFINITION 7.2

A *type II error* is committed if the null hypothesis is accepted when it is false. The *probability* of a type II error is denoted by the Greek letter *β (beta)*.

The probability α of making a type I error is the probability of rejecting the null hypothesis when it is true. The probability β of a type II error is the probability of accepting the null hypothesis when it is false. Ideally we would like both α and β to be zero, but this is impossible. As an alternative we might try to make both α and β small. Although it is usually much more difficult to determine β than α, there is a relationship between these two probabilities. For a given sample size, α and β are inversely related; that is, as one goes up, the other goes down, as shown in figure 7.3.

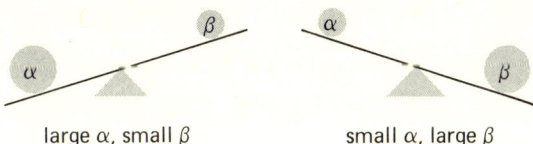

large α, small β   small α, large β

Figure 7.3 Relation Between α and β

As we will show subsequently, the experimenter must specify a tolerable value for α prior to running the statistical test. Thus he or she may choose α to be .01, .05, and so on. Specification of a value for α locates the rejection region. Determination of β is more complicated and is beyond the scope of this text. **Hence in this text, if the observed value of the test statistic does not fall in the rejection region, we will withhold judgment by specifying that there is insufficient evidence to reject the null hypothesis, thus eliminating the possibility of committing a type II error.**

Let us now see how the choice of α locates the rejection region. Returning to our investigation involving the antihypertensive drug product, we would reject the null hypothesis ($\mu = 10$) for large values of the sample mean $\bar{y}$.

Suppose that we decide to take a random sample of $n = 45$ hypertensive patients. The average drop in standing diastolic blood pressure two hours after administration of a single dose of the product is $\bar{y} = 15.3$, and the standard deviation is $s = 5.1$. Can we reject $H_0: \mu = 10$ and conclude that $\mu > 10$?

Before answering this question we must specify α. If we are willing to take the risk that one time in 40 we will incorrectly reject the null hypothesis, than $\alpha = 1/40 = .025$. An appropriate rejection region can be specified for this (or any other) value of α by referring to the sampling distribution of $\bar{y}$. If the null hypothesis is true, then $\bar{y}$ is normally distributed with mean $\mu_{\bar{y}} = \mu = 10$ and standard deviation $\sigma_{\bar{y}} \approx s/\sqrt{n} = 5.1/\sqrt{45} = .76$. Since the shaded area in figure 7.2 corresponds to α, we must locate the rejection region so that an area $\alpha = .025$ lies in the upper tail (above $\mu = 10$). From our knowledge of the normal curve, we know that the rejection region is located at a distance of 1.96 standard deviations ($1.96\sigma_{\bar{y}}$) above the mean $\mu = 10$ (see figure 7.4). If the observed value of $\bar{y}$ is more than 1.96 standard deviations above $\mu = 10$, we reject the null hypothesis; otherwise there is insufficient evidence to reject H_0. This is shown in figure 7.4.

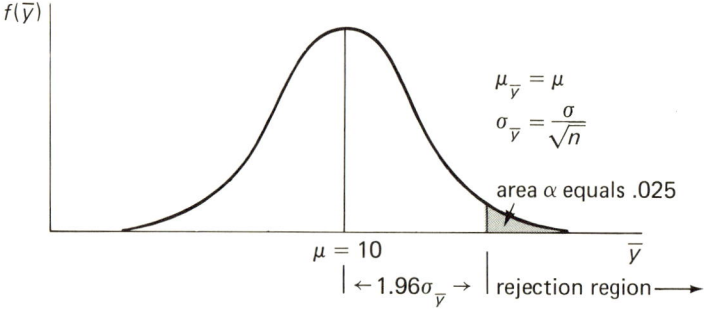

Figure 7.4 Rejection Region for the Antihypertensive Drug Example When $\alpha = .025$

Example 7.1

Set up all parts of the statistical test for the antihypertensive drug example. Use the sample data ($\bar{y} = 15.3$, $s = 5.1$, $n = 45$) to reach a decision when $\alpha = .025$.

Solution

The four parts of the statistical test are as follows:

Null hypothesis: $\mu = 10$.

Alternative hypothesis: $\mu > 10$.

Test statistic: $\bar{y}$.

Testing an Hypothesis About a Population Mean

Rejection region: For $\alpha = .025$ reject H_0: $\mu = 10$ if $\bar{y}$ is more than 1.96 standard deviations above $\mu = 10$.

To determine the number of standard deviations $\bar{y} = 15.3$ lies above $\mu = 10$, we compute a z score, using the formula

$$z = \frac{\bar{y} - \mu}{\sigma_{\bar{y}}}$$

Substituting $\bar{y} = 15.3$, $\mu = 10$, and $\sigma_{\bar{y}} = .76$, we obtain

$$z = \frac{15.3 - 10}{.76} = 6.97$$

Since the observed value of $\bar{y}$ is 6.97 standard deviations above the hypothesized mean ($\mu = 10$), we reject H_0: $\mu = 10$ and conclude that the mean drop in standing diastolic blood pressure is greater than 10 millimeters Hg at the two-hour, postdrug time period.

one-tailed test

The statistical test we conducted in example 7.1 is called a <u>one-tailed test</u>, because the rejection region is located in only one tail of the sampling distribution for $\bar{y}$. If our alternative hypothesis had been H_a: $\mu < 10$, small values of $\bar{y}$ would have indicated rejection of the null hypothesis. This situation would also have been a one-tailed test, but the rejection region would have been located in the lower tail of the sampling distribution of $\bar{y}$. Figure 7.5 shows the rejection region for the alternative hypothesis H_a: $\mu < 10$ when $\alpha = .025$.

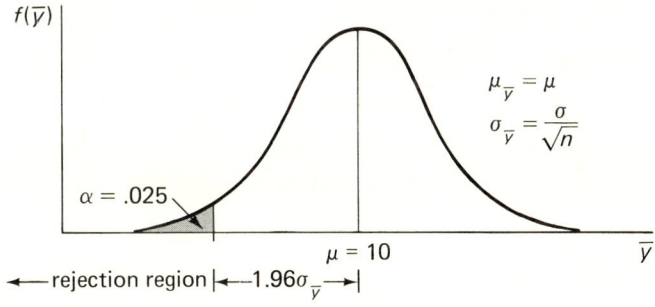

Figure 7.5 Rejection Region for H_a: $\mu < 10$ When $\alpha = .025$; for the Antihypertensive Drug Example

two-tailed test

We can also formulate a <u>two-tailed test</u> when, for example, the alternative hypothesis is of the form H_a: $\mu \neq 10$. Here we would be interested in detecting whether μ is larger or smaller than $\mu = 10$, the value specified in the null hypothesis. For a two-tailed test we locate the rejection region in both tails of the sampling distribution of $\bar{y}$. The rejection region for a two-tailed test of

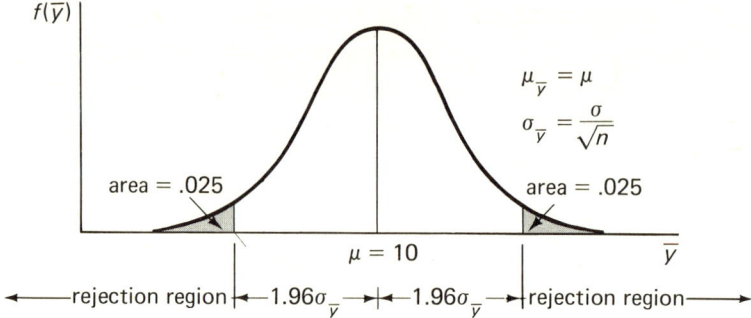

Figure 7.6 Two-Tailed Rejection Region for $H_a: \mu \neq 10$ When $\alpha = .05$; for the Antihypertensive Drug Example

$H_a: \mu \neq 10$ is shown in figure 7.6 when the probability of a type I error is set at $\alpha = .05$.

Example 7.2

A study was conducted to determine whether the average amount of money expended per household per week for food in a particular community differed from the national average ($58.00 per week). A random sample of $n = 100$ households in the community gave a mean and a standard deviation of $62.00 and $10.28, respectively. Do these data provide sufficient evidence to indicate that the mean weekly household expenditure for the community is different from the national average? Use $\alpha = .05$.

Solution

The four parts of the statistical test for this example are as follows:

Null hypothesis: $\mu = 58$, where μ is the average amount of money expended per household per week for food in this community.

Alternative hypothesis: $\mu \neq 58$.

Test statistic: $\bar{y}$.

Rejection region: Reject $H_0: \mu = 58$ if $\bar{y}$ is more than 1.96 standard deviations from $\mu = 58$.

For our example $\bar{y} = 62$ and $\sigma_{\bar{y}} = \sigma/\sqrt{n}$. Substituting s for σ and 100 for n, we obtain

$$\sigma_{\bar{y}} = \frac{10.28}{\sqrt{100}} = 1.03$$

Testing an Hypothesis About a Population Mean

The z score corresponding to $\bar{y} = 62$ is

$$z = \frac{\bar{y} - \mu}{\sigma_{\bar{y}}} = \frac{62 - 58}{1.03} = 3.88$$

Since $\bar{y} = 62$ is 3.88 standard deviations above $\mu = 58$, we reject H_0 and conclude that the average food expenditure per week in this community is greater than \$58.00. See figure 7.7.

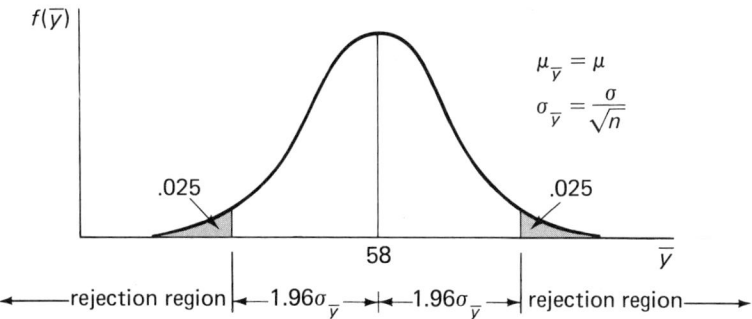

Figure 7.7 Rejection Region for Example 7.2

Now that we know how to conduct a statistical test for μ, we can simplify the mechanics of the test by making the test statistic z (the z score) rather than $\bar{y}$. For example:

Null hypothesis: $\mu = \mu_0$ (where μ_0 is some specific value).

Alternative hypothesis: $\mu > \mu_0$.

Test statistic: $z = \dfrac{\bar{y} - \mu_0}{\sigma_{\bar{y}}}$.

When H_a is $\mu > \mu_0$ and $\alpha = .025$, we reject the null hypothesis if the computed value of z is greater than 1.96—that is, if $\bar{y}$ is more than 1.96 standard deviations above the hypothesized mean μ_0. Similarly, for $H_0: \mu = \mu_0$, $H_a: \mu \neq \mu_0$, and $\alpha = .05$, we reject the null hypothesis if the computed value of z is greater than 1.96 or less than -1.96. Stated more simply, we reject H_0 if $|z| > 1.96$.

The statistical test for a population mean can now be summarized. The summary follows this paragraph. For $H_0: \mu = \mu_0$, three different alternative hypotheses (two one-sided alternatives and one two-sided alternative) are shown. The corresponding rejection regions are shown also for $\alpha = .05$ or $\alpha = .01$. Although theoretically the experimenter may choose any value for α, values of .05 or .01 are the most common and are used here. In conducting a

Making Inferences: Testing Hypotheses

statistical test for μ you must choose one of the three alternative hypotheses with its associated rejection region and set α equal to .05 or .01.

Summary of a Large-Sample Statistical Test for μ ($n \geq 30$)

Null hypothesis: $\mu = \mu_0$ (μ_0 is specified).

Alternative hypothesis: For a one-tailed test:
1. $\mu > \mu_0$.
2. $\mu < \mu_0$.

For a two-tailed test:
3. $\mu \neq \mu_0$.

Test statistic: $z = \dfrac{\bar{y} - \mu_0}{\sigma_{\bar{y}}}$ where $\sigma_{\bar{y}} = \dfrac{\sigma}{\sqrt{n}}$

Rejection region: For $\alpha = .05$ (or .01) and for a one-tailed test:
1. Reject H_0 if $z > 1.645$ (or 2.33).
2. Reject H_0 if $z < -1.645$ (or -2.33).

For $\alpha = .05$ (or .01) and for a two-tailed test:
3. Reject H_0 if $|z| > 1.96$ (or 2.58).

Note: When $n \geq 30$ you may substitute s for σ in the formula for $\sigma_{\bar{y}}$. You must use the methods of chapter 9 when $n < 30$.

EXERCISES

7.1. The mean and the standard deviation of a random sample of $n = 50$ measurements are $\bar{y} = 63.7$ and $s = 14.2$. Conduct a statistical test of H_0: $\mu = 68$ against the alternative H_a: $\mu < 68$, using $\alpha = .05$.

7.2. Refer to exercise 7.1. Would your conclusion be different if you had selected $\alpha = .01$? Explain.

7.3. To evaluate the success of a one-year experimental program designed to increase the mathematical achievement of underprivileged high school seniors, the mathematics scores for a sample of $n = 100$ underprivileged seniors were obtained for comparison with the previous year's statewide average of 525 for underprivileged seniors. You wish to examine whether there has been an increase in the mean achievement level over last year's statewide average. Discuss whether you would use a one-tailed or a two-tailed test. Set up all parts of the statistical test for μ, using $\alpha = .05$.

7.4. Refer to exercise 7.3. Suppose you wish to examine whether the mean achievement has changed (up or down) over the past year. Would you use a two-tailed test? Explain. Set up all parts of the statistical test for μ, using $\alpha = .01$.

7.5. To study the effectiveness of a weight-reducing agent, a clinical trial was conducted in which 35 overweight males were placed on a fixed diet. After a two-week period each male was weighed and then given a supply of the weight-reducing agent. The diet was to be maintained and, in addition, a single dose of the weight-reducing agent was to be taken each day. At the end of the next two-week period, weights were again obtained. Set up all parts of the statistical test for the alternative hypothesis that μ, the average weight loss, is greater than 0. Why is a one-tailed test appropriate? Use $\alpha = .05$.

7.6. Refer to exercise 7.5.
(a) The average weight loss for the second two-week period was $\bar{y} = 10.3$ pounds, and the standard deviation was $s = 4.6$. Perform a statistical test and draw conclusions. Use $\alpha = .05$.
(b) Based on the results for part (a), can you conclude that the weight-reducing agent is effective? Explain.

7.7. Transportation, getting people to their destination and home again, is a national problem. One aspect of this problem currently being studied by the Federal Highway Administration is how to merge successfully automobiles entering at high speed with congested interstate traffic. To study this problem, an automobile merging system has been installed on the entrance to I-75 at Tampa, Florida. Through the use of a series of display lights, a driver is told whether or not he is traveling at an appropriate speed to merge successfully into the existing traffic on the highway. Prior to the installation of the system, investigators measured the stress levels of many drivers merging onto the highway during the 4 to 6 P.M. rush hour period. Similar testing on a random sample of 50 drivers is to be conducted now that the merging system has been installed.

For the purposes of illustration, suppose that the average stress level prior to the installation of the system is 8.2 (measured on a 10-point scale). Set up appropriate null and alternative hypotheses to test the research hypothesis that the average stress level for drivers under the merging system is less than that observed prior to the installation of the system. Is this a one- or two-tailed test?

7.8. Refer to exercise 7.7. The sample mean and standard deviation for the

50 drivers tested are, respectively, 7.6 and 1.8. Use these data to test the alternative hypothesis of exercise 7.7. Use α = .05.

 7.9. Tooth decay generally develops first on those teeth that have irregular shapes (typically molars). The most susceptible surfaces on these teeth are the chewing surfaces. Usually the enamel on these surfaces contains tiny pockets that tend to hold food particles. Bacteria begin to eat the food particles to create an environment in which the tooth surface will decay.

Of particular importance in the decay rate of teeth, in addition to the natural hardness of the teeth, is the form of the food eaten by the individual. Some forms of carbohydrates are particularly detrimental to dental health. Many studies have been conducted to verify these findings and we can imagine how the study might have been run. A random sample of 60 adults was obtained from a given locale. Each person was examined and then maintained on a diet supplemented with a sugar solution at all meals. At the end of a one-year period the average number of newly decayed teeth for the group was .70, and the standard deviation was .4. Do these data present sufficient evidence to indicate that the mean number of newly decayed teeth for people whose diet includes a sugar solution is greater than .30, a rate that had been shown to apply to a person whose diet did not contain the sugar solution supplement? Why would a two-tailed test be inappropriate? Use α = .05.

7.3 TESTING AN HYPOTHESIS ABOUT A BINOMIAL PARAMETER

Many surveys are conducted to make inferences about the proportion of people favoring a particular issue. A random sample of n people is selected from the total population of interest to the researcher and each is interviewed to determine his or her position, pro or con, on the issue. The number y in favor divided by the sample size n represents the sample proportion, and this quantity should come close to the true (unknown) population proportion. We use the letter p to denote the population proportion and we let $\hat{p} = y/n$ denote the corresponding sample proportion.

When the number of people in the group surveyed is large relative to the sample size, this sampling procedure is a binomial experiment. The true proportion p in the population that are "in favor," then, represents the probability that the first person interviewed favors the issue. As additional people are selected for the sample, the probability of interviewing a person "in favor" remains constant, for all practical purposes. A random selection of the sample produces near independence for the binomial trials and ensures the validity of the statistical procedures that follow. Surveys that satisfy these conditions may be viewed as binomial experiments.

The general test procedure for p is identical to that for μ, with p replacing μ and $\hat{p}$ replacing $\bar{y}$. A summary follows:

Testing an Hypothesis About a Binomial Parameter

Summary of a Large-Sample Statistical Test for p

Null hypothesis: $p = p_0$ (p_0 is specified).

Alternative hypothesis: For a one-tailed test:
1. $p > p_0$.
2. $p < p_0$.

For a two-tailed test:
3. $p \neq p_0$.

Test statistic: $z = \dfrac{\hat{p} - p_0}{\sigma_{\hat{p}}}$ where $\sigma_{\hat{p}} = \sqrt{\dfrac{p_0 q_0}{n}}$

Rejection region: For $\alpha = .05$ (or .01) and for a one-tailed test:
1. Reject H_0 if $z > 1.645$ (or 2.33).
2. Reject H_0 of $z < -1.645$ (or -2.33).

For $\alpha = .05$ (or .01) and for a two-tailed test:
3. Reject H_0 if $|z| > 1.96$ (or 2.58).

Note: $q_0 = 1 - p_0$. This test is valid when np_0 and nq_0 are both 10 or more.

Example 7.3

According to recent marketing research reports, 12% of potential customers (businesses) purchase a given brand of computer equipment for small computers manufactured by a company. An extensive advertising and promotional campaign for the same brand is conducted over a broad market area. At the end of the campaign a sample of 300 potential new customers is polled to determine if the customer was favorably inclined to buy the advertised equipment. If p denotes the proportion of all potential businesses that will purchase the specified brand of computer equipment, then it would be desirable from the company's standpoint to increase p through the campaign.

(a) Set up all parts of a statistical test for p to determine if the campaign was successful. Use $\alpha = .05$.

(b) Suppose 45 of the 300 sampled companies expressed an interest in the advertised brand of computers. Conduct a statistical test for these data.

Solution

(a) The statistical test is as follows:

Null hypothesis: $p = .12$.

Alternative
hypothesis: $p > .12$ (i.e., the campaign was successful).

Test statistic: $z = \dfrac{\hat{p} - .12}{\sigma_{\hat{p}}}$ where $\sigma_{\hat{p}} = \sqrt{\dfrac{(.12)(.88)}{300}} = .0188$

Rejection region: For $\alpha = .05$ and for a one-tailed test with H_a: $p > .12$, we will reject H_0 if $z > 1.645$.

(b) For 45 "successes" from the sample of 300, the sample proportion is $\hat{p} = 45/300 = .15$. Substituting into z, we have

$$z = \dfrac{.15 - .12}{.0188} = 1.60$$

Since the computed value of z does not exceed 1.645, we have insufficient evidence to show that the advertising campaign was successful.

EXERCISES

7.10. A professor wishes to determine whether a student's guessing ability on a true-false test differs from the results that could be obtained by flipping a coin to answer each question. An examination is composed of 200 questions. State the alternative (research) and null hypotheses for this statistical test.

7.11. Refer to exercise 7.10. If the student answers $y = 110$ of the questions correctly, conduct the statistical test indicated in exercise 7.10. Use $\alpha = .05$.

7.12. For $n = 400$, $y = 180$, and $\alpha = .05$, perform the calculations necessary to test H_0: $p = .4$ versus H_a: $p \neq .4$.

7.13. In a random sample of $n = 1000$ voters, $y = 560$ favored the passage of a controversial tax issue. Let p denote the proportion of all registered voters who favor the passage of the tax issue. Use these data to test H_0: $p = .5$ against the alternative H_a: $p > .5$. Use $\alpha = .05$.

7.14. The accounting department of a large manufacturing firm is concerned about the number of errors that are detected during routine audits. In a sample of 10,000 single-digit entries, the auditor detects 20 errors. Is there evidence to indicate that the error rate is greater than one in 1,000? Use $\alpha = .05$.

7.15. Airlines have kept accurate records over the past few years concerning the number of persons who purchased tickets for a flight but then did not show up at the scheduled departure time. The no-show problem has led to the practice of overbooking flights. Suppose that the Civil Aeronautics Board runs a check of this practice by sampling 500 different flights to research the number of times people have been denied a seat on a scheduled flight after having purchased a ticket in advance. Suppose 40 of these sampled flights had one or more persons denied a seat. Use these data to test the alternative hypothesis that an overbooking policy has led to seat denial more than 5% of the time (i.e., $H_a: p > .05$). Use $\alpha = .05$.

 7.16. A leakage test has been used to determine whether a large shipment of chemicals supplied in 16-ounce polyvinyl containers should be accepted from the supplier. A shipment of containers is not acceptable if the proportion of defective containers is greater than .10. Set up all parts of a statistical test for determining whether a shipment is acceptable. Draw a conclusion if $n = 100$ containers are inspected and 12 are defective. Use $\alpha = .05$.

7.17. Refer to exercise 7.16. If 18 containers were defective, what would your conclusion have been?

7.4 THE LEVEL OF SIGNIFICANCE OF A STATISTICAL TEST

In the previous sections we presented an introduction to hypothesis testing along rather traditional lines: we defined the parts of a statistical test along with the two types of errors and their associated probabilities, α and β. In recent years many scientists and other users of statistics have objected to this decision-based approach to hypothesis testing. They argue that, rather than conducting a statistical test with a preset value of α, we should specify the alternative and null hypotheses, collect the sample data, and determine the weight of the evidence for rejecting the null hypothesis. This weight, given in terms of a probability, is called the *level of significance* (or *p*-value) of the statistical test. We illustrate the calculation of a level of significance in the following example.

level of significance

Example 7.4

Consider the previously discussed experimental situation related to the testing of an antihypertensive drug product (example 7.1). Rather than specifying a preset value for α, determine the level of significance for a test of $H_0: \mu = 10$ against $H: \mu > 10$ if $\bar{y} = 11.8$ and $\sigma_{\bar{y}} = .76$.

Solution

From the sample data given here, the computed z score is

$$z = \frac{\bar{y} - \mu_0}{\sigma_{\bar{y}}} = \frac{11.8 - 10}{.76} = 2.37$$

The level of significance for this test is the probability of observing a value of

Making Inferences: Testing Hypotheses

z greater than 2.37. This probability can be found by referring to table 1 in the Appendix. The level of significance for this test is $.5 - .4911 = .0089$ (see figure 7.8).

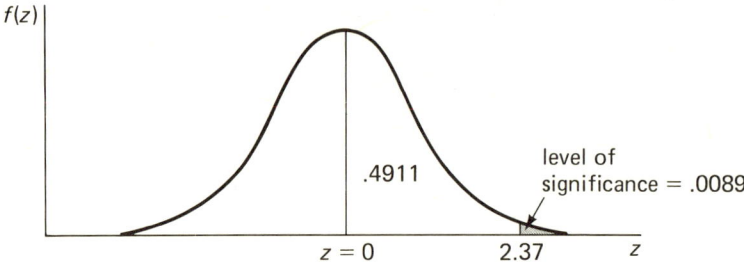

Figure 7.8 Level of Significance for Example 7.4

As shown in example 7.4, the level of significance represents the probability of observing a sample outcome more contradictory to H_0 than the observed sample result if, in fact, H_0 is true. **The smaller the value of this probability, the heavier is the weight of the sample evidence for rejecting H_0.** For example, a statistical test with a level of significance of .01 has more evidence for the rejection of H_0 than another statistical test with a level of significance of .20.

If the null and alternative hypotheses in example 7.4 had been

$$H_0: \mu = 10 \quad \text{and} \quad H_a: \mu < 10$$

and the computed value of z had been -2.37, the level of significance would still have been .0089 (see figure 7.9).

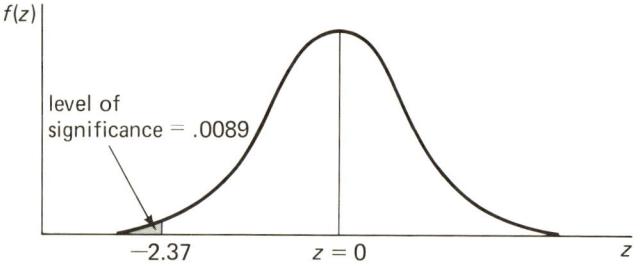

Figure 7.9 Level of Significance for $H_0: \mu = 10$, $H_a: \mu < 10$, and $z = -2.37$

The level of significance for a two-tailed test must take into account that contradictory values are located in both tails of the distribution. For example,

The Level of Significance of a Statistical Test

when testing $H_0: \mu = \mu_0$ against the alternative hypothesis $H_a: \mu \neq \mu_0$, if the computed value of z is 2.33, the level of significance is determined by finding the probability of observing a z value greater than 2.33 or less than -2.33. This value is .0198. We illustrate such a calculation in the next example.

Example 7.5

Give the level of significance for the data of example 7.4 if the null and alternative hypotheses are

$$H_0: \mu = 10 \quad \text{and} \quad H_a: \mu \neq 10$$

and the computed value of z is 2.37.

Solution

Values of z that are at least as contradictory to H_0 as the observed z (assuming H_0 is true) are those that are too large (i.e., $z > 2.37$) or too small (i.e., $z < -2.37$). Since the probability of observing a value of z greater than 2.37 or less than -2.37 is .0178, the level of significance for this test is .0178.

There is much to be said in favor of using the level of significance to summarize the results for any statistical test. Rather than reaching a decision directly, the statistician (or person performing the statistical test) presents the experimenter with the weight of evidence for rejecting the null hypothesis. The experimenter can then draw his or her own conclusions. Many professional journals have followed this approach by reporting the results of a statistical test in terms of its level of significance. Thus we might read that a particular test was significant at the .05 level or perhaps the .01 level.

One word of warning should be voiced. The .05 level of significance has become a magic level, and many seem to feel that a particular null hypothesis should not be rejected unless the test achieves a p-value of .05 or less. This, of course, has resulted in part from attitudes held by people who for many years have used the decision-based approach with α preset at .05. Keep this in mind when reading journal articles or reporting the results of statistical tests. **After all, statistical significance at a particular level does not dictate practical significance. Hence after determining the level of significance of a test, the experimenter should always consider the practical significance of the finding.**

Throughout the text we will conduct statistical tests from both the decision-based approach and from the standpoint of reporting levels of significance, to familiarize you with both avenues of thought.

EXERCISES

7.18. Sample data for a statistical test of H_0: $\mu = 40$ yielded a z score of 1.86.
(a) Determine the level of significance for a test of H_a: $\mu > 40$.
(b) Determine the level of significance for a test of H_a: $\mu \neq 40$.

7.19. Suppose a statistical test about a binomial proportion p had H_0: $p = .70$ and a value of the test statistic z equal to -2.32.
(a) Give the level of significance of the test for the alternative hypothesis H_a: $p < .70$.
(b) Give the level of significance of the test for the alternative hypothesis H_a: $p \neq .70$.

7.20. A random sample of 36 cigarettes of a certain brand was tested for nicotine content. The sample mean and standard deviation (in milligrams) are, respectively, 15.1 and 3.8. Give the level of significance of the statistical test of H_0: $\mu = 14$ (the claimed nicotine content) against the alternative hypothesis H_a: $\mu > 14$.

SUMMARY

Recall that the objective of statistics is to make an inference about a population based on information contained in a sample. Chapters 2 and 3 showed how we describe a set of measurements, thus providing us with a way to phrase an inference about a population. Chapters 4 and 5 introduced the concepts of probability, probability distributions, and sampling distributions, providing us with the mechanism for making inferences.

Since populations are described by parameters, we can make inferences about them in two ways: we can test hypotheses about their values or we can estimate them. Chapter 6 introduced point and interval estimation of a population mean and a binomial proportion based on large samples. Statistical tests related to these parameters were presented in this chapter.

A statistical test is composed of four parts: a null hypothesis, an alternative (research) hypothesis, a test statistic, and a rejection region. The rejection region is a set of values of the test statistic that are contradictory to the null hypothesis. Contradictory values are those that lie too many standard deviations away from the hypothesized value of the mean.

The rejection region for most tests is found by first specifying the value for α, the probability of a type I error. Then from knowledge of the sampling distribution for the test statistic, we can locate the rejection region. One-sided alternatives require one-sided rejection regions. Two-sided alternative hypotheses have the rejection region located in both tails of the sampling distribution.

Note that the sample size n plays an important role in testing hypotheses, because it measures the amount of data (and hence information) upon which we base a decision. If the data are quite variable and n is too small, it is unlikely that we will reject the null hypothesis even when the null hypothesis is false. That is, the probability β of making a type II error will be large. This is an important point. One frequently hears that so-called experts, panels, and high-level government commissions have reached conclusions that will vitally affect our society. These reports are sometimes based on pitifully small quantities of extremely variable data that do not support the experts' conclusions. In fact, these people rarely reveal their sample sizes.

An alternative to the standard decision-based test procedure is also presented in this chapter. Rather than using a preset value of α, the experimenter computes the level of significance for the test. This p-value measures the weight of evidence for rejection of the null hypothesis. The smaller the level of significance, the more contradictory are the sample data to the null hypothesis.

REFERENCES

Barr, A. J.; Goodnight, J. H.; Sall, J. P.; Helwig, J. T. *A User's Guide to SAS 76*. Raleigh, N.C.: SAS Institute, Inc., 1976.

Dixon, W. J. *BMDP: Biomedical Computer Programs*. Berkeley: University of California Press, 1975.

Freund, J. E. *Statistics: A First Course*. 2d ed. Englewood Cliffs, N.J.: Prentice-Hall, 1976.

Mendenhall, W. *Introduction to Probability and Statistics*. 5th ed. N. Scituate, Mass.: Duxbury Press, 1979. Chapter 8.

Nie, N.; Hull, C. H.; Jenkins, J. G.; Steinbrenner, K.; and Bent, D. H. *Statistical Package for the Social Sciences*. 2d ed. New York: McGraw-Hill, 1975.

Ryan, T. A.; Joiner, B. L.; and Ryan, B. F. *Minitab Student Handbook*. N. Scituate, Mass.: Duxbury Press, 1976.

Service, J. *A User's Guide to the Statistical Analysis System*. Raleigh, N.C.: Student Supply Stores, North Carolina State University, 1972.

Walpole, R. E. *Introduction to Statistics*. 2d ed. New York: Macmillan, 1974. Chapter 10.

SUPPLEMENTARY EXERCISES

 7.21. Suppose that we wish to test the hypothesis that a voter population is equally distributed in its preferences between two candidates, *A* and *B*, against the alternative hypothesis that one of the candidates is preferred to the other. A random sample of 100 voters is selected, and the number y favoring B is recorded. Specify the appropriate null hypothesis. Construct a test of the null hypothesis, being certain to identify each of the four parts of the test. Set the rejection region so that α is approximately equal to .05.

7.22. Refer to exercise 7.21. Suppose that we observe 61 voters ($y = 61$) favoring B. What would you conclude? When using this test, what is the probability that you will incorrectly reject the null hypothesis?

7.23. Refer to exercise 7.21. Give the level of significance for the test.

7.24. A random group of 300 homemakers was interviewed to determine the preference for one of two types of fabric softeners. Brand A was favored by 135 homemakers; the others favored B. Do these data provide sufficient evidence to indicate a difference in preference for the two fabric softeners? Test by using $\alpha = .05$.

7.25. To determine consistency in evaluating student behavior, two evaluators were presented with a random group of 200 students for examination. Each student was examined by both of the evaluators. The evaluators agreed on 133 of the evaluations. Does this indicate that their agreement is due to reasons other than pure chance? Give the level of significance for your test.

7.26. A group of 40 rats was selected for study. Each rat's heart rate was measured prior to receiving a single dose of the test preparation and then again two hours after administration of the drug. The sample mean drop in blood pressure from the predrug reading to the two-hour postdrug reading was 30.2, and the standard deviation was 10.0. Use these data to test the null hypothesis H_0: $\mu_{\text{drop}} = 0$ against the alternative H_a: $\mu_{\text{drop}} > 0$. Use $\alpha = .05$.

7.27. Refer to exercise 7.26. Give the level of significance for the test.

7.28. A psychological experiment was conducted to investigate the length of time (time delay) between the administration of a stimulus and the observation of a specified reaction. A random sample of 36 persons was subjected to the stimulus and observed for the time delay. The sample mean and standard deviation were 2.2 and .57 seconds, respectively. Test the null hypothesis that the mean time delay for the hypothetical set of all persons who may be subjected to the stimulus is $\mu = 1.6$ against the alternative hypothesis that the mean time delay differs from 1.6. Use $\alpha = .05$.

7.29. The diameter of extruded plastic pipe varies about a mean value that is controlled by a machine setting. A random sample of the diameters of 50 pieces of plastic pipe gave a mean and a standard deviation of 4.05 and .12 inches, respectively. Do the data present sufficient evidence to indicate that the mean diameter differs from 4 inches? Use $\alpha = .05$.

7.30. A hospital claims that the average length of patient confinement is five days. A study of the length of patient confinements for 36 people showed

$\bar{y} = 6.2$ and $s = 5.2$. Do these data present sufficient evidence to contradict the hospital's claim? Use $\alpha = .05$.

7.31. The manufacturer of an automatic control claims that the device will maintain a mean room humidity of 80%. The humidity in a controlled room was recorded for a period of 30 days, and the mean and the standard deviation were found to be 78.3 and 2.9, respectively. Do the data present sufficient evidence to contradict the manufacturer's claim?

7.32. A buyer wishes to determine whether the mean sugar content per orange shipped from a particular grove is less than .027 pounds. A random sample of 50 oranges produced a mean sugar content of .025 and a standard deviation of .003 pounds. Do the data present sufficient evidence to indicate that the mean sugar content is less than .027 pounds? Use $\alpha = .05$.

7.33. A manufacturer claimed that more than 20% of the public preferred its product. A random sample of 100 persons is taken to check the claim; $y = 25$ preferred the product. Give the level of significance for a test of H_0: $p = .20$ and H_a: $p > .20$.

7.34. Why is the z test known as a large-sample statistical test?

7.35. A white mouse is running a maze that has two doors of equal size at one end. One of the doors has a piece of cheese behind it, the other does not. If the mouse does not "learn" to choose the door with the cheese, he should choose either door with probability equal to ½. In 90 trials the mouse chooses the door with the cheese 62 times ($y = 62$). Can we conclude that the mouse has "learned" where the cheese is? Use $\alpha = .05$.

7.36. A manufacturer claims that, at most, 5% of the goods it produces are defective. Out of 200 items randomly selected from production, 14 are found to be defective. Is there enough evidence to indicate that more than 5% are defective? Use $\alpha = .01$.

7.37. As part of the early laboratory work that precedes the use of a drug in human patients, an experiment was conducted to determine the effect of Benzedrine on the heart rate of dogs. A random sample of 40 dogs was included in the study. At the beginning of the experiment the heart rate (in beats per minute) was measured for each dog prior to receiving a measured dose of Benzedrine. After waiting for a fixed period of time following the administration of the drug, each dog was again examined to determine its heart rate. The sample mean increase in beats per minute was 6.2, and the standard deviation was 7.0. Use these data to test the alternative hypothesis that the mean increase in heart rate for dogs following the administration of Benzedrine is greater than zero. Use $\alpha = .05$.

7.38. One method for solving the electrical power shortage makes use of floating nuclear power plants located a few miles offshore in the ocean. Because there is great concern about the possibility of a ship colliding with the floating (but anchored) power plant, navigation experts have stated that it would be desirable if the average number of ships per day passing within 10 miles of the proposed power site location were less than 7. To verify this hypothesis for the proposed site, a random sample of 60 days was used throughout the peak shipping months. For each day the number of ships passing within the 10-mile limit was recorded. The sample mean and standard deviation were 6.3 days and 2 days, respectively. Use these data to test the navigation experts' alternative hypothesis. Use $\alpha = .05$.

7.39. Administrative officials for a university are concerned that the freshman students taking advantage of off-campus housing facilities have significantly lower grade point averages (GPA) than all freshmen at the school. After the fall quarter the all-freshman average GPA was 2.1 (on a 4-point system). Since it was not possible to isolate grades for all students living in off-campus housing by using the university records, a random sample of 81 off-campus freshmen was obtained by tracing students through their permanent home address. The sample mean and standard deviation were found to be 1.92 and .2, respectively. Do these data present sufficient evidence to indicate that the average GPA for all off-campus freshmen is lower than the all-freshman average? Use $\alpha = .05$.

7.40. Industrial waste and sewage dumped into our rivers and streams absorb oxygen and thereby reduce the amount of dissolved oxygen available for fish and other forms of aquatic life. One state agency requires a minimum of 5 parts per million (ppm) of dissolved oxygen in order that the oxygen content be sufficient to support aquatic life. During the low-water season (July), 30 specimens taken from a river at a specific location gave a sample mean and standard deviation of 4.9 and .2, respectively, in parts per million of dissolved oxygen. Do the data provide sufficient evidence to indicate that the average dissolved oxygen content is less than 5 ppm? Use $\alpha = .05$.

7.41. A data processing department claimed that in converting data from handwritten pages to computer cards, no more than .1% of the data entries (keystrokes in keypunching) were in error. For a large study a sample of $n = 20,000$ keystrokes was checked against the handwritten copy; $y = 50$ errors were found. Conduct a statistical test of the data processing department's claim. Use $\alpha = .05$.

7.42. Refer to exercise 7.41. Give the level of significance for the statistical test.

7.43. In a standard dissolution test for tablets of a particular drug product, the manufacturer must obtain the dissolution rate for a batch of tablets prior to release of the batch. Suppose that the dissolution test consists of assays for 36 individual 25-milligram tablets. For each assay the tablet is suspended in an acid bath and then assayed after 30 minutes. The sample mean and standard deviation after 30 minutes are 19.8 and .42 milligrams, respectively. Use these data to test H_0: $\mu = 20$ (80% of the labeled amount in the tablets) against the alternative hypothesis H_a: $\mu < 20$. Use $\alpha = .05$.

7.44. Refer to exercise 7.43. Give the level of significance for the test when the alternative hypothesis is H_a: $\mu \neq 20$.

EXPERIENCES WITH REAL DATA

Select an issue that is of concern to your university, your college, or your local community. For example, at the time of this writing, our county commission is holding hearings preliminary to a decision to authorize the construction of a large shopping mall adjacent to a new elementary school. Some witnesses appear before the television camera and strongly support the mall; others denounce it. Evidence suggests that elected officials, such as our county commission, place great weight on the proportion of speakers who favor or oppose the proposed construction (all other things being equal). Indeed, we frequently observe similar behavior on the part of our congressmen who quote counts of letters supporting or opposing some issue, thereby implying public support or opposition.

Do the speakers who appear before a county commission (or the writers of letters to a senator) represent a random sample of an elected official's constituency? Why might the majority of these speakers (or letter writers) favor one side of an issue when the constituency they are supposed to represent favor the other side?

Conduct a survey of public opinion to test the theory that the public supports (or opposes) the issue. Obtain a listing of the public and select a random sample (see section 5.2 on how to draw a random sample) of 400 people from this group. Then test the hypothesis that the proportion p of people favoring the issue is equal to .5 against the alternative that p is less than .5. Collect the data and draw your conclusions. Do the public favor the issue? Why are the conclusions based on your sample survey more valid than an inference based on the proportion of speakers (or letters to a senator) favoring the issue?

8 Comparisons

CHAPTER OUTLINE

8.1 Introduction

8.2 The Sampling Distribution of the Difference Between Two Sample Statistics

8.3 Comparing Two Population Means

8.4 Comparing Two Binomial Proportions

8.1 INTRODUCTION

Chapters 2 and 3 showed how to describe a set of measurements by using graphical and numerical descriptive techniques. Chapters 4 and 5 dealt with probability, and probability distributions, and sampling distributions, showing how to reason from a known population to a sample, and thus provided the mechanism for making an inference about a population. Chapters 6 and 7 presented the two methods for making inferences, estimation and testing hypotheses. These techniques were illustrated with two very practical inferential problems: making inferences about a population mean μ and about a binomial parameter p based on large samples. Thus we have stated that the objective of statistics is to make inferences, have explained how inferences are made, and have given two practical illustrations. Where do we go now? To more practical applications. Now it is necessary to show exactly how statistics can make a worthwhile contribution to society and to you.

Rarely do you read one of the popular newsmagazines or the Sunday edition of a leading newspaper without confronting articles dealing with the comparison of two populations. We read that factory orders in July rose 1.7% over those in June, car production for September is scheduled to drop 4.5% in comparison with August, the public school teachers of a certain state receive salaries less than the national average, and the percentage of people suffering from arteriosclerosis is higher for individuals with cadmium in their water supply than for those whose water lacks the element. All these examples involve the comparison of two populations based on information contained in samples selected from each.

How can you tell whether the observed differences in the previous examples are real or whether they are due to random variation? People unfamiliar with statistics frequently answer this question by saying, "But you can see the difference, can't you? There's no question about it!" They forget—or do not know—that the difference they observe is based on samples, and hence they confuse this observed difference in the sample means or percentages with the corresponding population difference.

There are numerous examples that emphasize this point. For instance, recall that the Federal Trade Commission has placed pressure on numerous companies to restrain them from using misleading advertising. And research articles published in some professional journals can be very misleading. Typical of this practice is the reporting of results in a medical journal comparing two products, I and II, which, for purposes of illustration, might be drugs used to treat some rare disease. Suppose the author of the article reports that only 14 persons out of 21 treated with drug I recovered, while 19 out of 22 treated with drug II recovered. Do these data imply that drug II is more effective than drug I in treating the disease? If these data, along with a thorough and valid description of the actual experiments, are presented in condensed form to a practicing physician who is overburdened with patients and who has little time to read the hundreds of similar articles on new treatments, instruments, and drugs, what will he or she conclude?

The sample proportions of patients who recovered are substantially different for the two drug products: .67 versus .86. But keep in mind that these proportions are estimates of binomial parameters based on relatively small samples. As a consequence, the sample estimates can vary considerably about the true binomial parameters. It can be shown that the probability that these two sample proportions differ by as much as .19 would be rather large, even when there is no real difference in recovery rates for the two drug products. Hence the sample data do not provide sufficient evidence to indicate a difference in the recovery rates for patients treated with the two drugs. Drug I may be better than II, II may be better than I, or they may be equally effective.

In this chapter we will present large-sample techniques for comparing two population means or two population proportions. Small-sample methods for comparing two population means are discussed in chapter 9.

8.2 THE SAMPLING DISTRIBUTION OF THE DIFFERENCE BETWEEN TWO SAMPLE STATISTICS

In many sampling situations we will select independent random samples from two populations in order to compare the population means or proportions. The statistics used to make these inferences will, in many cases, be the difference between the corresponding sample statistics. For example, suppose we select independent random samples of n_1 observations from one population and n_2 observations from a second population. We will use the difference between the population means, $(\bar{y}_1 - \bar{y}_2)$, to make an inference about the difference between the population means, $(\mu_1 - \mu_2)$.

The following theorem is of help in finding the sampling distribution for the difference between sample statistics computed from independent random samples.

THEOREM 8.1

If two independent random variables y_1 and y_2 are normally distributed with means and variances (μ_1, σ_1^2) and (μ_2, σ_2^2), respectively, then the difference between the random variables will be normally distributed with mean equal to $(\mu_1 - \mu_2)$ and variance equal to $(\sigma_1^2 + \sigma_2^2)$.*

Theorem 8.1 can be applied directly to find the sampling distribution of the difference between two independent sample means or two independent sample proportions. The Central Limit Theorem (discussed in chapter 5) implies that if independent samples of sizes n_1 and n_2 are selected from two populations, 1 and 2, then, when n_1 and n_2 are large, the sampling distributions of $\bar{y}_1$ and $\bar{y}_2$ will be approximately normal, with means and variances $(\mu_1, \sigma_1^2/n_1)$ and $(\mu_2, \sigma_2^2/n_2)$, respectively. Consequently, since $\bar{y}_1$ and $\bar{y}_2$ are independent, normally distributed random variables, it follows from theorem 8.1 that the sampling distribution for the difference in the sample means, $(\bar{y}_1 - \bar{y}_2)$, will be approximately normal, with a mean

$$\mu_{\bar{y}_1 - \bar{y}_2} = \mu_1 - \mu_2$$

and a variance

$$\sigma^2_{\bar{y}_1 - \bar{y}_2} = \sigma^2_{\bar{y}_1} + \sigma^2_{\bar{y}_2} = \frac{\sigma_1^2}{n_1} + \frac{\sigma_2^2}{n_2}$$

and a standard deviation

$$\sigma_{\bar{y}_1 - \bar{y}_2} = \sqrt{\frac{\sigma_1^2}{n_1} + \frac{\sigma_2^2}{n_2}}$$

The sampling distribution of the difference between two independent, normally distributed sample means is shown in figure 8.1.

*Note: The sum $(y_1 + y_2)$ of the random variables will also be normally distributed with mean $(\mu_1 + \mu_2)$ and variance $(\sigma_1^2 + \sigma_2^2)$.

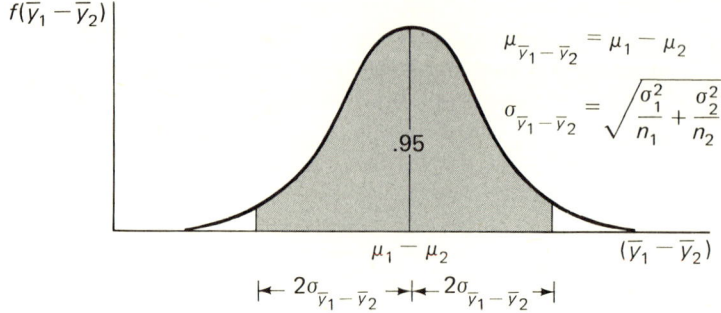

Figure 8.1 Sampling Distribution for the Difference Between Two Sample Means

---

Properties of the Sampling Distribution for the Difference Between Two Sample Means, $(\bar{y}_1 - \bar{y}_2)$

1. The sampling distribution of $(\bar{y}_1 - \bar{y}_2)$ is approximately normal for large samples.
2. The mean of the sampling distribution, $\mu_{\bar{y}_1 - \bar{y}_2}$, is equal to the difference between the population means, $(\mu_1 - \mu_2)$.
3. The standard deviation of the sampling distribution is

$$\sigma_{\bar{y}_1 - \bar{y}_2} = \sqrt{\frac{\sigma_1^2}{n_1} + \frac{\sigma_2^2}{n_2}}$$

Similarly, theorem 8.1 enables us to find the sampling distribution for the difference in sample proportions, $(\hat{p}_1 - \hat{p}_2)$, computed from independent random samples of n_1 and n_2 observations, respectively, selected from two binomial populations. From chapter 5 we know that for large samples $\hat{p}_1$ and $\hat{p}_2$ will be approximately normally distributed, with means and variances $(p_1, p_1 q_1/n_1)$ and $(p_2, p_2 q_2/n_2)$, respectively. Hence it follows from theorem 8.1 that the sampling distribution for the difference in sample proportions, $(\hat{p}_1 - \hat{p}_2)$, will be approximately normally distributed, with a mean

$$\mu_{\hat{p}_1 - \hat{p}_2} = p_1 - p_2$$

and a variance

$$\sigma^2_{\hat{p}_1 - \hat{p}_2} = \sigma^2_{\hat{p}_1} + \sigma^2_{\hat{p}_2} = \frac{p_1 q_1}{n_1} + \frac{p_2 q_2}{n_2}$$

and a standard deviation

$$\sigma_{\hat{p}_1-\hat{p}_2} = \sqrt{\frac{p_1 q_1}{n_1} + \frac{p_2 q_2}{n_2}}$$

The sampling distribution for $(\hat{p}_1 - \hat{p}_2)$ is as shown in figure 8.2.

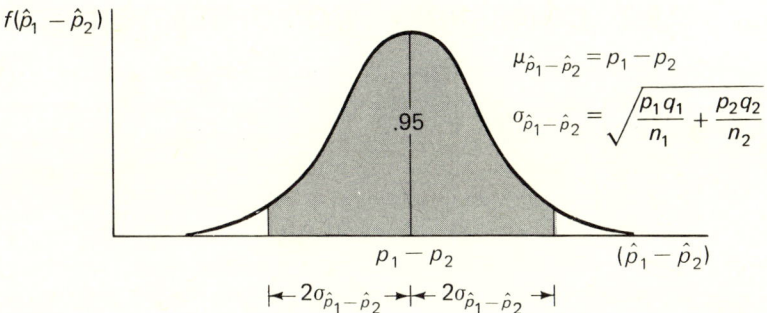

Figure 8.2 Sampling Distribution for the Difference Between Two Sample Proportions

---

Properties of the Sampling Distribution for the Difference Between Two Binomial Sample Proportions, $(\hat{p}_1 - \hat{p}_2)$

1. The sampling distribution of $(\hat{p}_1 - \hat{p}_2)$ will be approximately normal for large samples.
2. The mean of the sampling distribution is $\mu_{\hat{p}_1-\hat{p}_2} = (p_1 - p_2)$.
3. The standard deviation of the sampling distribution is

$$\sigma_{\hat{p}_1-\hat{p}_2} = \sqrt{\frac{p_1 q_1}{n_1} + \frac{p_2 q_2}{n_2}}$$

The sampling distributions for the difference between two sample means, $(\bar{y}_1 - \bar{y}_2)$, or the difference between two sample proportions, $(\hat{p}_1 - \hat{p}_2)$, can be used to answer the same types of questions as were asked about the sampling distributions for $\bar{y}$ or $\hat{p}$ in chapter 5. Since sample statistics are used to make inferences about corresponding population parameters, we can use the sampling distribution of a statistic to calculate the probability that the statistic will be within a specified distance of the population parameter. For example, we could use the sampling distribution of the difference in sample means to calculate the probability that $(\bar{y}_1 - \bar{y}_2)$ will be within a specified distance of the unknown difference in population means, $(\mu_1 - \mu_2)$. Inferences (estimations or tests) about $(\mu_1 - \mu_2)$ or $(p_1 - p_2)$ based on large samples will be discussed in succeeding sections of this chapter.

8.3 COMPARING TWO POPULATION MEANS

Populations are frequently compared by examining the difference in their means. For example, we might want to compare the mean strengths of two different mixes of concrete before proceeding with the construction of a large bridge, or the mean assessed values of homes in two different subdivisions of a city, or the mean response of humans to two different stimuli. For each of these problems we assume that we are sampling from two populations, the first with unknown parameters mean μ_1 and variance σ_1^2, the second with unknown parameters mean μ_2 and variance σ_2^2. Independent random samples of n_1 and n_2 measurements are then drawn from populations 1 and 2, respectively. Finally, the estimates $\bar{y}_1, s_1^2, \bar{y}_2$, and s_2^2 of the corresponding population parameters are computed from the sample data. (See table 8.1.)

Table 8.1 Sampling from Two Populations

	Population 1	Population 2
Population Mean	μ_1	μ_2
Population Variance	σ_1^2	σ_2^2
Sample Size	n_1	n_2
Sample Mean	$\bar{y}_1$	$\bar{y}_2$
Sample Variance	s_1^2	s_2^2

A logical estimate for the difference in population means is the difference in sample means, $(\bar{y}_1 - \bar{y}_2)$. And as we have shown in section 8.2, the difference in sample means, $(\bar{y}_1 - \bar{y}_2)$, has a sampling distribution that is approximately normal with mean $\mu_{\bar{y}_1 - \bar{y}_2} = (\mu_1 - \mu_2)$ and standard deviation

$$\sigma_{\bar{y}_1 - \bar{y}_2} = \sqrt{\frac{\sigma_1^2}{n_1} + \frac{\sigma_2^2}{n_2}}$$

standard error of the difference

The quantity $\sigma_{\bar{y}_1 - \bar{y}_2}$ is often referred to as the standard error of the difference.

Knowing the properties of the sampling distribution for $(\bar{y}_1 - \bar{y}_2)$, we can determine point and interval estimates for $(\mu_1 - \mu_2)$, which have the same form as point and interval estimates for μ or for p (see chapter 6).

Point Estimation of $(\mu_1 - \mu_2)$

point estimate of $(\mu_1 - \mu_2)$: $(\bar{y}_1 - \bar{y}_2)$
bound on error: $2\sigma_{\bar{y}_1 - \bar{y}_2}$

where

$$\sigma_{\bar{y}_1 - \bar{y}_2} = \sqrt{\frac{\sigma_1^2}{n_1} + \frac{\sigma_2^2}{n_2}}$$

Note: For $n_1 \geq 30$ and $n_2 \geq 30$, s_1^2 and s_2^2 can be substituted for σ_1^2 and σ_2^2.

Large-Sample Interval Estimation of $(\mu_1 - \mu_2)$, $n_1 \geq 30$ and $n_2 = \geq 30$

$$\bar{y}_1 - \bar{y}_2 \pm z\sigma_{\bar{y}_1 - \bar{y}_2}$$

where

$$\sigma_{\bar{y}_1 - \bar{y}_2} = \sqrt{\frac{\sigma_1^2}{n_1} + \frac{\sigma_2^2}{n_2}}$$

and $z = 1.645$, 1.96, or 2.58 for a 90%, a 95%, or a 99% confidence interval, respectively. Note: For $n_1 \geq 30$ and $n_2 \geq 30$, s_1^2 and s_2^2 can be substituted for σ_1^2 and σ_2^2.

Example 8.1

A study was conducted to determine if persons in suburban district I have a different mean income from those in district II. A random sample of 50 homeowners was taken in district I. Although 50 homeowners were to be interviewed in district II also, one person refused to provide the information requested, even though the researcher promised to keep the interview confidential. So only 49 observations were obtained from district II. The data, recorded in thousands of dollars, produced sample means and variances as shown in table 8.2. Use these data to construct a 95% confidence interval for $(\mu_1 - \mu_2)$.

Table 8.2 Income Data for Example 8.1

	District I	District II
Sample Size	50	49
Sample Mean	14.27	12.78
Sample Variance	8.74	6.58

Solution

The difference in the sample means is

$$\bar{y}_1 - \bar{y}_2 = 14.27 - 12.78 = 1.49$$

Since the sample sizes are both larger than 30, we can substitute s_1^2 and s_2^2 for the unknown population variances σ_1^2 and σ_2^2 in the formula for $\sigma_{\bar{y}_1-\bar{y}_2}$ to obtain

$$\sqrt{\frac{\sigma_1^2}{n_1} + \frac{\sigma_2^2}{n_2}} \approx \sqrt{\frac{8.74}{50} + \frac{6.58}{49}} = .56$$

Hence $1.96\sigma_{\bar{y}_1-\bar{y}_2} = 1.10$.

A 95% confidence interval for the difference in mean incomes for the two districts has a lower limit of

$$\bar{y}_1 - \bar{y}_2 - 1.96\sigma_{\bar{y}_1-\bar{y}_2} = 1.49 - 1.10 = .39$$

and an upper limit of

$$\bar{y}_1 - \bar{y}_2 + 1.96\sigma_{\bar{y}_1-\bar{y}_2} = 1.49 + 1.10 = 2.59$$

We are 95% confident that the difference in the population means is in the interval from .39 to 2.59; that is, we are quite confident that the difference in mean incomes lies between $390 and $2590.

The corresponding statistical test of an hypothesis about the difference between two population means μ_1 and μ_2 follows the logic developed for a test concerning a single population mean. The test procedure requires adequate sample sizes to estimate the population variances σ_1^2 and σ_2^2, because they will rarely be known. We suggest the requirement that n_1 and n_2 both be 30 or more. Then you can use s_1^2 and s_2^2 to approximate σ_1^2 and σ_2^2. What test can you use if one or more of the sample sizes is less than 30? You can employ the small-sample test described in section 9.4.

A summary of the four parts of a large-sample test for $(\mu_1 - \mu_2)$ is given next.

Comparing Two Population Means

Large-Sample Test for Comparing Two Population Means
($n_1 \geq 30$ and $n_2 \geq 30$)

Null hypothesis: $\mu_1 - \mu_2 = 0$ (i.e., $\mu_1 = \mu_2$).

Alternative hypothesis: For a one-tailed test:

1. $\mu_1 - \mu_2 > 0$.
2. $\mu_1 - \mu_2 < 0$.

For a two-tailed test:

3. $\mu_1 - \mu_2 \neq 0$.

Test statistic: $z = \dfrac{\bar{y}_1 - \bar{y}_2}{\sigma_{\bar{y}_1 - \bar{y}_2}}$ where $\sigma_{\bar{y}_1 - \bar{y}_2} = \sqrt{\dfrac{\sigma_1^2}{n_1} + \dfrac{\sigma_2^2}{n_2}}$

Rejection region: For $\alpha = .05$ (or .01) and for a one-tailed test:

1. Reject H_0 if $z > 1.645$ (or 2.33).
2. Reject H_0 if $z < -1.645$ (or -2.33).

For $\alpha = .05$ (or .01) and for a two-tailed test:

3. Reject H_0 if $|z| > 1.96$ (or 2.58).

Note: For $n_1 \geq 30$ and $n_2 \geq 30$, s_1^2 and s_2^2 can be substituted for σ_1^2 and σ_2^2.

We will illustrate with an example the use of the large-sample test for comparing means.

Example 8.2

Refer to the data of example 8.1 (see table 8.2). Test the alternative hypothesis that the mean incomes for the two districts are different. Use $\alpha = .05$.

Solution

The four parts of the statistical test for $(\mu_1 - \mu_2)$ are as follows:

Null hypothesis: $\mu_1 - \mu_2 = 0$.

Alternative hypothesis: $\mu_1 - \mu_2 \neq 0$.

Test statistic: $z = \dfrac{\bar{y}_1 - \bar{y}_2}{\sigma_{\bar{y}_1 - \bar{y}_2}}$.

Rejection region: For $\alpha = .05$ reject H_0 if $|z| > 1.96$.

Using the sample data, we obtain

$$\bar{y}_1 - \bar{y}_2 = 14.27 - 12.78 = 1.49$$

and

$$\sigma_{\bar{y}_1-\bar{y}_2} = \sqrt{\frac{8.74}{50} + \frac{6.58}{49}} = .56$$

Substituting into z, we have

$$z = \frac{1.49}{.56} = 2.66$$

Since 2.66 falls in the rejection region, we reject H_0 and conclude that $(\mu_1 - \mu_2) \neq 0$. Practically speaking, since $\bar{y}_1 > \bar{y}_2$, we conclude that the mean income for all homeowners in district I is higher than that for district II.

EXERCISES

8.1. A random sample of $n_1 = 36$ measurements was drawn from a population with mean $\mu_1 = 600$ and variance $\sigma_1^2 = 900$. A second sample of $n_2 = 49$ measurements was selected from a population with mean $\mu_2 = 400$ and variance $\sigma_2^2 = 400$. If the two samples are independent, describe the sampling distribution for the difference in sample means, $(\bar{y}_1 - \bar{y}_2)$.

8.2. Refer to exercise 8.1. What percentage of the sample differences $(\bar{y}_1 - \bar{y}_2)$ should be within $2\sigma_{\bar{y}_1-\bar{y}_2}$ of $(\mu_1 - \mu_2)$?

8.3. The sample data for a study to compare two population means are shown in the accompanying table. Test the alternative hypothesis $H_a: \mu_1 - \mu_2 > 0$. Use $\alpha = .05$.

	Population	
	1	2
Sample Mean	580.3	576.8
Sample Variance	39.1	48.6
Sample Size	35	37

8.4. Refer to exercise 8.3. Give the level of significance for a test of the alternative hypothesis: $H_a: \mu_1 - \mu_2 > 0$.

8.5. Two samples of 30 standard metropolitan statistical areas were collected, one from the North and the other from the South. Use the sample data shown here to construct a 95% confidence interval for the difference between the average percentages of adults who have completed four years of high school or more.

	North	South
Sample Mean	58.2	49.2
Sample Variance	43.6	38.5

8.6. Construct a 99% confidence interval for $(\mu_1 - \mu_2)$, using the data from exercise 8.5.

8.7. Faculty members at many universities throughout the nation have been encouraged to join forces to engage in collective bargaining for better wages, additional fringe benefits, and better working conditions. One university enlisted the help of a statistician to determine if faculty members on campuses with collective bargaining received higher salary increases last year, on the average, than those who did not have union support. A random sample of 60 full professors was obtained from universities with collective bargaining, while another random sample of 60 full professors was obtained from those universities without collective bargaining. Each faculty member was interviewed to determine his or her increase in salary over the past year. Those increases (based on a 9-month salary) are summarized in the accompanying table. Do these data present sufficient evidence to indicate that the average annual increase last year in salaries for full professors was higher for universities with collective bargaining? Use $\alpha = .05$.

	Universities with Collective Bargaining	Universities without Collective Bargaining
Sample Mean	1,522	1,350
Sample Variance	256,000	312,000
Sample Size	60	60

8.8. In dentistry a commonly used measure of the extent of tooth decay for a person is the DMF count, where D refers to the number of permanent teeth decayed, M the number of missing, and F the number filled. The sum of these three components is the DMF count. Using the DMF count, researchers have shown that for any given age and diet, the inclusion of small amounts of fluoride reduces the amount of tooth decay significantly. To substantiate this phenomenon, a random sample of 100 children (ages 5–9) is selected from areas containing low fluoride levels in their drinking water. A similar sample of 100 children from the same age group is obtained from areas with adequate fluoride levels in the drinking water. For each child included in the study, a DMF count is obtained. The sample data are shown in the accompanying table. Use these data to test the research hypothesis that children (ages 5–9) in areas with adequate fluoride in the drinking water have, on the average, a lower DMF count than children from low-fluoride areas. Let $\alpha = .05$.

	Low-Fluoride Areas	Adequate-Fluoride Areas
Sample Mean	1.20	.15
Sample Variance	.10	.08
Sample Size	100	100

8.4 COMPARING TWO BINOMIAL PROPORTIONS

Many practical problems require the comparison of two binomial parameters. We might wish to compare the proportions of housewives who utilize prenatal health services before and after a campaign to publicize the services, or the proportions of households in two states that are entirely supported by welfare, or the proportions of voters favoring the Democratic candidate in a suburban and a rural area.

We assume that independent random samples are drawn from two binomial populations with unknown parameters p_1 and p_2. We further assume that the samples contain n_1 and n_2 observations, respectively. If y_1 represents the number of successes in the n_1 trials for sample 1 and y_2 the number of successes in n_2 trials for sample 2, then the sample proportions

$$\hat{p}_1 = \frac{y_1}{n_1} \quad \text{and} \quad \hat{p}_2 = \frac{y_2}{n_2}$$

are point estimates of p_1 and p_2, respectively. These results are summarized in table 8.3.

Table 8.3 Sampling from Two Binomial Populations

	Population 1	Population 2
Population Proportion	p_1	p_2
Sample Size	n_1	n_2
Number of Successes	y_1	y_2
Sample Proportion	$\hat{p}_1 = y_1/n_1$	$\hat{p}_2 = y_2/n_2$

As discussed in section 8.2, the sampling distribution of $(\hat{p}_1 - \hat{p}_2)$ is approximately normal for large values of n_1 and n_2 (see figure 8.3), with mean and standard deviation (standard error of the difference between proportions) given by

$$\mu_{\hat{p}_1 - \hat{p}_2} = p_1 - p_2 \quad \text{and} \quad \sigma_{\hat{p}_1 - \hat{p}_2} = \sqrt{\frac{p_1 q_1}{n_1} + \frac{p_2 q_2}{n_2}}$$

Comparing Two Binomial Proportions

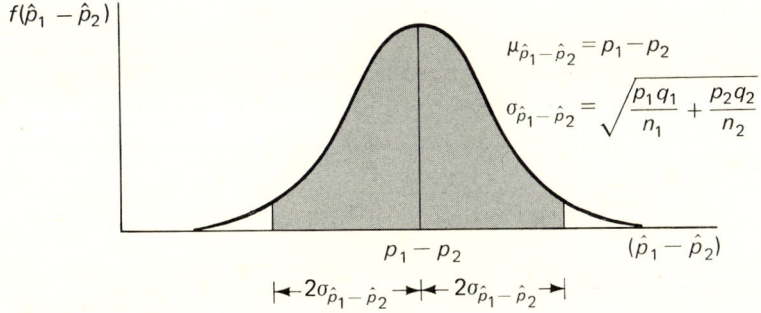

Figure 8.3 Sampling Distribution of $(\hat{p}_1 - \hat{p}_2)$

The sample sizes are considered large enough when the sample size and the value of p for each population satisfy the requirement that both np and nq are 10 or more. Since p and q are never known, we use $n\hat{p}$ and $n\hat{q}$ for each sample.

We summarize point and interval estimation of $(p_1 - p_2)$, based on large samples, next.

Point Estimation of $(p_1 - p_2)$

$$\text{point estimate of } (p_1 - p_2): \quad \hat{p}_1 - \hat{p}_2$$
$$\text{bound on error:} \quad 2\sigma_{\hat{p}_1 - \hat{p}_2}$$

where

$$\sigma_{\hat{p}_1 - \hat{p}_2} = \sqrt{\frac{p_1 q_1}{n_1} + \frac{p_2 q_2}{n_2}}$$

Note: Substitute $\hat{p}_1$ and $\hat{p}_2$ for the unknown parameters p_1 and p_2 in the formula for the bound on error. Very little error will result provided the sample sizes are large.

Large-Sample Confidence Interval for $(p_1 - p_2)$

$$\hat{p}_1 - \hat{p}_2 \pm z\sigma_{\hat{p}_1 - \hat{p}_2}$$

where

$$\sigma_{\hat{p}_1 - \hat{p}_2} = \sqrt{\frac{p_1 q_1}{n_1} + \frac{p_2 q_2}{n_2}}$$

with $q_1 = (1 - p_1)$ and $q_2 = (1 - p_2)$. Substitute $z = 1.645$, 1.96, or 2.58 for a 90%, a 95%, or a 99% confidence interval, respectively. Note: Use $\hat{p}_1$ and $\hat{p}_2$ for the unknown parameters p_1 and p_2 in the formula for $\sigma_{\hat{p}_1-\hat{p}_2}$. Very little error will result provided the sample sizes are large.

We illustrate these ideas with an example.

Example 8.3

In a survey to analyze the cost of funeral expenditures for various social classes, a random sample of 162 families from the lower and working classes was interviewed to determine the funeral expenses for a recent family death. Of the 162 families contacted, 61 spent over $800 on the funeral. In a sample of 189 middle- and upper-class families who had experienced a recent family death, 106 spent over $800 on the funeral. Estimate $(p_1 - p_2)$, the difference in the proportions of families paying over $800 for funeral expenses, for the two social classifications. Place a bound on the error of estimation.

Solution

The point estimate of $(p_1 - p_2)$ is the difference in the sample proportions, $(\hat{p}_1 - \hat{p}_2)$:

$$\hat{p}_1 - \hat{p}_2 = \frac{y_1}{n_1} - \frac{y_2}{n_2} = \frac{61}{162} - \frac{106}{189} = .376 - .561 = -.185$$

To determine the bound on error, we must evaluate $\sigma_{\hat{p}_1-\hat{p}_2}$; for this calculation we may substitute $\hat{p}_1$ and $\hat{p}_2$ for p_1 and p_2:

$$\sigma_{\hat{p}_1-\hat{p}_2} \approx \sqrt{\frac{\hat{p}_1 \hat{q}_1}{n_1} + \frac{\hat{p}_2 \hat{q}_2}{n_2}} = \sqrt{\frac{(.376)(.624)}{162} + \frac{(.561)(.439)}{189}} = .052$$

A bound on the error of estimation for $(p_1 - p_2)$ is, therefore,

$$2\sigma_{\hat{p}_1-\hat{p}_2} = 2(.052) = .104$$

We can readily formulate a statistical test for the equality of two binomial parameters. A logical test statistic is one that makes use of the difference $(\hat{p}_1 - \hat{p}_2)$ in the point estimates. For fixed sample sizes the greater the difference between $\hat{p}_1$ and $\hat{p}_2$, the greater is the evidence to indicate that p_1 does not equal p_2. The test statistic is

$$z = \frac{\hat{p}_1 - \hat{p}_2}{\sqrt{pq\left(\frac{1}{n_1} + \frac{1}{n_2}\right)}}$$

Comparing Two Binomial Proportions

You will need to use the data to approximate p in the formula for the test statistic. The best estimate of p, the proportion of successes common to both populations, is

$$\hat{p} = \frac{\text{total number of successes}}{\text{total number of trials}} = \frac{y_1 + y_2}{n_1 + n_2}, \quad \text{with} \quad \hat{q} = 1 - \hat{p}$$

We summarize the test procedure next.

Large-Sample Statistical Test for Comparing Two Binomial Proportions

Null hypothesis: $p_1 - p_2 = 0$.

Alternative hypothesis: For a one-tailed test:
1. $p_1 - p_2 > 0$.
2. $p_1 - p_2 < 0$.

For a two-tailed test:

3. $p_1 - p_2 \neq 0$.

Test statistic: $\quad z = \dfrac{\hat{p}_1 - \hat{p}_2}{\sigma_{\hat{p}_1 - \hat{p}_2}} \quad$ where $\quad \sigma_{\hat{p}_1 - \hat{p}_2} = \sqrt{pq \left(\dfrac{1}{n_1} + \dfrac{1}{n_2} \right)}$

and p is approximated by

$$\hat{p} = \frac{y_1 + y_2}{n_1 + n_2} \quad \text{with} \quad \hat{q} = 1 - \hat{p}$$

Rejection region: For $\alpha = .05$ (or .01) and for a one-tailed test:

1. Reject H_0 if $z > 1.645$ (or 2.33).
2. Reject H_0 if $z < -1.645$ (or -2.33).

For $\alpha = .05$ (or .01) and for a two-tailed test:

3. Reject H_0 if $|z| > 1.96$ (or 2.58).

Note: $n\hat{p}$ and $n\hat{q}$ must be greater than or equal to 10 for both populations.

Example 8.4

Two sets of $n_1 = n_2 = 60$ ninth graders were taught high school algebra by two different methods. Group I used a programmed learning text and had no formal lectures; group II was given formal lectures by a teacher. At the conclusion of a four-month period a comprehensive test was given to both groups

Comparisons

to determine the proportion of students in each group who obtained a score of 85 (out of 100) or better. The results are given in table 8.4.

Table 8.4 Data for Example 8.4

Group I	Group II
$n_1 = 60$	$n_2 = 60$
$y_1 = 41$	$y_2 = 24$

Let p_1 represent the population proportion of students taught with a programmed text who would achieve a score of 85 or more on the comprehensive test. Similarly, let p_2 represent the population proportion of students taught by formal lectures who would score 85 or more on the test. Test the research hypothesis that the two population proportions p_1 and p_2 are different. Use $\alpha = .01$.

Solution

For all practical purposes, sampling from each population satisfies the requirements of a binomial experiment. The four parts of the statistical test are as follows:

Null hypothesis: $p_1 - p_2 = 0$.

Alternative hypothesis: $p_1 - p_2 \neq 0$.

Test statistic: $$z = \frac{\hat{p}_1 - \hat{p}_2}{\sqrt{pq\left(\frac{1}{n_1} + \frac{1}{n_2}\right)}}.$$

Rejection region: For $\alpha = .01$ and a two-tailed test, we will reject H_0 if $|z| > 2.58$ (see figure 8.4).

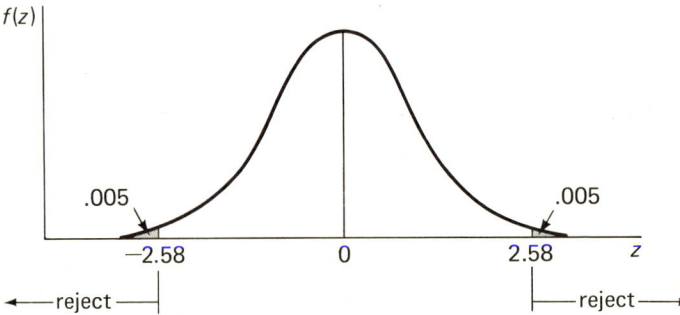

Figure 8.4 Rejection Region for Example 8.4

Exercises

From the sample data we have

$$\hat{p}_1 = \frac{41}{60} = .68 \qquad \hat{p}_2 = \frac{24}{60} = .40$$

$$\hat{p} = \frac{41 + 24}{60 + 60} = .54 \qquad \hat{q} = 1 - \hat{p} = .46$$

Substituting into the test statistic, we obtain

$$z = \frac{.68 - .40}{\sqrt{(.54)(.46)\left(\frac{1}{60} + \frac{1}{60}\right)}} = 3.08$$

Since the computed value of z exceeds 2.58, we reject H_0 and conclude that the two proportions are different. In fact, since $\hat{p}_1 = .68$ is greater than $\hat{p}_2 = .40$, we may conclude that the programmed teaching technique produces a higher proportion of students scoring 85 or more.

EXERCISES

8.9. A random sample of $n_1 = 1000$ measurements was obtained from a binomial population with $p = .4$. Another random sample (independent of the first sample) of $n_2 = 1000$ measurements was selected from a second binomial population with $p = .2$. Describe the sampling distribution of $(\hat{p}_1 - \hat{p}_2)$.

8.10. In a study to compare two binomial proportions, $n_1 = 50$, $n_2 = 40$, $y_1 = 18$, and $y_2 = 10$. Use these data to construct a 99% confidence interval for the difference between the corresponding population proportions.

8.11. Refer to exercise 8.10. Use the same data to test H_0: $p_1 - p_2 = 0$ against the alternative hypothesis H_a: $p_1 - p_2 > 0$. Use $\alpha = .01$.

8.12. In a recent survey of county high school students ($n_1 = 100$ males and $n_2 = 100$ females), 58 of the males sampled said they consume alcohol (beer included) on a regular basis and 46 of the females sampled responded in the same way. Use these data to test the alternative hypothesis H_a: $p_1 - p_2 > 0$, that is, that the proportion of county high school males who consume alcohol on a regular basis is greater than the corresponding proportion of females. Give the level of significance for your test.

8.13. A law student believes that the proportion of Republicans in favor of the unrestricted right of executive privilege is greater than the proportion of Democrats in favor of the unrestricted right. She acquired independent random samples of 200 Republicans and 200 Democrats and found 46 Republicans and 37 Democrats in favor of the unrestricted right of executive privilege. Construct a 95% confidence interval for the difference in proportions for Democrats and Republicans.

8.14. In a comparison of the incidence of tumor potential in two strains of rats, 100 rats (50 males, 50 females) were selected from each of two strains and were examined for a period of one year. All the rats were approximately the same age and were housed and fed under comparable conditions. Use the accompanying one-year sample data to construct a 95% confidence interval for the difference in the proportions of rats exhibiting tumor potential for the two strains.

One-Year Results

	Strain A	Strain B
Sample Size	100	100
Number Exhibiting Tumor Potential	25	15

 8.15. The traffic congestion in cities has become a major cause for concern not only from an environmental standpoint but also from an engineering viewpoint. The Federal Highway Administration in conjunction with the Urban Mass Transit Authority contracted with the University of Florida Transportation Center to test and evaluate the effectiveness of an exclusive bus lane constructed on a 10-mile (north-south) section of 7th Avenue in Miami and a parallel section of Interstate 95. Initially a parking facility and bus terminal were built at the Golden Glades exit to provide commuters with ready access to an inexpensive, fast form of transportation to their destinations in downtown Miami. Each bus was equipped with a device that enabled the driver to preempt traffic signals along the 10-mile stretch of highway. Thus a driver could flash ahead to a traffic signal if the bus was going to arrive at the intersection during the red phase of the traffic signal cycle and the light would either remain green or switch as quickly as possible to allow uninterrupted bus travel. One major task of the University of Florida Transportation Center was to compare various preempting strategies (how much leeway was given to the individual bus driver in controlling the green phase of the signal) on

(a) the average travel for buses from the terminal to each of the four destination areas
(b) the effect of these preempting strategies on congestion at intersections along the 10-mile section of 7th Avenue

In addition to these considerations, a major objective involved a comparison of the auto occupancies throughout various phases of the three-and-a-half-year project. For example, it was important to compare the proportion of cars with two or more persons through each of the test phases in order to determine if there were shifts in commuter car usage as a result of or in conjunction with changes in bus usage.

Suppose that initial results were as follows: During phase 0 (which ran for approximately three months) the buses were not given a preempting option at the intersections. Phase 1 allowed local preemption of a traffic signal by a bus but no other signal coordination. Of 10,000 cars observed during the sample period in phase 0, 10% had two or more persons in a car. Similarly, for 10,000 cars sampled during phase 1, 15% had two or more persons in a car. Do these data present sufficient evidence to indicate whether there is a difference in the proportion of cars carrying two or more persons for the two phases? Use $\alpha = .05$.

 8.16. A pharmaceutical company compared two frequently used antibiotics, erythromycin and tetracycline, in a bacterial sensitivity study. Bacterial cultures from a random sampling of patients who had received injections of one of the two drugs were analyzed to ascertain antibacterial activity against specific bacteria. A summary of the survey data is given in the accompanying table for erythromycin and tetracycline for the streptococcus pyogenes bacteria. Use these sample data to test whether there is evidence to indicate a difference in the proportions of patient cultures showing antibacterial activity against the streptococcus pyogenes bacteria for the two test drug products. Use $\alpha = .05$.

	Erythromycin	*Tetracycline*
Number of Patient Cultures Analyzed	528	481
Number of Patient Cultures Showing Antibacterial Activity	515	394

8.17. The accompanying article alludes to a study of the effects of orange juice in reducing symptoms of respiratory infections. Examine this article and comment on the published statistics contained in it. Be sure to comment on the adequacy of the sample, the inferences that are drawn, and additional material that might have been included in the article to help you reach a conclusion.

Researchers Study Effects of Orange Juice

Gainesville, Fla.—A quart of orange juice, taken every day, has significant effects in reducing the symptoms of respiratory infections caused by the rubella virus, report medical scientists from the University of Florida.

The Florida study involved 55 human volunteers. Many of those drinking orange juice produced antibodies earlier than those not drinking orange juice. The researchers said that the early antibody production may or may not be the means by which the symptoms were reduced. However, the appearance of this antibody production, they pointed out, may indicate that citrus has a localized effect in stimulating the immune system to fight respiratory infection.

Many of the volunteers who drank fresh frozen orange juice did not experience any sore throat or runny nose. Other volunteers, forbidden to drink orange juice or eat citrus fruit, did show these mild symptoms. Prior to the study all participants were free of any evidence of infection and had normal antibody levels.

The volunteers were subjected to a weakened, nontransmissible form of rubella virus. The rubella virus was chosen because it can induce either respiratory or systemic infections, depending on how the virus is administered. Introduced in the nasal passages, the virus causes respiratory infection; introduced under the skin, the virus causes systemic infection.

Half of the 55 volunteers were given the rubella virus by nose drops; the other half received the same virus by injection. Each group was again divided. Half of them were on a quart of orange juice a day; their controls were on regular diets with no orange juice or citrus fruit. Each volunteer was followed for 21 days after the virus was administered.

In the groups who received the virus by nose drops and were not allowed to drink orange juice, 77 percent showed symptoms of respiratory infection. Among the counterparts who received the same type of virus and drank orange juice, only 27 percent had respiratory symptoms.

"This represents a significant reduction in symptoms," one researcher said.

Volunteers in the study who received the virus by injection showed no difference in symptoms, whether they were drinking orange juice or not.

SUMMARY

Several different estimation and test procedures are available for comparing two populations. In particular, in this chapter we were concerned with the following question: If two sample means $\bar{y}_1$ and $\bar{y}_2$ differ, does this imply that the corresponding population means μ_1 and μ_2 differ? Similarly, if two sample proportions $\hat{p}_1$ and $\hat{p}_2$ differ, do we have sufficient evidence to indicate a difference in the corresponding population proportions? Or, rather than testing to see whether there is evidence of a difference between two population parameters, we might wish to estimate the difference by using a confidence interval. The confidence intervals and statistical tests from this chapter are summarized in tables 8.5 and 8.6.

Table 8.5 Large-Sample Confidence Intervals for $(\mu_1 - \mu_2)$ and $(p_1 - p_2)$

Parameter	Confidence Interval*
$\mu_1 - \mu_2$	$\bar{y}_1 - \bar{y}_2 \pm z\sigma_{\bar{y}_1 - \bar{y}_2}$, where $\sigma_{\bar{y}_1 - \bar{y}_2} \approx \sqrt{\dfrac{s_1^2}{n_1} + \dfrac{s_2^2}{n_2}}$ provided $n_1, n_2 \geq 30$
$p_1 - p_2$	$\hat{p}_1 - \hat{p}_2 \pm z\sigma_{\hat{p}_1 - \hat{p}_2}$, where $\sigma_{\hat{p}_1 - \hat{p}_2} \approx \sqrt{\dfrac{\hat{p}_1 \hat{q}_1}{n_1} + \dfrac{\hat{p}_2 \hat{q}_2}{n_2}}$

*The values of z for a 90%, a 95%, or a 99% confidence interval are 1.645, 1.96, or 2.58, respectively.

Table 8.6 Large-Sample Statistical Tests for $(\mu_1 - \mu_2)$ and $(p_1 - p_2)$

Parameter	Test Statistic
$\mu_1 - \mu_2$	$z = \dfrac{\bar{y}_1 - \bar{y}_2}{\sigma_{\bar{y}_1 - \bar{y}_2}}$, where $\sigma_{\bar{y}_1 - \bar{y}_2} \approx \sqrt{\dfrac{s_1^2}{n_1} + \dfrac{s_1^2}{n_2}}$ provided $n_1, n_2 \geq 30$
$p_1 - p_2$	$z = \dfrac{\hat{p}_1 - \hat{p}_2}{\sigma_{\hat{p}_1 - \hat{p}_2}}$, where $\sigma_{\hat{p}_1 - \hat{p}_2} \approx \sqrt{\hat{p}\hat{q}\left(\dfrac{1}{n_1} + \dfrac{1}{n_2}\right)}$ and $\hat{p} = \dfrac{y_1 + y_2}{n_1 + n_2}$

REFERENCES

Freund, J. E. *Statistics: A First Course*. 2d ed. Englewood Cliffs, N.J.: Prentice-Hall, 1976.

Mendenhall, W. *Introduction to Probability and Statistics*. 5th ed. N. Scituate, Mass.: Duxbury Press, 1979. Chapter 8.

Ott, L. *An Introduction to Statistical Methods and Data Analysis*. N. Scituate, Mass.: Duxbury Press, 1977. Chapter 11.

Siegel, S. *Nonparametric Statistics for the Behavioral Sciences*. New York: McGraw-Hill, 1956.

SUPPLEMENTARY EXERCISES

8.18. A survey is conducted to determine whether a difference exists between the proportion of married persons and the proportion of single persons in the 20–29 age group who smoke. A sample of 200 persons from each group is polled, and 64 married persons and 80 single persons are found to smoke. Do the data provide sufficient evidence to indicate a difference in the proportions of smokers for the two populations? Use $\alpha = .05$.

8.19. Sixty of 87 couples sampled from the United States prefer a certain contraceptive technique; 40 of 100 couples interviewed from a European country prefer the same technique. Estimate the difference in the population proportions, using a 95% confidence interval.

8.20. Refer to exercise 8.19. Test the null hypothesis that couples from the United States and from the European country do not differ in their preference for the contraceptive technique. Use $\alpha = .01$.

8.21. There are some persons who experience chronic depression, which ultimately may require hospitalization. The data in the accompanying table summarize the reductions in depression (as measured by a commonly used rating scale) for patients treated with one of two marketed antidepressive agents. Use these data to construct a 99% confidence interval for $(\mu_1 - \mu_2)$, the difference in mean reductions for the two products.

	Product 1	Product 2
Sample Mean	22.3	21.6
Sample Variance	58	45
Sample Size	30	33

8.22. Refer to exercise 8.21. A third group of patients treated with a placebo (control) showed $\bar{y}_3 = 15.8$, $s_3^2 = 40.1$, and $n_3 = 35$. Use these data to test H_0: $\mu_1 - \mu_3 = 0$ against the alternative H_a: $\mu_1 - \mu_3 > 0$. Give the level of significance for this test.

8.23. Independent random samples of 800 Republicans and 800 Democrats showed 40% and 32%, respectively, in favor of the death sentence for major crimes. Do these data provide sufficient evidence to indicate a difference in the population proportions favoring the death sentence? Use $\alpha = .01$.

8.24. For the accompanying news article, comment on the appropriateness of the sample and the inferences that are drawn. Is there additional information that would have been helpful for you in drawing your own inferences?

Study of Divided Families Shows Positive Attitudes

Chicago—A study of divorced mothers and their children has revealed some positive attitudes among members of divided families. Perhaps a broken home is not the psychological disaster for family members that society has long suspected.

The study, involving 20 mothers with one or more children between the ages of 6 and 18, was conducted to determine the basic concerns of divorced mothers and their children. There were 20 mothers and 35 children involved in the study.

All the women were working full time. Most of them had made plans toward bettering their earning power. The women had been divorced from 3 months to 15 years. The educational level of the women in the study was high, compared to the national average: 12 years to 18 years of education.

A key aim of the study was to determine the feelings of the women and their children about their acceptance in society.

Eighty-six percent of the children felt that at school they were treated the same as children whose parents were married. Children aged 10 through 12 especially preferred that teachers and friends be told about the home situation. They wanted news of the divorce not to come as a surprise to others or to be a source of embarrassment for them.

In general, the children were doing well in school and even excelled in some areas.

Although the trend among most of the women was to socialize mainly with single persons, eighty percent of them felt accepted in their neighborhoods. Half of them said they felt accepted at church.

Among the children, ninety-one percent indicated they were treated no differently at Sunday school. Ninety percent of the sample were active church members.

Most of the women, eighty-five percent, said that after their divorces their attitudes toward divorce had shifted from negative to positive. The same proportion saw advantages for their children, in terms of understanding life and people, as a result of the divorce.

Supplementary Exercises

8.25. Seventy-two French students were randomly divided into two equal groups and subjected to one of two teaching techniques. Achievement test scores for the two classes of French students possessed the sample means and variances shown in the accompanying table. Use a 99% confidence interval to estimate the difference in mean achievement for the two teaching techniques.

	Group 1	Group 2
Sample Size	36	36
Sample Mean	260	294
Sample Variance	3600	4300

8.26. An auditor wished to determine whether a difference existed in the mean amount of accounts receivable (per person) for two utility companies. Random samples of 50 customers were selected from each company. The accounts receivable produced the sample means and standard deviations given in the accompanying table. Do the data provide sufficient evidence to indicate a difference in the mean amount of accounts receivable for the two utility companies? Use $\alpha = .05$.

	Company 1	Company 2
Sample Mean	$39.52	$47.10
Sample Standard Deviation	$24.98	$27.37

8.27. The lengths of time (in hours) to the first repair were recorded for thirty new lawn mowers for each of two brands. The means and variances for the two samples are shown in the accompanying table. Do the data present sufficient evidence to indicate a difference in the mean time to the first repair for the two different brands of lawn mowers? Use $\alpha = .05$.

	Brand 1	Brand 2
Sample Mean	137	115
Sample Variance	420	595

8.28. An ornithologist was firmly convinced that the number of fleas on a blue warbler was different, on the average, from the number on the grey variety. He trapped and inspected 45 blue warblers and 39 grey warblers in a one-week period and recorded the number of fleas per bird. The sample averages and variances are shown in the accompanying table. Do the data provide sufficient evidence to indicate a difference in the mean number of fleas for blue and grey warblers? Use $\alpha = .05$.

	Blue Warbler	Grey Warbler
Sample Mean	23.2	19.1
Sample Variance	24.1	29.6

212   Comparisons

8.29. The weight gains for two groups of rats, one group fed an ordinary diet (group 1) and the other fed a diet supplemented with a growth hormone (group 2), were recorded over a period of four weeks. These data are summarized in the accompanying table. Do the data present sufficient evidence to indicate a higher mean weight gain for rats with a diet supplement? Use $\alpha = .05$.

	Group 1	Group 2
Sample Mean	2.3 grams	2.7 grams
Sample Variance	1.12	1.05
Sample Size	40	40

8.30. The tax assessor for two counties had the unenviable task of trying to adjust property tax assessments. To help in assigning realistic values to properties in the two counties, he first compared the selling prices for homes in the two counties. The selling prices of houses were recorded for randomly selected bills of sale obtained from two county court houses. Each sample contained 60 bills of sale. The sample means and variances are shown in the accompanying table. Estimate the difference in mean selling prices for the two counties, using a 95% confidence interval.

	County I	County II
Sample Mean	59,840	57,520
Sample Variance	25,000,000	19,400,000

8.31. On a question concerning library hours on a college campus, 42 of 50 men interviewed favored the proposal to increase the hours of operation, and 78 of 80 women favored the proposal. Is there evidence of a difference in the proportions of all men and women favoring the proposal? Give the level of significance for your test.

8.32. The advent of teacher evaluations by students has seemed to change the criteria on which students are judged. While this might be arguable, it is a fact that at many universities the percentages of students awarded A's and B's has increased substantially in the past five years. Administrators who are concerned about this trend have taken a dim view of the excessively high percentages of A's and B's. The percentages of A's and B's awarded by two college philosophy professors were duly noted by the dean. Professor I achieved a rate of 53% as opposed to 40% for professor II, based upon 200 and 180 students, respectively. Do the data indicate a difference in the rates of awarding A's and B's by the two philosophy professors? Use $\alpha = .05$.

8.33. As part of the investigation of a new drug product developed for the relief of angina pectoris, a heart disease characterized by spasms of pain in

Supplementary Exercises

the chest, researchers examined the amount of time persons suffering from the disease could tolerate a specific exercise activity. A total of 70 persons were randomly divided into two groups, with those in group I receiving the new drug while those in group II received an identical-appearing placebo. Use the summary data (in minutes) to determine whether there is an increase in mean exercise times for the group I patients. Use $\alpha = .05$.

	Group I	Group II
Sample Mean	30.5	22.3
Sample Standard Deviation	10.6	9.8
Sample Size	35	35

8.34. An experiment was conducted to compare two different rations of feed for baby chicks. From a total of 200 baby chicks, 100 were randomly assigned to group I and fed ration A. The other 100 chicks were assigned to group II and fed ration B. One study of interest was a comparison of the proportions of chicks that died while being fed the two rations. Assuming all factors other than the rations were the same for the two groups, use the accompanying sample data to conduct a statistical test to determine whether the proportions of chickens that die are different for the two groups. Use $\alpha = .05$.

	Ration A	Ration B
Sample Size	100	100
Number That Died	14	8

8.35. A large federally funded project related to air pollution is underway to compare the lung capacities of individuals over a three-year period. Several different cities with varying degrees of air pollution have been chosen to participate in the study. Within each city a sample of adults and schoolchildren will be closely watched over the three-year study period to determine any differences in lung capacities. Suppose the data in the accompanying table summarize the changes in adult lung capacities over three years for samples from two of the participating cities. Set up an appropriate statistical test for comparing mean lung capacities in the two cities. Give the level of significance for your test.

	City 1 (Low Pollution Level)	City 2 (High Pollution Level)
Sample Size	100	91
Sample Mean	15.3	20.6
Sample Standard Deviation	8.16	9.82

Comparisons

8.36. A marketing research firm is interested in comparing the proportions of potential buyers of a soon-to-be-released product for chain stores and for independent stores. Use the accompanying sample data to construct a 95% confidence interval for the difference in proportions.

	Chain Stores	Independent Stores
Sample Size	110	650
Number of Potential Buyers	26	275

EXPERIENCES WITH REAL DATA

From your campus library, and with the help of your professor, choose one of the major journals from each of two subject matter areas (e.g., sociology and education).

1. Peruse each issue from both journals over the past calendar year to estimate the proportion of published articles that employ inferential statistical techniques. Use these data to construct a 95% confidence interval for the difference in proportions of articles using inferential statistics for the two journals.

2. While scanning those journal articles, also record the time from submission to actual publication for each article. Use these data to test the hypothesis that the average length of time from submission to publication is different for the two journals.

9
Inference Based on Small Samples

CHAPTER OUTLINE

9.1 Introduction
9.2 A Small-Sample Test of an Hypothesis About the Population Mean μ
9.3 A Small-Sample Confidence Interval for μ
9.4 A Small-Sample Test of a Difference in Means
9.5 A Small-Sample Confidence Interval for $(\mu_1 - \mu_2)$

9.1 INTRODUCTION

The techniques presented in chapters 6, 7, and 8 for estimating and testing hypotheses concerning a population mean and the difference between two population means demonstrate the concepts of statistical inference. And they provide some elementary statistical tools to help you evaluate experimental data more critically. But not all problems you will encounter will have large enough sample sizes so that you can use the techniques of chapters 6, 7, and 8. Hence you will need small-sample methods for testing and estimating a population mean and the difference between two means, and these methods are presented in this chapter. Thus chapter 9 is concerned with methodology rather than with basic concepts.

9.2 A SMALL-SAMPLE TEST OF AN HYPOTHESIS ABOUT THE POPULATION MEAN μ

A large-sample method for testing an hypothesis about the mean μ was presented in section 7.2. We noted there that this method is inappropriate for samples containing fewer than 30 measurements when σ is unknown. However, because of cost, time, or other restrictions, we must frequently use small

Inferences Based on Small Samples

samples. Hence we need a small-sample procedure to test an hypothesis about the mean μ. We illustrate the method in the following example.

Example 9.1

An experiment was performed to determine if the use of pictures would facilitate or impede a child's ability to learn the meanings of words. A random sample of 10 kindergarten children was assigned to a class that used pictures to assist in learning. Each child was tested after the experimental period and the number of words correctly identified from a total of 20 words chosen for the test was recorded. This test was repeated for five successive days, and the score assigned to each student was the average of the five separate tests. These data are shown below.

$$12.0 \quad 13.6 \quad 15.2 \quad 14.4 \quad 17.8 \quad 8.2 \quad 9.6 \quad 16.0 \quad 12.2 \quad 18.8$$

We would like to compare the achievement of this experimental group of children with the performance of others who have not used pictures to aid (or impede) their ability to learn the meaning of words. Fortunately, sample test data have been collected over a long period of time for a very large group of children who have not employed pictures in their learning process. The mean for their tests was found to be 17.1.

Suppose we regard 17.1 to be the true mean achievement for students who have learned without the use of pictures. Do the data above provide sufficient evidence to indicate that the use of pictures either improves or impedes word learning? That is, do the data provide sufficient evidence to show that the mean μ for the test population differs from 17.1? Note that we cannot use the z test of chapter 7 because the sample size is too small (less than 30). Consequently, we will defer the solution of this example until we explain how to test hypotheses about a population mean based on small samples.

Problems similar to example 9.1 were encountered by experimenters early in the twentieth century, when the only available test statistic was

$$z = \frac{\bar{y} - \mu}{\sigma/\sqrt{n}}$$

What statistical test did they use? Faced with small samples, early experimenters computed the sample standard deviation s to approximate σ and substituted it into the formula for z. Although aware that their procedure might be invalid, they took the only course open to them—they used the z test of section 7.2.

Many of these experimenters, however, were quite concerned that their

A Small-Sample Test of an Hypothesis About the Population Mean μ

statistical tests, based on z, were leading them to incorrect decisions. One of these experimenters, W. S. Gosset, translated his concern into action. Gosset was a chemist for the Guinness Breweries, and, as you might suspect, he was only provided with small samples for use in tests of quality.

Gosset was faced with the problem of finding the sampling distribution for

$$t = \frac{\bar{y} - \mu}{s/\sqrt{n}}$$

a statistic similar to z but with s substituted for σ. Gosset found that this test statistic, which he called t, possessed a probability distribution similar in appearance to the normal distribution but with a much wider spread. In fact, the smaller the sample size n, the greater was the spread in the probability distribution for t.

You will recall that the quantity

$$z = \frac{\bar{y} - \mu}{\sigma/\sqrt{n}}$$

possesses a standard normal distribution ($\mu = 0$, $\sigma = 1$). A standard normal (z) distribution and a t distribution based on a sample of six measurements are shown in figure 9.1. Note that t is more "spread out" than z.

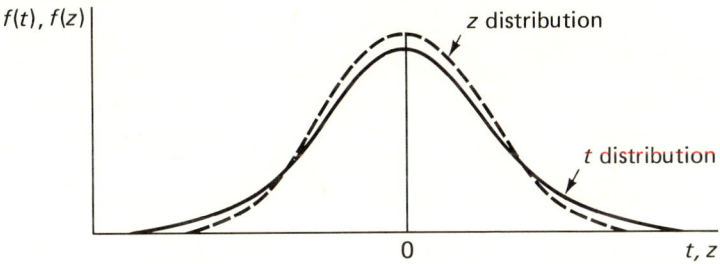

Figure 9.1 A z Distribution (Standard Normal Distribution) and a t Distribution Based on $n = 6$ Measurements

Gosset published his work on the t distribution in 1908; because of company policy, he used the pen name Student. His publication included the exact form of the distribution as well as the tail-end values of t, which are helpful in locating rejection regions. Gosset's statistic has many other applications in statistical decision making and has achieved a position of major importance in the field of statistics. His unique choice of a pen name has caused the statistic to be called Student's t.

Student's t

The t distribution possesses the following characteristics.

Characteristics of the *t* Distribution

1. Like z, it is mound-shaped and symmetrical about 0 (see figure 9.1).
2. It is more variable than z since both $\bar{y}$ and s change for each sample that is drawn from a population.
3. There are many t distributions. We determine a particular one by specifying a quantity called the <u>degrees of freedom</u>, which is directly related to the sample size.
4. As the degrees of freedom (or, equivalently, the sample size n) become large, the t distribution becomes a standard normal (z) distribution. This change is reasonable because s provides a better estimate of σ as n increases.

degrees of freedom

The values of t used to locate the rejection region for a statistical test are presented in table 2 in the Appendix. Since the t distribution is symmetrical about $t = 0$, we give only right-tail values. A value in the left tail is simply the negative of the corresponding right-tail value. **An entry in the table specifies a value of t, say t_a, such that an area a lies to its right** (see the shaded portion of figure 9.2). For example, $t_{.05}$ is the value of t such that an area equal to .05 lies to its right.

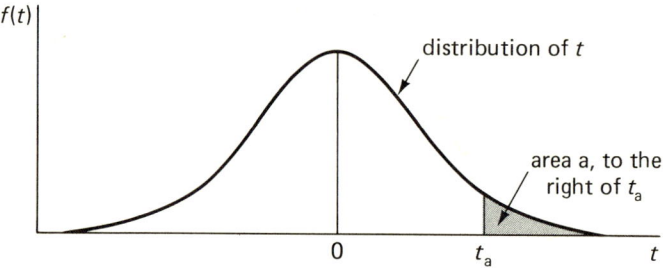

Figure 9.2 Use of the *t* Table

The format of table 2 is shown in table 9.1. Columns corresponding to various values of a are shown at the top of the table for a = .100, .05, .025, .010, and .005.

The symbols df, denoting "degrees of freedom," are shown in the first and last columns of the table. We will not explain the meaning of this term except to note that it is related to the sample size and, particularly, to the amount of

A Small-Sample Test of an Hypothesis About the Population Mean μ

Table 9.1 Format of the t Table, Table 2 in the Appendix

df	.100	.05	.025	.010	.005	df
1	3.078	6.314	12.706	31.821	63.657	1
2	1.886	2.920	4.303	6.965	9.925	2
.	.	.	.	.	.	.
.	.	.	.	.	.	.
9	1.383	1.833	2.262	2.821	3.250	9

(column header spanning .100 through .005: a)

information available to estimate the unknown quantity σ. For the test statistic

$$t = \frac{\bar{y} - \mu}{s/\sqrt{n}}$$

the degrees of freedom will always be one less than the sample size, that is, df $= (n - 1)$.

The numbers recorded in the table give the values of t_a. For example, suppose we have a sample of $n = 10$ measurements and wish to use a one-tailed t test, rejecting in the upper tail of the t distribution with a probability of a type I error of $\alpha = .05$. Then we would want to find the value of t_a corresponding to an area a $= .05$. (See figure 9.3.)

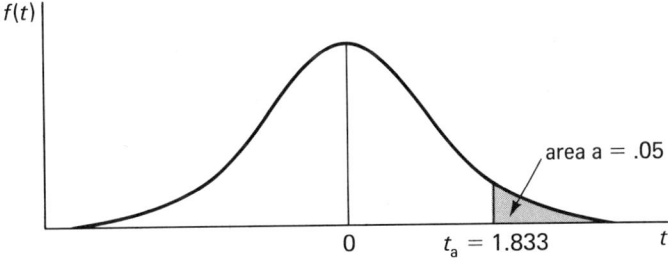

Figure 9.3 Rejection Region for a One-Tailed t Test for $n = 10$ and a $= .05$

To find the value of t corresponding to a $= .05$ and $n = 10$, look at table 2 in the Appendix. Select the column corresponding to a $= .05$ and proceed down the column to df $= (n - 1) = 9$. The value of t_a is 1.833. That value is shown in figure 9.3.

Inferences Based on Small Samples

You will recall that we deferred the solution of example 9.1 until we had a small-sample statistical test for an hypothesis about a population mean. Now let us return to testing the hypothesis that the mean number of words correctly identified in a posttest for children learning to identify words using illustrative pictures differs from 17.1, the average for children learning without such pictures. Use $\alpha = .05$.

Solution

The appropriate null and alternative hypotheses are

$H_0: \mu = 17.1$ (the population mean for the plan with pictures is 17.1; i.e., it is the same as the population mean for the plan with no pictures)

$H_a: \mu \neq 17.1$

The test statistic is

$$t = \frac{\bar{y} - \mu_0}{s/\sqrt{n}}$$

where $\bar{y}$ and s are the sample mean and standard deviation, respectively, and μ_0 is the hypothesized value of the population mean (in this case, 17.1).

Our next step is to calculate $\bar{y}$ and s for the data.

$$\Sigma y = 12.0 + 13.6 + \cdots + 18.8 = 137.8$$
$$\Sigma y^2 = (12.0)^2 + (13.6)^2 + \cdots + (18.8)^2 = 2001.88$$

Hence

$$\bar{y} = \frac{\Sigma y}{n} = \frac{137.8}{10} = 13.78$$

$$s^2 = \frac{\Sigma y^2 - (\Sigma y)^2/n}{n-1} = \frac{2001.88 - (137.8)^2/10}{9} = 11.444$$

and

$$s = \sqrt{11.444} = 3.383$$

The test statistic for this small-sample test is

$$t = \frac{\bar{y} - \mu_0}{s/\sqrt{n}}$$

where μ_0 is the hypothesized value of μ, that is, $\mu_0 = 17.1$. Substituting, we obtain

$$t = \frac{13.78 - 17.1}{3.383/\sqrt{10}} = -3.10$$

A Small-Sample Test of an Hypothesis About the Population Mean μ

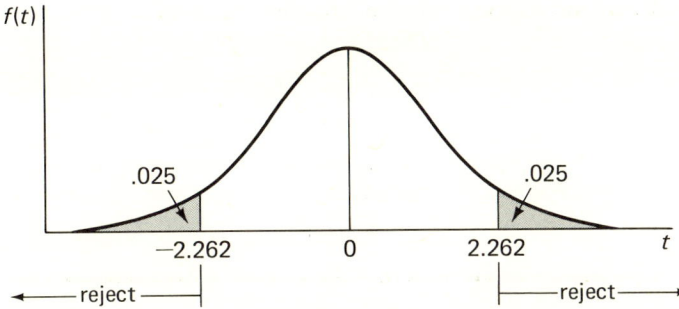

Figure 9.4 Rejection Region for Testing the Hypothesis $\mu = 17.1$ Against the Alternative Hypothesis $\mu \neq 17.1$ for df = 9 and a = .025

To locate the rejection region for our test, we must specify the degrees of freedom. The sample size is $n = 10$ and hence df = 9. Then for a two-tailed test with $\alpha = .05$, we can locate the rejection region from table 9.2 in the Appendix for df = 9 and a = .025. The table value of t is 2.262 (see figure 9.4). Since the calculated value of t does fall in the rejection region, we reject H_0 and conclude that the mean number of correctly identified words for the hypothetical population of children under the experimental program is different from the mean for all children not involved in the new program. Because the computed value of t falls in the lower tail of the t distribution, it appears that the average is less than 17.1. Indeed, the use of pictures appears to be interfering with a child's ability to learn words.

We summarize the parts of the small sample t test next.

Summary of Student's t Test for a Population Mean

Null hypothesis: $\mu = \mu_0$ (μ_0 is specified).

Alternative hypothesis: For a one-tailed test:
1. $\mu > \mu_0$.
2. $\mu < \mu_0$

For a two-tailed test:
3. $\mu \neq \mu_0$.

Test statistic: $t = \dfrac{\bar{y} - \mu_0}{s/\sqrt{n}}.$

Rejection region: For a specified value of α, df $= (n - 1)$, and for a one-tailed test:

1. Reject H_0 if $t > t_a$, where a $= \alpha$.
2. Reject H_0 if $t < -t_a$, where a $= \alpha$.

For a specified value of α, df $= n - 1$, and for a two-tailed test:

3. Reject H_0 if $|t| > t_a$, where a $= \alpha/2$.

Let us try a second small-sample test of an hypothesis concerning a population mean μ.

Example 9.2

An experiment was conducted to determine the abrasion resistance of a new type of automobile paint. Twelve different strips of metal were painted with the new paint. The abrasion resistance of each strip was then tested on a machine. These results are given in table 9.2. Determine whether there is evidence to indicate that the new paint has a mean resistance to abrasion greater than 2.9, the mean for the paint that is now being used. Use $\alpha = .05$.

Table 9.2 Abrasion Resistance of Paint

Strip	Abrasion Resistance	Strip	Abrasion Resistance
1	2.1	7	3.2
2	4.3	8	4.0
3	4.2	9	2.9
4	3.3	10	2.5
5	3.7	11	2.3
6	2.8	12	3.1

Solution

The four parts of the t test are as follows:

Null hypothesis: $\mu = 2.9$, where μ is the mean of the hypothetical population of abrasion resistances for the new paint.

Alternative hypothesis: $\mu > 2.9$.

A Small-Sample Test of an Hypothesis About the Population Mean μ

Test statistic:
$$t = \frac{\bar{y} - 2.9}{s/\sqrt{n}}.$$

Rejection region: For a one-tailed test with $\alpha = .05$, and $n = 12$, we look up the t value for a = .05 and df = 11. From table 2 this value is 1.796. So we will reject H_0 if $t > 1.796$.

Before computing t, we must first obtain $\bar{y}$ and s for the sample data. We find that

$$\Sigma y = 2.1 + 4.3 + \cdots + 3.1 = 38.40$$
$$\Sigma y^2 = (2.1)^2 + (4.3)^2 + \cdots + (3.1)^2 = 128.76$$

Hence

$$\bar{y} = \frac{\Sigma y}{n} = \frac{38.4}{12} = 3.2$$

$$s^2 = \frac{\Sigma y^2 - (\Sigma y)^2/n}{n-1} = \frac{128.76 - (38.4)^2/12}{11}$$
$$= \frac{128.76 - 122.88}{11} = .535$$

and

$$s = \sqrt{.535} = .731$$

Substituting into the test statistic, we have

$$t = \frac{3.2 - 2.9}{.731/\sqrt{12}} = \frac{(.3)(3.464)}{.731} = 1.42$$

Since the computed value of t does not fall in the rejection region, we have insufficient evidence to indicate that the new paint is more resistant to abrasion.

Keep in mind that the exact probability distribution of the quantity

$$\frac{\bar{y} - \mu}{s/\sqrt{n}}$$

depends on the form of the distribution from which the sample observations y were obtained. **Only if these observations come from a normal distribution with mean μ and standard deviation σ will $(\bar{y} - \mu)/(s/\sqrt{n})$ possess a t distribution based on $(n - 1)$ degrees of freedom.** Gosset's results, and hence the tail values in table 2, are based on this assumption (that the original sample was selected from a population that possesses a normal probability distribution).

How restrictive is this assumption? Fortunately, the probability distribution of $(\bar{y} - \mu)/(s/\sqrt{n})$ is relatively stable even for probability distributions that are not normal but are mound-shaped. This property of the t distribution and the common occurrence of mound-shaped distributions in practice make the Student's t invaluable for use in statistical inference.

EXERCISES

9.1. Why is the z test of chapter 7 inappropriate for testing $H_0: \mu = \mu_0$ when $n < 30$?

9.2. Set up the rejection region for $H_0: \mu = \mu_0$ when $\alpha = .05$ and for the following conditions:
(a) $H_a: \mu < \mu_0$, $n = 15$
(b) $H_a: \mu \neq \mu_0$, $n = 23$
(c) $H_a: \mu > \mu_0$, $n = 6$

9.3. Repeat exercise 9.2 with $\alpha = .01$.

9.4. The sample data for a t test of $H_0: \mu = 15$ and $H_a: \mu > 15$ are $\bar{y} = 16.2$, $s = 3.1$, and $n = 18$. Use $\alpha = .05$ to draw your conclusions.

9.5. The voter turnout in upper-middle-class areas of a city has averaged 650 for every 1000 registered voters in past years. This year a random sample of five precincts shows a turnout of 635, 655, 640, 643, and 620 per 1000 registered voters. Do these data indicate an overall average of less than 650? Use $\alpha = .05$.

9.6. Measurements of water intake, obtained from a sample of 17 rats that had been injected with a sodium chloride solution, produced a mean and standard deviation of 31.0 and 6.2 cubic centimeters, respectively. The historical average water intake for noninjected rats observed over a comparable period of time is 22.0 cubic centimeters. Do the data indicate that injected rats drink more water than noninjected rats? Test the hypothesis by using $\alpha = .05$.

9.7. Industrial wastes and sewage dumped into our rivers and streams absorb oxygen and thereby reduce the amount of dissolved oxygen available for fish and other forms of aquatic life. One state agency requires a minimum of 5 parts per million (ppm) of dissolved oxygen in order that the oxygen content be sufficient to support aquatic life. Six water specimens taken from a river at a specific location during the low-water season (July) gave readings of 4.9, 5.1, 4.9, 5.0, 5.0, and 4.7 ppm of dissolved oxygen. Do the data provide sufficient evidence to indicate that the dissolved oxygen content is less than 5 ppm? Test by using $\alpha = .05$. (You can verify that the sample standard deviation is .137.)

9.3 A SMALL-SAMPLE CONFIDENCE INTERVAL FOR μ

A small-sample confidence interval for μ follows directly from the large-sample formula. Instead of using the interval $(\bar{y} \pm z\sigma_{\bar{y}})$, where $\sigma_{\bar{y}} = \sigma/\sqrt{n}$, we can substitute t for z and replace σ by s to form the small-sample confidence interval for μ. The only restriction that we have in using this small-sample confidence interval is that again we must assume the sample was drawn from a population that is approximately normal.

General Small-Sample Confidence Interval for μ

$$\bar{y} \pm \frac{ts}{\sqrt{n}}$$

The value of t corresponding to a 90%, a 95%, or a 99% confidence interval is found in table 2 of the Appendix for df $= (n - 1)$ and a $= .05, .025,$ or $.005$, respectively.

We illustrate the use of the small-sample confidence interval with an example.

Example 9.3

A pharmaceutical firm has been conducting highly restricted studies on small groups of people to determine the effectiveness of a measles vaccine. The measurements below are readings on the antibody strength of five individuals injected with the vaccine:

$$1.2 \quad 3.0 \quad 2.5 \quad 2.4 \quad 1.9$$

Use the sample data to estimate μ, the population mean antibody strength for individuals vaccinated with the new drug. Use a 95% confidence interval.

Solution

For this example we have

$$\Sigma y = 11.00 \quad \text{and} \quad \Sigma y^2 = 26.06$$

Thus

$$\bar{y} = \frac{11}{5} = 2.2$$

$$s^2 = \frac{\Sigma y^2 - (\Sigma y)^2/n}{n-1} = \frac{26.06 - (11)^2/5}{4}$$
$$= \frac{26.06 - 24.20}{4} = .465$$
$$s = \sqrt{.465} = .682$$

The general form of the confidence interval is

$$\bar{y} \pm \frac{ts}{\sqrt{n}}$$

From table 2 in the Appendix, for a = .025 and df = $(n-1) = 4$, we find that $t = 2.776$. Substituting into the formula, we obtain a lower and an upper confidence limit of

$$2.2 - \frac{2.776(.682)}{\sqrt{5}} = 2.2 - .847 = 1.353$$
$$2.2 + \frac{2.776(.682)}{\sqrt{5}} = 2.2 + .847 = 3.047$$

The 95% confidence interval is then 1.353 to 3.047. We note that this interval either does or does not enclose the true mean μ. However, in repeated sampling, 95% of the intervals calculated in this way will enclose μ. Hence we are 95% confident that the true mean antibody strength for the measles vaccine is in the interval 1.353 to 3.047.

The small-sample confidence interval, like the corresponding t test, is based on the assumption that the sample was selected from a normal population. If we had to adhere strictly to this assumption, the procedure would be of little value because we rarely know the exact form of the population probability distribution. Fortunately, the confidence interval will work satisfactorily as long as the population probability distribution is mound-shaped. Thus our small-sample confidence interval for μ has wide applicability.

EXERCISES

9.8. Refer to example 9.3. Construct a 90% confidence interval for μ.

9.9. The sample mean and standard deviation from a random sample of $n = 5$ measurements were $\bar{y} = 38.6$ and $s = 5.7$. Construct a 95% confidence interval for μ.

9.10. Organic chemists often purify organic compounds by a method known as fractional crystallization. An experimenter desired to prepare and purify

4.85 grams of aniline. Ten 4.85-gram quantities of aniline were individually prepared and purified to acetanilide. The following dry yields were recorded:

$$
\begin{array}{ccccc}
3.85 & 3.90 & 3.72 & 3.85 & 4.01 \\
3.88 & 3.62 & 3.80 & 3.36 & 3.83
\end{array}
$$

Estimate the mean number of grams of acetanilide that could be recovered from an initial amount of 4.85 grams of aniline. Use a 95% confidence interval. (You can verify that the sample standard deviation is .181.)

9.11. An experimenter, interested in determining the mean thickness of the cortex of the sea urchin egg, employed an experimental procedure developed by Sakai. The thickness of the cortex was measured for 10 sea urchin eggs. The following measurements were obtained:

$$
\begin{array}{ccccc}
4.5 & 3.2 & 4.7 & 2.6 & 4.6 \\
6.1 & 3.9 & 5.2 & 3.7 & 4.1
\end{array}
$$

Estimate the mean thickness of the cortex, using a 95% confidence interval.

9.4 A SMALL-SAMPLE TEST OF A DIFFERENCE IN MEANS

The small-sample test of an hypothesis concerning the difference between two population means is similar to a large-sample test, except that we make some additional assumptions concerning the nature of the sampled populations. We will comment on the restrictive nature of these assumptions at the end of this section.

As with the large-sample test, the most logical test statistic for testing the hypothesis $(\mu_1 - \mu_2) = 0$ is based on the difference in the sample means, $(\bar{y}_1 - \bar{y}_2)$. If the sample difference is too far away from $(\mu_1 - \mu_2) = 0$, we reject the null hypothesis.

If we assume that both populations are normal and that the populations possess a common variance, say σ^2—that is, if we assume that both populations possess roughly the same amount of variation—the test statistic,

$$ t = \frac{\bar{y}_1 - \bar{y}_2}{s\sqrt{\dfrac{1}{n_1} + \dfrac{1}{n_2}}} $$

possesses a t distribution.

The quantity s in the test statistic is an estimate of the common standard deviation σ for the two populations and is formed by combining information from the two samples.

$$ s = \sqrt{\frac{(n_1 - 1)s_1^2 + (n_2 - 1)s_2^2}{n_1 + n_2 - 2}} $$

pooled estimate of variance

In fact, s^2 is a weighted average of the sample variances s_1^2 and s_2^2 and is sometimes called a pooled estimate of the variance, and $s = \sqrt{s^2}$. For the special case where the sample sizes are the same ($n_1 = n_2$), the formula for s^2 reduces to $s^2 = (s_1^2 + s_2^2)/2$, the mean of the two sample variances.

The rejection region for the t test is selected in the same manner as is the rejection region for the test of an hypothesis about a single mean (section 9.2). The degrees of freedom for the test are the sum of the degrees of freedom for the two samples:

$$df = (n_1 - 1) + (n_2 - 1) = n_1 + n_2 - 2$$

Thus for a two-tailed test with $\alpha = .05$, we look up in table 2 of the Appendix the t value corresponding to $a = .025$ and $(n_1 + n_2 - 2)$ degrees of freedom.

We illustrate this two-sample t test with an example.

Example 9.4

An experiment was conducted to investigate the effect of two diets on weight gain of 14-year-old children suffering from malnutrition. Ten children were subjected to diet I and nine to diet II. The gains in weight over a nine-month period are shown in table 9.3. Determine if the data indicate a difference between the mean gains in weight for children fed on the two diets. Use $\alpha = .05$.

Table 9.3 Data for Example 9.4

	Weight Gain
Diet I	14.0 12.5 10.2 9.8 10.5 11.2 15.0 22.0 13.0 9.6
Diet II	14.4 18.2 19.5 21.2 15.3 11.6 12.8 13.1 11.3

Solution

First we assume that weight gains for the two diets are normally distributed, with unknown means μ_1 and μ_2 and common unknown variance σ^2. The sample means and variances can be shown to be as given in table 9.4.

Table 9.4 Means and Variances for Data

	Diet I	Diet II
Sample Mean	12.78	15.27
Sample Variance	13.88	12.81
Sample Size	10	9

A Small-Sample Test of a Difference in Means

The estimate of the common population standard deviation is

$$s = \sqrt{\frac{(n_1 - 1)s_1^2 + (n_2 - 1)s_2^2}{n_1 + n_2 - 2}}$$

$$= \sqrt{\frac{9(13.88) + 8(12.81)}{10 + 9 - 2}} = \sqrt{\frac{124.92 + 102.48}{17}} = \sqrt{13.38}$$

and so

$$s = \sqrt{13.38} = 3.66$$

Let μ_1 and μ_2 be the means for the hypothetical populations of weight gains associated with diets I and II, respectively; then the null and alternative hypotheses are

$$H_0: \mu_1 - \mu_2 = 0 \quad H_a: \mu_1 - \mu_2 \neq 0$$

The value of the test statistic t for this test is

$$t = \frac{\bar{y}_1 - \bar{y}_2}{s\sqrt{\frac{1}{n_1} + \frac{1}{n_2}}} = \frac{12.78 - 15.27}{3.66\sqrt{\frac{1}{10} + \frac{1}{9}}} = \frac{-2.49}{3.66(.459)} = -1.48$$

The rejection region for $\alpha = .05$ utilizes a t value corresponding to a $= .025$ and df $= (10 + 9 - 2) = 17$. From table 2 of the Appendix, this value is 2.110. Thus we reject the null hypothesis if the computed value of t is greater than 2.110 or less than -2.110. This rejection region is shown in figure 9.5.

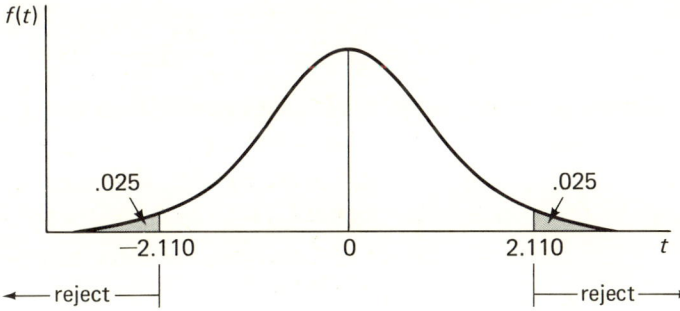

Figure 9.5 Two-Tailed Rejection Region for a $= .025$ and df $= 17$

Noting that the computed value of t, $t = -1.48$, does not fall in the rejection region, we conclude that there is insufficient evidence to indicate a difference in the mean weight gains for the two diets.

Before concluding our discussion of the t test, we question whether the assumptions of normal populations and equal variances ($\sigma_1^2 = \sigma_2^2$) must hold

Inferences Based on Small Samples

in order for the t test to be valid. The test functions satisfactorily for populations possessing mound-shaped probability distributions. The assumption that $\sigma_1^2 = \sigma_2^2$ is more critical. But again, it does not seriously affect the properties of the test if n_1 and n_2 are approximately equal. Consequently, these assumptions are not too restrictive, giving the test wide applicability.

A summary of the elements of the t test for comparing two means is given next.

Summary of a Two-Sample t Test for Comparing Two Population Means

Null hypothesis: $\mu_1 - \mu_2 = 0$.

Alternative hypothesis: For a one-tailed test:
1. $\mu_1 - \mu_2 > 0$.
2. $\mu_1 - \mu_2 < 0$.

For a two-tailed test:

3. $\mu_1 - \mu_2 \neq 0$.

Test statistic:
$$t = \frac{\bar{y}_1 - \bar{y}_2}{s\sqrt{\frac{1}{n_1} + \frac{1}{n_2}}}. \quad \text{where}$$

$$s = \sqrt{\frac{(n_1 - 1)s_1^2 + (n_2 - 1)s_2^2}{n_1 + n_2 - 2}}$$

Rejection region: For a specified value of α, for df $= (n_1 + n_2 - 2)$, and for a one-tailed test:

1. Reject H_0 if $t > t_a$, where a $= \alpha$.
2. Reject H_0 if $t < -t_a$, where a $= \alpha$.

For a specified value of α, for df $= (n_1 + n_2 - 2)$, and for a two-tailed test:

3. Reject H_0 if $|t| > t_a$, where a $= \alpha/2$.

EXERCISES

9.12. Set up the rejection regions for testing $H_0: \mu_1 - \mu_2 = 0$ for the following conditions:
(a) $H_a: \mu_1 - \mu_2 \neq 0$, $n_1 = 12$, $n_2 = 14$, and $\alpha = .05$
(b) $H_a: \mu_1 - \mu_2 > 0$, $n_1 = n_2 = 8$, and $\alpha = .01$

(c) $H_a: \mu_1 - \mu_2 < 0$, $n_1 = 6$, $n_2 = 4$, and $\alpha = .05$
(d) What assumptions must be made prior to applying a two-sample t test?

9.13. Conduct a test of $H_0: \mu_1 - \mu_2 = 0$ against the alternative hypothesis $H_a: \mu_1 - \mu_2 < 0$ for the sample data shown here. Use $\alpha = .05$.

	Population 1	Population 2
Sample Size	16	13
Sample Mean	71.5	79.8
Sample Variance	68.35	70.26

9.14. Refer to the data of exercise 9.13. Give the level of significance for your test. (Hint: Since the t table is not constructed in a way that makes it possible to obtain an exact p value, the best we can do is approximate the level of significance, using entries in table 2. For example, a level of significance stated as $p < .10$ is perfectly acceptable.)

9.15. In an effort to link cold environments with hypertension in humans, a preliminary experiment was conducted to investigate the effect of cold on hypertension in rats. Two random samples of 6 rats each were exposed to different environments. One sample of rats was held in a normal environment at 26°C. The other sample was held in a cold 5°C environment. Blood pressures and heart rates were measured for rats for both groups. The blood pressures for the 12 rats are shown in the accompanying table. (These data represent a small portion of the data resulting from an experiment conducted by Dr. Garth Resch, University of Florida.) Do the data provide sufficient evidence to indicate that rats exposed to a 5°C environment have a higher mean blood pressure than rats exposed to a 26°C environment? Test by using $\alpha = .05$.

Environmental Temperature and Blood Pressure

26°		5°	
Rat	Blood Pressure	Rat	Blood Pressure
1	152	7	384
2	157	8	369
3	179	9	354
4	182	10	375
5	176	11	366
6	149	12	423

9.16. Refer to exercise 9.7, where we measured the dissolved oxygen content in river water to determine whether a stream possessed sufficient oxygen to support aquatic life. A pollution control inspector suspected that a riverside community was releasing semitreated sewage into a river and this, as a consequence, was changing the level of dissolved oxygen of the river. To check his theory, he drew five randomly selected specimens of river water at a location above the town and another five specimens below. The dissolved oxygen readings, in parts per million, are given in the accompanying table. Do the data provide sufficient evidence to indicate a difference in mean oxygen content between locations above and below the town? Use $\alpha = .05$.

Above Town	4.8 5.2 5.0 4.9 5.1
Below Town	5.0 4.7 4.9 4.8 4.9

9.17. The length of time to complete recovery was recorded for patients randomly assigned and subjected to two different surgical procedures. The data, in days, are given in the accompanying table. (Hint: Any time the number of degrees of freedom exceeds 29, use df $= \infty$ in table 2.) Do the data present sufficient evidence to indicate a difference in mean recovery time for the two surgical procedures? Test by using $\alpha = .05$.

	Procedure I	Procedure II
Sample Size	21	23
Sample Mean	7.3	8.9
Sample Variance	1.23	1.49

9.18. Refer to exercise 9.17. Give the level of significance for your test.

9.5 A SMALL-SAMPLE CONFIDENCE INTERVAL FOR $(\mu_1 - \mu_2)$

In addition to being able to test an hypothesis about the equality of two population means for small sample sizes, we can construct a small-sample confidence interval for $(\mu_1 - \mu_2)$. The small-sample confidence interval formula for $(\mu_1 - \mu_2)$ is similar to the large-sample confidence interval formula, but z is replaced by t. The same assumptions apply to a small-sample confidence interval for $(\mu_1 - \mu_2)$ as were required for a small-sample test of H_0: $\mu_1 - \mu_2 = 0$. (See section 9.4.)

Small-Sample Confidence Interval for $(\mu_1 - \mu_2)$

$$\bar{y}_1 - \bar{y}_2 \pm ts\sqrt{\frac{1}{n_1} + \frac{1}{n_2}}$$

where

$$s = \sqrt{\frac{(n_1 - 1)s_1^2 + (n_2 - 1)s_2^2}{n_1 + n_2 - 2}}$$

The value of t corresponding to a 90%, a 95%, or a 99% confidence interval is found in table 2 of the Appendix for df $= (n_1 + n_2 - 2)$ and a $= .05, .025,$ or $.005$, respectively.

We illustrate the use of the confidence interval with an example.

Example 9.5

A psychologist was interested in comparing the average length of time it takes individuals to complete two different psychological checklists. From a relatively homogeneous group of 20 individuals, 10 were randomly assigned to list 1 and the other 10 to list 2. The appropriate checklists were then administered, and the amount of time required to complete the task was recorded for each individual. These data are summarized in table 9.5. Find a 95% confidence interval for $(\mu_1 - \mu_2)$, the difference in the mean completion times.

Table 9.5 Data for Example 9.5

List 1	List 2
$\bar{y}_1 = 54.3$ minutes	$\bar{y}_2 = 48.1$ minutes
$n_1 = 10$	$n_2 = 10$
$s_1^2 = 16.0$	$s_2^2 = 12.2$

Solution

First we must compute the estimate of the common standard deviation σ. Using the sample data, we have

$$s = \sqrt{\frac{(n_1 - 1)s_1^2 + (n_2 - 1)s_2^2}{n_1 + n_2 - 2}} = \sqrt{\frac{9(16.0) + 9(12.2)}{18}} = 3.75$$

Inferences Based on Small Samples

Next, from table 2 of the Appendix we find the tabulated t value for df = $(n_1 + n_2 - 2) = 18$ and a = .025; it is 2.101. Substituting into the formulas for the lower and upper confidence limits, we have

$$\bar{y}_1 - \bar{y}_2 - ts\sqrt{\frac{1}{n_1} + \frac{1}{n_2}} = 54.3 - 48.1 - (2.101)(3.75)\sqrt{\frac{1}{10} + \frac{1}{10}}$$
$$= 6.2 - 3.52 = 2.68$$

$$\bar{y}_1 - \bar{y}_2 + ts\sqrt{\frac{1}{n_1} + \frac{1}{n_2}} = 54.3 - 48.1 + (2.101)(3.75)\sqrt{\frac{1}{10} + \frac{1}{10}}$$
$$= 6.2 + 3.52 = 9.72$$

Thus we estimate that the difference in the mean completion times, $(\mu_1 - \mu_2)$, is somewhere in the interval 2.68 to 9.72.

In conclusion, you might ask how this section is relevant to you. Perhaps you will be faced with a similar problem in your particular field of study. Or perhaps you are interested in keeping abreast of the times. Reading a newspaper or periodical, you may find a point estimate $(\bar{y}_1 - \bar{y}_2)$ for some comparison that interests you. We hope that you will wonder about the width of the confidence interval (which is likely to be missing) or about the corresponding bound on the error of estimation. Has the author used the appropriate statistical method to analyze the data?

EXERCISES

9.19. Refer to the data of example 9.5. Construct a 99% confidence interval for $(\mu_1 - \mu_2)$.

9.20. Use the data of exercise 9.15 to construct a 90% confidence interval for $(\mu_1 - \mu_2)$, the difference in mean blood pressure in rats subjected to the two environments.

9.21. An experiment was conducted to compare the mean lengths of time required for the bodily absorption of two drugs, A and B. Ten people were randomly selected and assigned to each drug treatment. Each of the ten persons in the sample received an oral dosage of the assigned drug and the length of time (in minutes) for the drug to reach a specified level in the blood was recorded. The means and variances for the two samples are given in the accompanying table. Fing a 95% confidence interval for the difference in mean times for absorption.

	Drug A	Drug B
Sample Mean	27.2	33.5
Sample Variance	16.36	18.92

References

SUMMARY

This chapter presents small-sample methods for estimating and testing hypotheses about a population mean μ and the difference between two population means, $(\mu_1 - \mu_2)$. All the methods are similar to those of the corresponding large-sample tests, and all make use of the Student's t statistic. These methods are needed to adjust for the use of the sample standard deviations to approximate the corresponding population parameters. Since the assumptions required for the use of the small-sample methods are not restrictive, the methods possess wide applicability.

The small-sample estimation and test procedures presented in this chapter are summarized in tables 9.6 and 9.7.

Table 9.6 Small-Sample Confidence Intervals for μ and $(\mu_1 - \mu_2)$

Parameter	Confidence Interval*
μ	$\bar{y} \pm \dfrac{ts}{\sqrt{n}}$, where df $= n - 1$
$\mu_1 - \mu_2$	$\bar{y}_1 - \bar{y}_2 \pm ts\sqrt{\dfrac{1}{n_1} + \dfrac{1}{n_2}}$, where $s = \sqrt{\dfrac{(n_1 - 1)s_1^2 + (n_2 - 1)s_2^2}{n_1 + n_2 - 2}}$ and df $= n_1 + n_2 - 2$

*The value of t for a 90%, a 95%, or a 99% confidence interval is found in table 2 of the Appendix for a = .05, .025, or .005, respectively.

Table 9.7 Small-Sample Statistical Tests for μ and $(\mu_1 - \mu_2)$

Parameter	Test statistic
μ	$t = \dfrac{\bar{y} - \mu_0}{s/\sqrt{n}}$
$\mu_1 - \mu_2$	$t = \dfrac{\bar{y}_1 - \bar{y}_2}{s\sqrt{\dfrac{1}{n_1} + \dfrac{1}{n_2}}}$, where $s = \sqrt{\dfrac{(n_1 - 1)s_1^2 + (n_2 - 1)s_2^2}{n_1 + n_2 - 2}}$ and df $= n_1 + n_2 - 2$

REFERENCES

Barr, A. J.; Goodnight, J. H.; Sall, J. P.; Helwig, J. T. *A User's Guide to SAS 76.* Raleigh, N.C.: SAS Institute, Inc., 1976.

Dixon, W. J. *BMDP: Biomedical Computer Programs.* Berkeley: University of California Press, 1975.

, and Brown, M. B., eds. *Biomedical Computer Programs*. Rev. ed. Los Angeles: University of California Press, 1978.

Freund, J. E. *Statistics: A First Course*. 2d ed. Englewood Cliffs, N.J.: Prentice-Hall, 1976.

Helwig, J. T. *SAS Supplementary Library User's Guide*. Raleigh, N.C.: SAS Institute, Inc., 1977.

Huntsberger, D. V., and Billingsley, P. *Elements of Statistical Inference*. 4th ed. Boston: Allyn and Bacon, 1977.

Mendenhall, W. *Introduction to Probability and Statistics*. 5th ed. N. Scituate, Mass.: Duxbury Press, 1979. Chapter 9.

Nie, N.; Hull, C. H.; Jenkins, J. G.; Steinbrenner, K.; and Bent, D. H. *Statistical Package for the Social Sciences*. 2d ed. New York: McGraw-Hill, 1975.

Ott, L. *An Introduction to Statistical Methods and Data Analysis*. N. Scituate, Mass.: Duxbury Press, 1977. Chapter 11.

Service, J. *A User's Guide to the Statistical Analysis System*. Raleigh, N.C.: Student Supply Stores, North Carolina State University, 1972.

Siegel, S. *Nonparametric Statistics for the Behavioral Sciences*. New York: McGraw-Hill, 1956.

SUPPLEMENTARY EXERCISES

9.22. Statistics has become a valuable tool for auditors, especially where large inventories are involved. It would be costly and time-consuming for an auditor to inventory each item in a large operation. Thus he frequently resorts to obtaining a random sample of items and using the sample results to check the validity of a company's financial statement. For example, a hospital financial statement claims an inventory that averages $300 per item. An auditor's random sample of 20 items yielded a mean and standard deviation of $160 and $90, respectively. Do the data contradict the hospital's claimed mean value per inventoried item and indicate that the average is less than $300? Use $\alpha = .05$.

9.23. Over the past five years the mean time for a warehouse to fill a buyer's order has been 25 minutes. Officials of the company believe that the length of time has increased recently, either due to a change in the work force or due to a change in customer purchasing policies. The processing time (in minutes) was recorded for a random sample of 15 orders processed over the past month.

28	25	27	31	10
26	30	15	55	12
24	32	28	42	38

The sample standard deviation for these data is 11.44 minutes. Do the data present sufficient evidence to indicate that the mean time to fill an order has increased? Use $\alpha = .05$.

Supplementary Exercises

9.24. Give the level of significance for the statistical test in exercise 9.23.

9.25. If a new process for mining copper is to be put into full-time operation, it must produce an average of more than 50 tons of ore per day. A five-day trial period gave the results shown in the accompanying table. Do these figures warrant putting the new process into full-time operation? Test by using $\alpha = .05$.

Day	1	2	3	4	5
Yield in Tons	50	47	53	51	52

9.26. A test was conducted to determine the length of time required for a student to read a specified amount of material. All students were instructed to read at the maximum speed at which they could still comprehend the material. Sixteen students took the test, with the following results (in minutes):

$$\begin{array}{cccccccc} 25 & 18 & 27 & 29 & 20 & 19 & 25 & 24 \\ 32 & 21 & 24 & 19 & 23 & 28 & 31 & 22 \end{array}$$

Estimate the mean length of time required for all students to read the material, using a 95% confidence interval.

9.27. A random sample of 8 students participated in a psychological test of depth perception. Two markers, one labeled A and the other B, were arranged a fixed distance apart at the far end of the laboratory. One by one the students were ushered into the room and asked to judge the distance between the two markers at the other end of the room. The sample data (in feet) were as follows:

$$\begin{array}{cccc} 2.1 & 2.2 & 2.6 & 2.3 \\ 1.8 & 2.3 & 2.4 & 2.5 \end{array}$$

Construct a 90% confidence interval for μ, the mean judged distance for all students for which this sample is representative.

9.28. The lifetimes of 10 automobile batteries of a certain brand (in years) are

$$2.4 \quad 1.9 \quad 2.0 \quad 2.1 \quad 1.8 \quad 2.3 \quad 2.1 \quad 2.3 \quad 1.7 \quad 2.0$$

Estimate the mean lifetime, using a 95% confidence interval.

9.29. A drug antibiotic manufacturer randomly sampled 12 different locations in the fermentation vat to determine average potency for the batch of antibiotic being prepared. Readings were as follows:

$$\begin{array}{cccc} 8.9 & 9.0 & 9.1 & 8.9 \\ 9.1 & 9.0 & 9.0 & 9.0 \\ 8.9 & 8.8 & 9.1 & 8.8 \end{array}$$

240 Inferences Based on Small Samples

Estimate the mean potency for the batch, using a 95% confidence interval. Interpret the interval.

 9.30. The accompanying computer output gives the drops in blood pressure for three groups of six rats from a strain of hypertensive rats. The six rats in the first group were treated with a low dose of an antihypertensive product, the second group with a higher dose of the same antihypertensive product, and the third group with an inert control. Note the great variability in blood pressure decreases, even for rats in the control group.

(a) Draw conclusions for a comparison of the mean drop for the high-dose group and the control group.

(b) Is there evidence to indicate a difference between the low- and high-dose groups? Explain.

```
******************************************************************
DESCRIPTIVE STATISTICS
                FILE: Low-Dose Group
                    -51.00000
                     15.00000
                     48.00000
                     65.00000
                    -20.00000
                     75.00000

Low-Dose Group
NUMBER:      6
  MEAN:     22.00000
STD DEV:    49.95198
******************************************************************
DESCRIPTIVE STATISTICS
                FILE: High-Dose Group
                     69.00000
                     24.00000
                     63.00000
                     87.50000
                     77.50000
                     40.00000

High-Dose Group
NUMBER:      6
  MEAN:     60.16667
STD DEV:    23.86769
******************************************************************
DESCRIPTIVE STATISTICS
                FILE: Control Group
                      9.00000
                     12.00000
```

Supplementary Exercises

```
                        36.00000
                        77.50000
                        -7.50000
                        32.50000
```

Control Group

NUMBER: 6
  MEAN: 26.58333
STD DEV: 29.66381

****

UNPAIRED T TEST

Low-Dose Group		High-Dose Group
-51.00000		69.00000
15.00000		24.00000
48.00000		63.00000
65.00000		87.50000
-20.00000		77.50000
75.00000		40.00000

NO. OF OBSERVATIONS	6.	6.
MEAN	22.00000	60.16667
STANDARD DEVIATION	49.95198	23.86769
STANDARD ERROR	20.39281	9.74394

RATIO OF MEANS (2ND/1ST)		2.73485
DIFFERENCE OF MEANS (2ND-1ST)		38.16667
STANDARD ERROR OF DIFFERENCE		22.60113
95% CONFIDENCE INTERVAL FOR DIFFERENCE OF MEANS	-12.18865,	88.521991
RATIO OF VARIANCES (2ND/1ST)		0.22831
T STATISTIC (EQUAL VARIANCES)		1.68871
DEGREES OF FREEDOM		10
PROBABILITY		0.12216

****

UNPAIRED T TEST

  FILES

High-Dose Group		Control Group
69.00000		9.00000
24.00000		12.00000
63.00000		36.00000
87.50000		77.50000
77.50000		-7.50000
40.00000		32.50000

NO. OF OBSERVATIONS	6.	6.
MEAN	60.16667	26.58333
STANDARD DEVIATION	23.86769	29.66381
STANDARD ERROR	9.74394	12.11020

RATIO OF MEANS (2ND/1ST)	0.44183
DIFFERENCE OF MEANS (2ND-1ST)	-33.58333

STANDARD ERROR OF DIFFERENCE	15.54353
95% CONFIDENCE INTERVAL FOR DIFFERENCE OF MEANS	−68.21432, 1.047661
RATIO OF VARIANCES (2ND/1ST)	1.54466
T STATISTIC (EQUAL VARIANCES)	−2.16060
DEGREES OF FREEDOM	10
PROBABILITY	0.05605

9.31. Use the data of exercise 9.30 to construct a 95% confidence interval for $(\mu_1 - \mu_3)$, the difference in population means for the low-dose group and the control group.

9.32. A petroleum corporation was interested in running some preliminary tests to compare the performance of a new gasoline mixture to one currently on the market. Ten identical new automobiles were randomly assigned, five to gasoline A and five to gasoline B. Gasoline A contained a mileage additive, and gasoline B was regular gasoline. Each automobile was filled with 10 gallons of gasoline and driven over a test course until it stopped. The mileage was recorded for each; see the accompanying table. Use a t test for the alternative hypothesis that there is a difference in mean mileage for the two brands of gasoline. Use $\alpha = .05$.

Gasoline A	Gasoline B
182	184
179	185
180	186
178	177
175	183
$\bar{y}_1 = 178.80$	$\bar{y}_2 = 183.00$
$s_1^2 = 6.70$	$s_2^2 = 12.40$

9.33. Use the data of example 9.5 to construct a 98% confidence interval for $(\mu_1 - \mu_2)$.

9.34. An experiment was conducted to compare the mean number of tape worms in the stomachs of sheep that had been treated for worms versus the mean number in those that were untreated. A sample of 14 worm-infected lambs was randomly divided into two groups. Seven were injected with the drug, and the remainder were left untreated. After a six-month period the lambs were slaughtered and the worm counts were recorded; see the accompanying table. Use the sample data to attempt to verify the research hypothesis that the population mean worm count for drug-treated lambs is less than the mean worm count for untreated lambs. Use $\alpha = .05$.

Drug Treated	18	43	28	50	16	32	13
Untreated	40	54	26	63	21	37	39

 9.35. The elasticity of plastic can vary depending on the process by which the plastic is prepared. To compare the elasticity of plastic produced by two different processes, six samples from each process were analyzed for elasticity. These data are shown in the accompanying table. Do the data present sufficient evidence to indicate a difference in the mean elasticities for the two processes? Use $\alpha = .05$.

Process A	Process B
6.1	9.1
9.2	8.2
8.7	8.6
8.9	6.9
7.6	7.5
7.1	7.9
$\bar{y}_1 = 7.93$	$\bar{y}_2 = 8.03$
$s_1^2 = 1.46$	$s_2^2 = .61$

 9.36. The purity of ore can vary greatly from one location to another. One determining factor, then, in choosing a site for mining would be the metal content of the ore. Two prospective locations were to be compared. Three ore samples were obtained from each location and analyzed to determine the metal content of the ore; see the accompanying table. Do the data provide sufficient evidence to indicate a difference in mean metal content for the two locations? Use $\alpha = .01$.

Location 1	50.1	49.6	51.2
Location 2	47.0	46.0	46.4

9.37. Refer to exercise 9.36. Give the level of significance for your test.

9.38. The amount of work accomplished on a construction job is frequently approximated by a visual estimate of the amount of material utilized per day. Six experienced men were employed to approximate the number of bricks utilized on two different jobs. Three were randomly assigned to job 1 and three were assigned to job 2. Each man, independent of the others, approximated the number of bricks utilized. The approximations (in thousands of bricks) are shown in the accompanying table. Assume that the men have been randomly selected from a very large set of experienced people. Thus μ_1 is the mean of the large set of approximations produced by people who visually estimate the number of bricks in job 1. Similarly, μ_2 is a corresponding mean of a large set of approximations that could be acquired for job 2. Do these data provide evidence to indicate that the mean number of bricks approximated for job 1 differs from the mean approximation for job 2? Use $\alpha = .05$.

Job 1	Job 2
107.2	103.2
108.1	105.9
105.7	104.1
$\bar{y}_1 = 107.00$	$\bar{y}_2 = 104.40$
$s_1^2 = 1.47$	$s_2^2 = 1.89$

9.39. Refer to exercise 9.35. Estimate the difference in the mean elasticities of the two processes, using a 95% confidence interval.

9.40. Refer to exercise 9.36. Estimate the difference in the mean metal content of the two locations, using a 90% confidence interval.

9.41. Refer to exercise 9.38. Construct a 95% confidence interval for the difference in the mean estimates for the two jobs.

9.42. An experiment was conducted to investigate the effect of the drug propranolol in reducing hypertension in rats. Two groups of rats were studied. One group received the drug and the other group served as the control group. Hypertension was induced in the rats by exposure to a cold environment. The extent of the induced hypertension in a given rat was measured by monitoring its blood pressure. After six weeks of cold exposure the sample blood pressure data were summarized for the two groups; see the accompanying table. Use these data to determine whether there is evidence to indicate that rats treated with propranolol have less hypertension, on the average, than those that are untreated. Use $\alpha = .05$.

	Group 1 (Received Propranolol)	Group 2 (Control)
Sample Size	7	5
Sample Mean	129.43	167.60
Sample Variance	583.95	249.30

EXPERIENCES WITH REAL DATA

Design an experiment for comparing two population means, using independent random samples selected from the two populations. You may wish to utilize data you can personally collect from a laboratory class in biology, chemistry, geology, psychology, or physics. For example, you may wish to compare the average length of time it takes mice to successfully travel through a maze for mice treated with a mild tranquilizer and others that are untreated. To do this, select perhaps 8 mice and randomly assign 4 to the group receiving the tranquilizer and the remaining 4 to the untreated group. Inject the mice to be treated with a mild dose of tranquilizer and, after a short delay, begin timing the mice (one at a time) to determine the amount of time it takes each of the 8 mice to thread the maze. Compare the mean times for the two groups of mice.

Calculation of $\bar{y}$ and s^2 can be accomplished most easily by using an electronic desk calculator, but you can also use packaged computer programs and an electronic computer to perform the calculations. Useful packaged programs are available in the Biomed (Biomedical Programs), SAS (Statistical Analysis System), and SPSS (Statistical Package for the Social Sciences) program libraries (see the References).

10 Regression and Correlation

CHAPTER OUTLINE

10.1 Introduction
10.2 Scatter Diagrams and the Freehand Regression Line
10.3 Method of Least Squares
10.4 Inferences Concerning the Slope of the Least Squares Regression Line
10.5 The Coefficient of Linear Correlation
10.6 Multiple Regression (Optional)

10.1 INTRODUCTION

An estimation problem of considerable interest to high school seniors, freshmen entering college, their parents, and university administrations concerns the expected academic achievement of a particular student after he or she has enrolled in a university. For example, we might wish to estimate what a student's grade point average (GPA) will be at the end of the freshman year before the student has been accepted or enrolled in the university. At first glance this task seems difficult. However, as you will learn in this chapter, there is a method that statisticians can use to find such estimates.

The statistical approach to this problem is, in many respects, a formalization of the procedure we might follow intuitively. Suppose data were available giving the high school academic grades, psychological and sociological information, as well as the grades attained at the end of the college freshman year for a large number of students. Then we might categorize the students into groups possessing similar characteristics. For example, highly motivated students who earned a high rank in their high school class, graduated from a high school with superior academic standards, and so forth, should achieve, on the average, a higher GPA at the end of their college freshman year than students who lack motivation and who achieved only moderate success in high school. Carrying this line of thought a little further, we would expect the GPA of a student to be related to many variables that define the individual's psychological and physical characteristics as well as to those that define the academic and social environment to which he or she will be exposed. Ideally we would like to obtain a mathematical equation that relates a student's GPA to all these variables, so it could be used for prediction.

Regression and Correlation

You will observe that the problem we have defined is of a very general nature. We are interested in some random variable y that is related to a number of variables $x_1, x_2, x_3, \ldots$. Generally, the random variable y is called the *dependent variable* and the x variables are designated as *independent variables*. We are interested in obtaining an equation that relates y to the independent variables. The variable y for our example is the student's grade point average at the end of the freshman year. The independent variables might be

x_1 = rank in high school class
x_2 = score on a mathematics achievement test
x_3 = score on a verbal achievement test

prediction equation

and so on. The ultimate objective is to measure $x_1, x_2, x_3, \ldots$ for a particular student, substitute these values into the mathematical equation, and thereby predict the student's grade point average. In order to accomplish this, we must first determine the related variables $x_1, x_2, x_3, \ldots$ and measure the strength of their relationship to y. Then we must construct a good prediction equation that will express y in terms of the selected independent variables.

Other practical examples of our prediction problem are numerous throughout business, industry, and the sciences. To illustrate this point, we look at several more examples in the following paragraphs.

Executives of a large oil company would like to relate the performance of a gasoline blend to a number of key ingredients in the gasoline mixture. By varying the proportions of the ingredients in the overall blend and using the performance information from these many different blends, it would be possible to obtain a prediction equation relating performance to the proportions of the blend ingredients. By substituting different values of the independent variables (proportions of the key ingredients) into the prediction equation, the company could determine the proportions that would provide the best blend in terms of gasoline performance.

A biologist would like to relate physical characteristics such as height, weight, blood pressure, pulse, age, and so forth to the amount of secretion from a gland in humans. By observing these variables (characteristics) and the glandular secretion for many different people, he could obtain a prediction equation relating the amount of secretion to these physical characteristics.

A political analyst may wish to predict the success of a candidate in a political primary based on a number of important variables. Success would probably be measured by the number of votes cast for the candidate. Variables affecting the outcome of the primary might be the amount of money spent on television advertising, the size of the campaign organization, and the amount of money spent advertising in papers and magazines. Ideally the political analyst would like to study the effects of these variables on the outcomes of previous elections to aid in predicting the best strategy for a future campaign.

We consider first the problem of predicting y based on a single independent variable x. Then we observe that the solution for a *multivariable problem*, where y is related to more than one independent variable, is based on a generalization of our technique. Since the methodology for the multivariable predictor is fairly complex, we will use computer programs for solutions to these problems.

10.2 SCATTER DIAGRAMS AND THE FREEHAND REGRESSION LINE

Consider the prediction of a student's GPA at the end of the freshman year based on his or her high school GPA. The objective is to obtain an equation that will predict the achievement of college freshmen and hence will be of assistance to admissions officers in identifying potentially successful students. The GPA data for a sample of 11 students are shown in table 10.1.

Table 10.1 Data for High School and College GPAs (Based on a 4.0 System)

Student	High School GPA, x	Freshman GPA, y
1	2.00	1.60
2	2.25	2.00
3	2.60	1.80
4	2.65	2.80
5	2.80	2.10
6	3.10	2.00
7	2.90	2.65
8	3.25	2.25
9	3.30	2.60
10	3.60	3.00
11	3.25	3.10

scatter diagram

A first approach to analyzing the GPAs in table 10.1 is to plot the data, using a scatter diagram. To construct a scatter diagram, we make vertical and horizontal axes of approximately equal length. It is generally agreed that the independent variable x is labeled along the horizontal axis, and the dependent variable y is labeled along the vertical axis. For our example the independent variable is high school GPA.

Having labeled the axes, we then draw scales along the axes in such a way that all measurements can easily be plotted along the appropriate axis. For our example both the independent and dependent variables range between 0 and 4.0. After the axes are drawn, labeled, and scaled, we plot the data of table 10.1. Each dot on the figure represents the information concerning one student and can be obtained by plotting the freshman GPA, y, against the corresponding high school value, x. The dot circled in figure 10.1 corresponds to student 1. Note that the dot is placed at a point on the graph corresponding to $x = 2.00$ and $y = 1.60$.

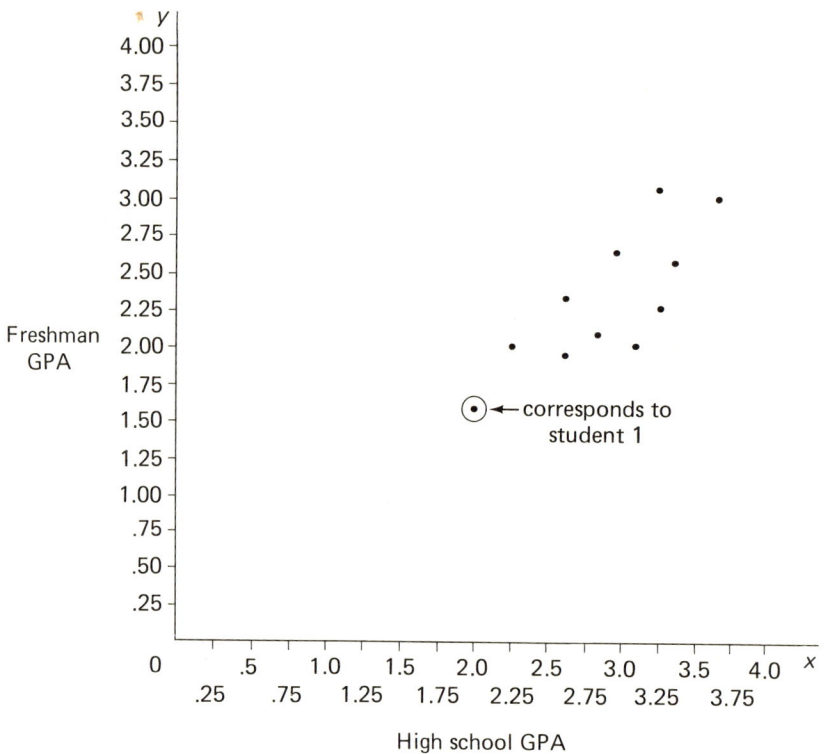

Figure 10.1 High School and College GPA Data

regression line

It appears from figure 10.1 that the freshman GPA (y) increases as the high school GPA (x) increases. In fact, many of the dots seem to be on a straight line. We call a line running through the dots of a scatter diagram a trend line, or a regression line. When the regression line is a straight line, we say there is a *linear relationship* between x and y. See figures 10.2(a) and (b). However, not all regression lines are linear. In some cases the trend line is curved, and then we say there is a *curvilinear relationship* between x and y. See figures 10.2(c) and (d).

Scatter Diagrams and the Freehand Regression Line

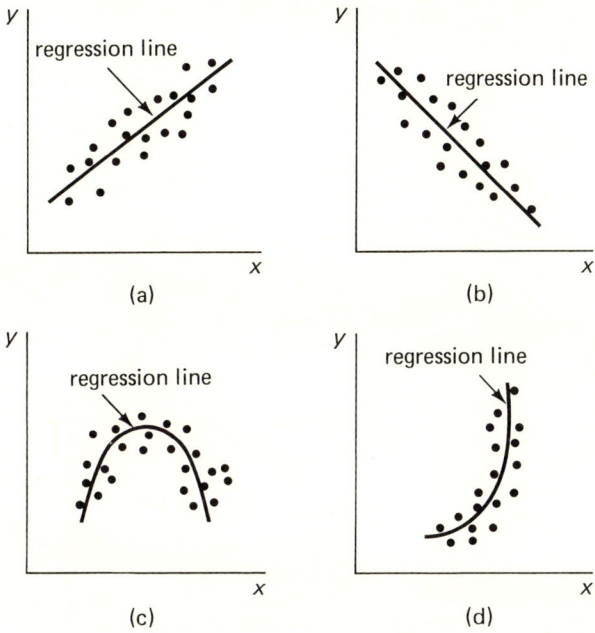

Figure 10.2 Different Types of Regression Lines. (a) Linear Relationship Between x and y; (b) Linear Relationship Between x and y; (c) Curvilinear Relationship Between x and y; (d) Curvilinear Relationship Between x and y

freehand regression line

There are many methods for obtaining a prediction equation relating y to x. The first is called an "eyeball fit," or a freehand regression line, which can be obtained by placing a ruler on the graph (figure 10.1) and moving it about until we think we have minimized the distances from the points to the fitted line. This has been done in figure 10.3. The resulting line can be used to make a prediction about a freshman's GPA based on high school achievement. To predict y when $x = 2.5$, refer to the graph and note that the y coordinate for the point corresponding to $x = 2.5$ is $y = 2.05$ (see the arrows on figure 10.3).

The freehand regression line shown in figure 10.3 can be represented by a linear equation of the form

linear equation

$$y = \beta_0 + \beta_1 x$$

The two constants β_0 and β_1 in the equation determine the location and the slope of the line. The constant β_0 is the y-intercept, that is, the value of y when the line crosses the vertical axis (y-axis). The constant β_1 is the slope of the line, that is, the change in y that corresponds to a one-unit increase in x. See figure 10.4.

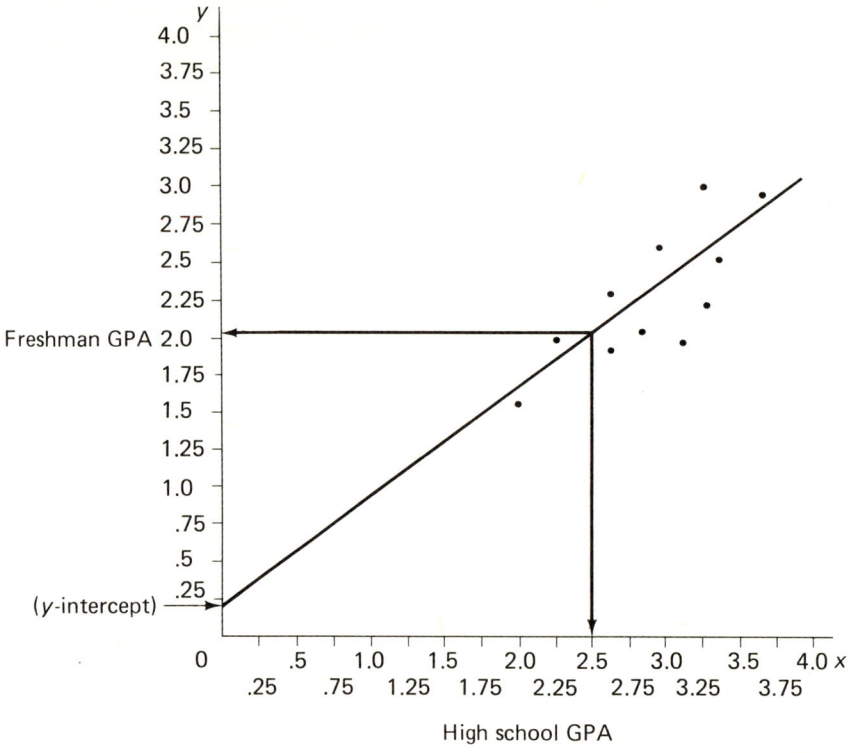

Figure 10.3 Freehand Regression Line for High School and College GPA Data

---

DEFINITION 10.1

y-intercept

The y-intercept β_0 for the straight line

$$y = \beta_0 + \beta_1 x$$

is the value of y at the point where the line crosses the y-axis.

DEFINITION 10.2

slope

The slope β_1 of the straight line

$$y = \beta_0 + \beta_1 x$$

Scatter Diagrams and the Freehand Regression Line

is the change (an increase or a decrease) in the value of y for a one-unit increase in x.

From figure 10.4 it appears that the y-intercept is approximately $\beta_0 = .20$. When x increases from 0 to 1.0, y increases from about .20 to .95. This change $(.95 - .20 = .75)$ represents the slope (β_1) of the straight line. Thus the equation corresponding to the freehand regression line of figure 10.4 is $y = .20 + .75x$. So to predict y when $x = 2.5$, we substitute $x = 2.5$ into our prediction equation, $y = .20 + .75x$, to obtain $y = .20 + .75(2.5) = 2.08$, which is about the same as our guessed value of 2.05 (figure 10.3). Thus you see that we can predict y by using either the graph, as in figure 10.3, or the prediction equation, $y = .20 + .75x$

Although the freehand regression line provides us with a prediction equation, there could be many different prediction equations from different eyeball

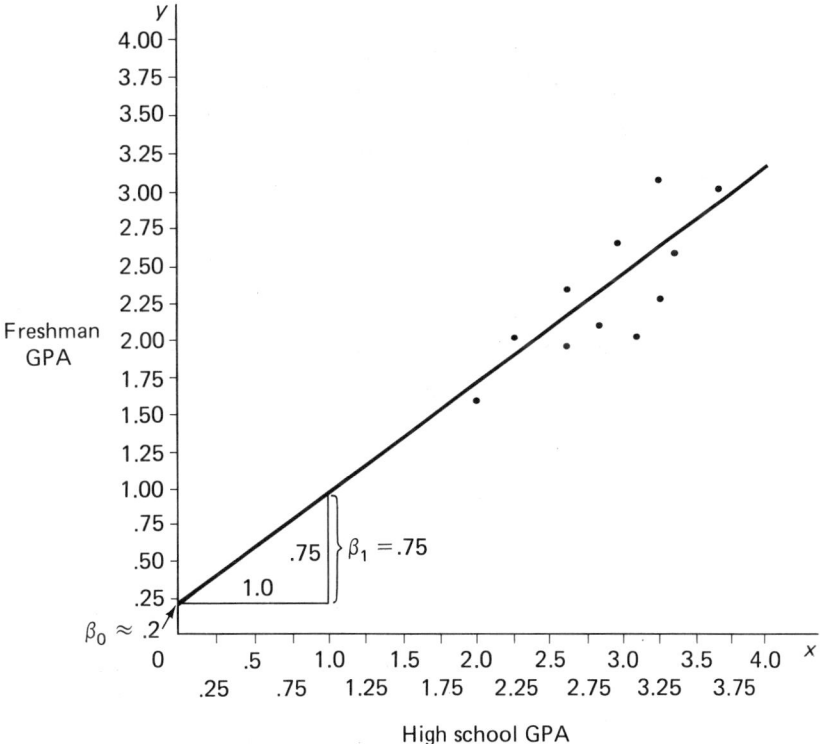

Figure 10.4 Slope β_1 and Intercept β_0 for the Freehand Regression Line

Regression and Correlation

fits to the same data. What we seek is a precise procedure for estimating the values of the constants β_0 and β_1 in our prediction equation $y = \beta_0 + \beta_1 x$. The procedure that we use for estimating β_0 and β_1 is called the *method of least squares*. We will discuss this method in the next section.

EXERCISES

10.1. Plot the data shown here in a scatter diagram.

x	5	10	12	15	18	24
y	10	19	21	28	34	40

10.2. Refer to the data of exercise 10.1.
(a) Use an eyeball fit to construct a freehand regression line.
(b) Identify the intercept and slope for your regression line.
(c) Predict y when $x = 20$.

10.3. Use the equation $y = 1.8 + 2.0x$.
(a) Predict y when $x = 3$.
(b) Plot the equation on a graph with the horizontal axis scaled from 0 to 5 and the vertical axis scaled from 0 to 12.

10.4. Results of a pharmaceutical pricing survey were reported recently relating the influence of wholesale discounts and other factors on prescription ingredient costs. One problem considered in the study was the relationship between the annual prescription volume y and the percentage of the drug ingredient purchases that were made directly from the drug manufacturers. Suppose that a sample of 10 independent pharmacies yielded the results shown in the accompanying table. Plot these data on a scatter diagram.

Annual Volume (× $1000), y	Percentage of Purchases, x	Annual Volume (× $1000), y	Percentage of Purchases, x
25	10	138	63
55	18	90	42
50	25	60	30
75	40	10	5
110	50	100	55

10.5. Refer to the data of exercise 10.4.
(a) Use an eyeball fit to determine a freehand regression line for predicting the annual prescription volume of an independent pharmacy based on the percentage of ingredients purchased directly from drug manufacturers.

(i) What is the y-intercept for your regression line?
(ii) What is the slope of the line?
(b) Predict the annual prescription volume, using the freehand regression line from part (a), when a pharmacy buys 45% of its ingredients directly from drug manufacturers.

10.6. Using data from the 1970 U.S. census, a sample of nine states showed the per capita income, x, and public education expenditure per student (in dollars), y, given in the table here.

State	Per Capita Income, x	Public Education Expenditure per Student, y
Arkansas	2520	535
California	4272	922
Colorado	3568	695
Michigan	3944	842
Mississippi	2194	476
North Carolina	2890	609
New York	4421	1237
Rhode Island	3779	904
South Dakota	3051	657

(a) Plot the data on a scatter diagram, with the horizontal axis labeled from 0 to 5000 and the vertical axis labeled from 0 to 1400.
(b) Using an eyeball fit, draw a freehand regression line for the data.
(c) Identify the slope and intercept for your regression line. Write the equation corresponding to your regression line.

10.3 METHOD OF LEAST SQUARES

method of least squares

The statistical procedure for finding the prediction equation—the <u>method of least squares</u>—is, in many respects, an objective way to obtain an eyeball fit to the points. For example, when we fit a line by eye to a set of points, we move the ruler until we think that we have minimized the distances from the points to the fitted line. This same minimizing technique is used in the statistical procedure.

predicted value of y

We denote the <u>predicted value of y</u> for a given value of x (obtained from the fitted line) as $\hat{y}$. The prediction equation obtained from the method of least squares is denoted by

$$\hat{y} = \hat{\beta}_0 + \hat{\beta}_1 x$$

The vertical distance from a point to the prediction line represents the deviation of a point from the predicted value of y. See figure 10.5. Symbolically, this deviation is written as $(y - \hat{y})$. To find the best prediction equation

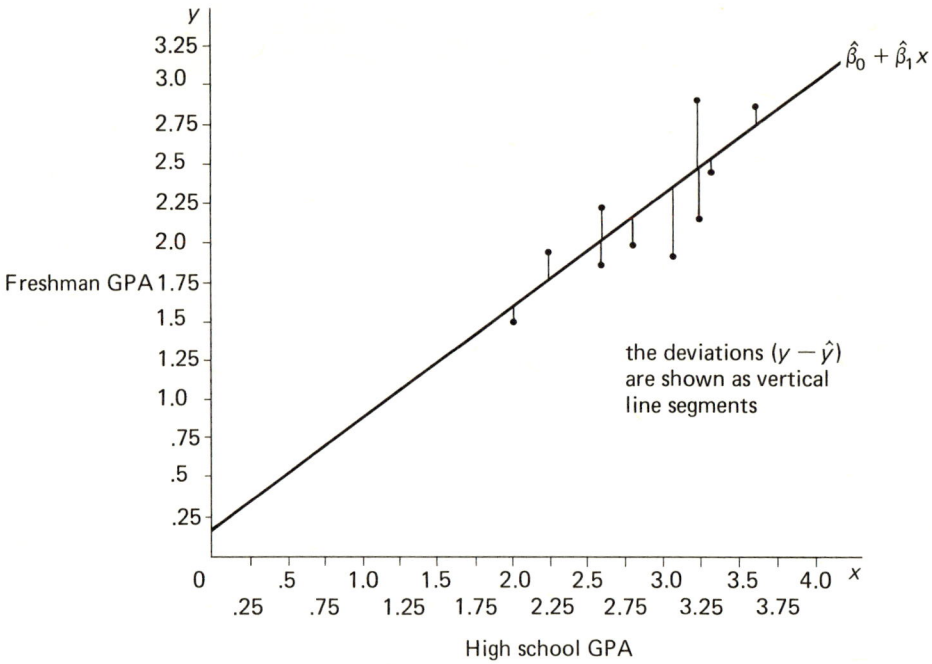

Figure 10.5 Least Squares Fit to the Data in Table 10.1

(regression line), we can work with deviations from the regression line. The method of least squares chooses the regression line that minimizes the sum of squares of the deviations of the observed values from the predicted values of y. If the line "fits" the data well, the deviations will be small and hence the sum of squares of the deviations will be small. A line that does not fit well will have some or all deviations large, making the sum of squares of the deviations large.

The sum of the squares of the deviations, which is also referred to as the **sum of squares for error** and is denoted by SSE, can be written as

$$\text{SSE} = \Sigma\, (y - \hat{y})^2$$

where y is an observed response and $\hat{y}$ is a point on the fitted line

$$\hat{y} = \hat{\beta}_0 + \hat{\beta}_1 x$$

Substituting for $\hat{y}$ in SSE, we have

$$\text{SSE} = \Sigma\, [y - (\hat{\beta}_0 + \hat{\beta}_1 x)\,]^2$$

The method of least squares chooses those values for the estimates $\hat{\beta}_0$ and $\hat{\beta}_1$ that make SSE a minimum. Derivation of these values is beyond the scope of this text, but they can be found by using the formulas that follow.

Method of Least Squares

Least Squares Estimates for the Regression Line $\hat{y} = \hat{\beta}_0 + \hat{\beta}_1 x$

$$\hat{\beta}_1 = \frac{S_{xy}}{S_{xx}}$$

and

$$\hat{\beta}_0 = \bar{y} - \hat{\beta}_1 \bar{x}$$

where

$$S_{xx} = \Sigma x^2 - \frac{(\Sigma x)^2}{n}$$

$$S_{xy} = \Sigma xy - \frac{(\Sigma x)(\Sigma y)}{n}$$

and $\bar{y}$ and $\bar{x}$ denote the sample means for the y values and x values, respectively; n is the number of y values.

The use of these formulas for finding $\hat{\beta}_0$, $\hat{\beta}_1$, and the least squares line is illustrated in the next example.

Example 10.1

Obtain the regression line $\hat{y} = \hat{\beta}_0 + \hat{\beta}_1 x$ for the GPA data in table 10.1 by using the method of least squares.

Solution

The calculation of the least squares estimates $\hat{\beta}_0$ and $\hat{\beta}_1$ is greatly simplified by using table 10.2.

Substituting the values in table 10.2 into the formulas, we obtain

$$S_{xx} = \Sigma x^2 - \frac{(\Sigma x)^2}{n} = 93.68 - \frac{(31.7)^2}{11} = 2.326$$

$$S_{xy} = \Sigma xy - \frac{(\Sigma x)(\Sigma y)}{n} = 76.3325 - \frac{(31.7)(25.9)}{11} = 1.693$$

$$\bar{y} = \frac{\Sigma y}{n} = \frac{25.9}{11} = 2.355$$

$$\bar{x} = \frac{\Sigma x}{n} = \frac{31.7}{11} = 2.882$$

Table 10.2 Calculations for the GPA Data

x	y	x^2	xy	y^2
2.00	1.60	4.0000	3.2000	2.5600
2.25	2.00	5.0625	4.5000	4.0000
2.60	1.80	6.7600	4.6800	3.2400
2.65	2.80	7.0225	7.4200	7.8400
2.80	2.10	7.8400	5.8800	4.4100
3.10	2.00	9.6100	6.2000	4.0000
2.90	2.65	8.4100	7.6850	7.0225
3.25	2.25	10.5625	7.3125	5.0625
3.30	2.60	10.8900	8.5800	6.7600
3.60	3.00	12.9600	10.8000	9.0000
3.25	3.10	10.5625	10.0750	9.6100
Total 31.70	25.90	93.6800	76.3325	63.5050

Hence

$$\hat{\beta}_1 = \frac{S_{xy}}{S_{xx}} = \frac{1.693}{2.326} = .728$$
$$\hat{\beta}_0 = \bar{y} - \hat{\beta}_1 \bar{x} = 2.355 - .728(2.882) = .257$$

The least squares prediction equation relating the freshman GPA, y, to the corresponding high school value, x, is then

$$\hat{y} = .257 + .728x$$

Notice that the least squares prediction equation is similar to the equation for the freehand line for the same data set, obtained on page 253.

EXERCISES

10.7. Use the accompanying data to determine the least squares prediction equation.

x	1	2	3	4	5
y	2	4	6	7	9

10.8. Use the accompanying data.

x	1	3	5	7	9
y	1	4	8	9	12

Method of Least Squares

(a) Determine the least squares prediction equation.
(b) Use the least squares prediction equation to predict y when $x = 6$.

10.9. Refer to the data of exercise 10.1. Find the least squares prediction equation and compare it to the freehand regression line you found in exercise 10.2.

10.10. A computer solution for the least squares prediction equation to the data of exercise 10.4 is shown here.
(a) Determine the least squares prediction equation.
(b) Predict y (annual prescription volume) for $x = 35$.

LINEAR REGRESSION

	ESTIMATE	95% CONFIDENCE LIMITS	
SLOPE:	1.97048	[1.61409,	2.32687]
INTERCEPT:	4.69785	[−9.02751,	18.42321]

X VALUE	OBSERVED Y	PREDICTED Y
10.00000	25.00000	24.40263
18.00000	55.00000	40.16645
25.00000	50.00000	53.95980
40.00000	75.00000	83.51696
50.00000	110.00000	103.22174
63.00000	138.00000	128.83795
42.00000	90.00000	87.45792
30.00000	60.00000	63.81218
5.00000	10.00000	14.55024
55.00000	100.00000	113.07413

10.11. Refer to exercise 10.6. If $S_{xx} = 4{,}881{,}194$, $S_{xy} = 1{,}369{,}845.67$, $\Sigma x = 30{,}639$, and $\Sigma y = 6{,}877$, obtain the least squares prediction equation relating a state's public education expenditure per student to the per capita income of the state. Compare the least squares prediction equation to your prediction equation of exercise 10.6. Use the least squares equation to predict the expenditure of a state having a per capita income of $3,600.

10.12. Compare your least squares prediction equation for exercise 10.11 to the computer output shown here for the same problem.

LINEAR REGRESSION

	ESTIMATE	95% CONFIDENCE LIMITS	
SLOPE:	0.28064	[0.17704,	0.38424]
INTERCEPT:	−191.27220	[−552.12069,	169.57630]

X VALUE	OBSERVED Y	PREDICTED Y
2520.00000	535.00000	515.93409
4272.00000	922.00000	1007.61084
3568.00000	695.00000	810.04210
3944.00000	842.000000	915.56177
2194.00000	476.00000	424.44629
2890.00000	609.00000	619.76993

4421.00000	1237.00000	1049.42582
3779.00000	904.00000	869.25660
3051.00000	657.00000	664.95256

10.4 INFERENCES CONCERNING THE SLOPE OF THE LEAST SQUARES REGRESSION LINE

In the material presented thus far in chapter 10 we have used the least squares regression line in a strictly descriptive sense. That is, for a given set of data, we were interested in determining the prediction equation, and no statistical inferences were made. Suppose, however, that the data represent a random sample from a larger body of data, the population. Then the intercept and slope of the least squares regression line represent estimates of the corresponding population intercept and slope. We let $\hat{\beta}_0$ and $\hat{\beta}_1$ denote the estimated intercept and slope computed from the sample data by using the method of least squares, and we let β_0 and β_1 denote the corresponding unknown intercept and slope for the population.

We may wish to test the null hypothesis $H_0: \beta_1 = 0$ (i.e., the slope of the population regression line is 0). If β_1 is different from zero and the population regression line slopes upward (or downward), as shown in figure 10.6(a), then knowledge of x will help us to predict values of y. When x is large, we predict large values of y; when x is small, we predict small values of y.

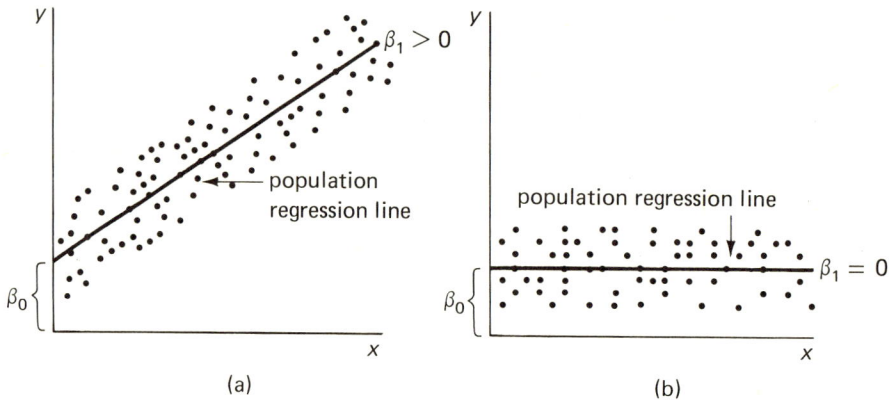

Figure 10.6 Can x Be Used to Predict y?

However, suppose that the slope of the population regression line is equal to 0 and the line, therefore, appears as shown in figure 10.6(b). When $\beta_1 = 0$, knowledge of x will be of no help in predicting y.

In our statistical test for β_1 we ask the question, "Does x contribute infor-

Inferences Concerning the Slope of the Least Squares Regression Line

mation for the prediction of y?" Or, restated, "Does the slope β_1 differ from 0?" Or could the apparent linear arrangement of the data points (e.g., figure 10.1) have occurred solely by chance?

A statistical test of H_0: $\beta_1 = 0$ is summarized next. In order to use this test, we assume that for each value of x there is a population of y values that can be represented by a normal distribution. Regardless of the value of x, the population of y values will have the same variance.

A Statistical Test Concerning the Slope of the Population Regression Line

Null hypothesis: $\beta_1 = 0$.

Alternative hypothesis:

For a one-tailed test:
1. $\beta_1 > 0$.
2. $\beta_1 < 0$.

For a two-tailed test:
3. $\beta_1 \neq 0$.

Test statistic: $t = \dfrac{\hat{\beta}_1}{\sqrt{s^2/S_{xx}}}$ where $s^2 = \dfrac{S_{yy} - \hat{\beta}_1 S_{xy}}{n-2}$ and $S_{yy} = (\Sigma y^2 - \Sigma y)^2$

Rejection region: For a given value of α, for df $= (n-2)$, and for a one-tailed test:
1. Reject H_0 if $t > t_a$, with a $= \alpha$.
2. Reject H_0 if $t < -t_a$, with a $= \alpha$.

For a given value of α, for df $= (n-2)$, and for a two-tailed test:
3. Reject H_0 if $|t| > t_a$, with a $= \alpha/2$.

We can also form a confidence interval for β_1 by using the sample data. This interval is as shown next.

General Confidence Interval for β_1

$$\hat{\beta}_1 \pm t \sqrt{\dfrac{s^2}{S_{xx}}} \quad \text{where} \quad s^2 = \dfrac{S_{yy} - \hat{\beta}_1 S_{xy}}{n-2}$$

Regression and Correlation

The value of t corresponding to a 90%, a 95%, or a 99% confidence interval is found in table 2 of the Appendix for df = $(n - 2)$ and a = .05, .025, or .005, respectively.

We illustrate the test procedure with an example.

Example 10.2

Use the data in table 10.1 to determine whether there is evidence to indicate that β_1, the population slope between high school and college freshman GPAs, is positive. Use $\alpha = .05$.

Solution

The four parts of the statistical test are as follows:

Null hypothesis: $\beta_1 = 0$.

Alternative hypothesis: $\beta_1 > 0$.

Test statistic: $t = \dfrac{\hat{\beta}_1}{\sqrt{s^2/S_{xx}}}$.

Rejection region: For $\alpha = .05$ and df = $(n - 2)$, reject H_0 if $t > t_a$, where a = .05.

In example 10.1 we computed

$$\hat{\beta}_1 = .728 \quad S_{xx} = 2.326 \quad S_{xy} = 1.693$$

In order to calculate s^2 we must first compute S_{yy}. Using table 10.2, we find

$$S_{yy} = \Sigma y^2 - \frac{(\Sigma y)^2}{n} = 63.505 - \frac{(25.9)^2}{11} = 2.522$$

Substituting, we have

$$s^2 = \frac{S_{yy} - \hat{\beta}_1 S_{xy}}{n - 2} = \frac{2.522 - (.728)(1.693)}{9} = \frac{1.289}{9} = .1432$$

The test statistic is then

$$t = \frac{.728}{\sqrt{.1432/2.326}} = 2.93$$

The t value in table 2 of the Appendix for a = .05 and df = 9 is 1.833. Since the computed value of t exceeds 1.833, we reject H_0 and conclude that the population regression line has a slope greater than zero and hence that the high school GPA is useful in predicting a college freshman's GPA.

Example 10.3

A biologist is interested in studying the growth rate of a bacteria culture over a period of time. In a laboratory experiment five different bacterial cultures were chosen. One culture was randomly selected and assigned to an incubation time of 1 hour, one to an incubation time of 3 hours, and one each to the incubation times 5, 7, and 9 hours. The growth rate y was measured on each culture after the required incubation period. Let x denote the incubation time. Use the sample data of table 10.3 and the method of least squares to obtain the regression line.

$$\hat{y} = \hat{\beta}_0 + \hat{\beta}_1 x$$

Conduct a test of significance to determine if there is a linear relationship between the mean growth rate and time. Use $\alpha = .05$.

Table 10.3 Incubation Time and Growth Rate

Incubation Time, x	Growth Rate, y
1	10.0
3	10.3
5	12.2
7	12.6
9	13.9

Solution

Before fitting the line, we must obtain the following quantities:

$$\Sigma x = 1 + 3 + 5 + 7 + 9 = 25$$
$$\Sigma x^2 = (1)^2 + (3)^2 + \cdots + (9)^2 = 165$$
$$\Sigma y = (10.0) + (10.3) + \cdots + (13.9) = 59$$
$$\Sigma y^2 = (10.0)^2 + (10.3)^2 + \cdots + (13.9)^2 = 706.9$$
$$\Sigma xy = 1(10.0) + 3(10.3) + \cdots + 9(13.9) = 315.2$$

We then obtain

$$S_{xy} = \Sigma xy - \frac{(\Sigma x)(\Sigma y)}{n} = 315.2 - \frac{(25)(59)}{5} = 20.2$$

$$S_{xx} = \Sigma x^2 - \frac{(\Sigma x)^2}{n} = 165 - \frac{(25)^2}{5} = 40$$

Regression and Correlation

The sample means are

$$\bar{y} = \frac{\Sigma y}{n} = \frac{59}{5} = 11.8$$

$$\bar{x} = \frac{\Sigma x}{n} = \frac{25}{5} = 5.0$$

Hence

$$\hat{\beta}_1 = \frac{S_{xy}}{S_{xx}} = \frac{20.2}{40.0} = .51$$

$$\hat{\beta}_0 = \bar{y} - \hat{\beta}_1 \bar{x} = 11.8 - (.51)(5) = 9.25$$

The prediction equation relating growth rate to time is then

$$\hat{y} = 9.25 + .51x$$

To determine whether or not the growth rate y of the bacteria is linearly related to time x, we consider the statistical test:

Null hypothesis: $\beta_1 = 0$.

Alternative
hypothesis: $\beta_1 \neq 0$.

Test statistic: $t = \dfrac{\hat{\beta}_1}{\sqrt{s^2/S_{xx}}}$.

Rejection region: For $\alpha = .05$ and df $= 3$, reject H_0 if $|t| > t_a$, where $a = .025$.

To calculate the value of the test statistic, we must first determine s^2.

$$S_{yy} = \Sigma y^2 - \frac{(\Sigma y)^2}{n} = 706.9 - \frac{(59)^2}{5} = 10.7$$

$$S_{yy} - \hat{\beta}_1 S_{xy} = 10.7 - (.51)(20.2) = .40$$

Then

$$s^2 = \frac{S_{yy} - \hat{\beta}_1 S_{xy}}{n-2} = \frac{.40}{3} = .13$$

Substituting into the test statistic, we have

$$t = \frac{\hat{\beta}_1}{\sqrt{s^2/S_{xx}}} = \frac{.51}{\sqrt{.13/40}} = \frac{.51}{.06} = 8.5$$

The t value in the appendix table for $a = .025$ and df $= 3$ is 3.182. Thus $t = 8.5$ falls in the rejection region, and we conclude that there is evidence to indicate that the mean growth rate of a bacteria culture is linearly related to time.

EXERCISES

10.13. Refer to exercises 10.6 and 10.11. Use $S_{yy} = 449{,}996.89$, and conduct a statistical test that the slope of the population regression line relating public education expenditure per student for a state to the per capita income of the state is positive. Use $\alpha = .05$.

10.14. Recent research in dentistry has indicated that plaque from different locations in the mouth can differ in chemical composition. Since the quantity of plaque at a given site might be quite small, it is necessary to have a sensitive procedure in order to study the chemical composition of plaque. One such procedure proposed by Drummond and Donkersloot in the *Journal of Dental Research* (1974) relates the DNA content (one important chemical component) of plaque to the weight of plaque.

In order to study the relationship between weight of plaque and DNA content, ten male volunteers (ages 18–20) were selected at random from a group of volunteers. Over a four-day period each person consumed his normal diet supplemented by 30 grams of sucrose per day. No tooth brushing was allowed. The four-day accumulation of plaque for each person was weighed and analyzed for DNA content. These sample data are summarized in the accompanying table.

Person	Plaque Weight (milligrams), x	DNA (micrograms), y
1	42.7	260
2	52.3	303
3	24.6	175
4	33.4	214
5	41.8	226
6	36.7	246
7	27.0	181
8	47.3	251
9	31.4	154
10	33.9	247

(a) Plot these data in a scatter diagram.

(b) Use the computer output shown here to determine the least squares regression line.

LINEAR REGRESSION

	ESTIMATE	95% CONFIDENCE LIMITS	
SLOPE :	4.38949	[2.22624 ,	6.55274]
INTERCEPT :	62.80609	[−19.49671 ,	145.10889]

X VALUE	OBSERVED Y	PREDICTED Y
42.70000	260.00000	250.23724
52.30000	303.00000	292.37633
24.60000	175.00000	170.78750
33.40000	214.00000	209.41500
41.80000	226.00000	246.28670
36.70000	246.00000	223.90031
27.00000	181.00000	181.32227
47.30000	251.00000	270.42889
31.40000	154.00000	200.63602
33.90000	247.00000	211.60974

10.15. Refer to exercise 10.14. If the least squares prediction equation is $\hat{y} = 62.81 + 4.39x$, test to see if there is a significant linear relationship between the DNA content and the plaque weight. That is, use the sample data to test whether the population slope β_1 differs from zero. Use $\alpha = .05$. (Note: $S_{yy} = 18{,}504.1$, $S_{xy} = 3{,}087.43$, and $S_{xx} = 703.37$.)

10.16. Use the output of exercise 10.14 to compare the results of your test in exercise 10.15. (Hint: Does the confidence interval for the slope include 0?)

10.5 THE COEFFICIENT OF LINEAR CORRELATION

correlation coefficient

It is sometimes desirable to quantify the strength of the linear relationship between two variables y and x. The most common statistical measure employed for this purpose is the coefficient of linear correlation, or, simply, correlation coefficient. To be consistent with our usage in previous chapters, we use different symbols to distinguish between the sample statistic and the population parameter. The sample correlation coefficient is designated by $\hat{\rho}$ (Greek lowercase letter rho), and the corresponding population correlation coefficient is denoted by ρ. The quantity $\hat{\rho}$ does not depend on the units of the measurements and can be computed by using this formula:

$$\hat{\rho} = \frac{S_{xy}}{\sqrt{S_{xx}S_{yy}}}$$

We illustrate the computation of $\hat{\rho}$ with an example; then we will explain its practical significance.

Example 10.4

Determine the strength of the linear relationship between growth rate y and time x by computing the sample correlation coefficient for the data in table 10.3.

The Coefficient of Linear Correlation

Solution

In example 10.3 we obtained these values:

$$S_{xy} = 20.2 \qquad S_{xx} = 40.0 \qquad S_{yy} = 10.7$$

The formula for obtaining $\hat{\rho}$ is

$$\hat{\rho} = \frac{S_{xy}}{\sqrt{S_{xx}S_{yy}}}$$

Hence

$$\hat{\rho} = \frac{20.2}{\sqrt{(40)(10.7)}} = \frac{20.2}{20.69} = .98$$

How can we interpret this correlation coefficient? First, note that there is a similarity between the sample correlation coefficient and the slope of the regression line $\hat{y} = \hat{\beta}_0 + \hat{\beta}_1 x$:

$$\hat{\rho} = \frac{S_{xy}}{\sqrt{S_{xx}S_{yy}}} \qquad \text{and} \qquad \hat{\beta}_1 = \frac{S_{xy}}{S_{xx}}$$

In particular, both have the same numerator; and their denominators will always be positive (because they involve the sum of squares of numbers). Thus the correlation coefficient $\hat{\rho}$ will have the same sign as $\hat{\beta}_1$; and $\hat{\rho} = 0$ when $\hat{\beta}_1 = 0$. A negative $\hat{\rho}$ implies a negative slope; a positive value implies that y increases as x increases. A slope $\hat{\beta}_1 = 0$ (and hence $\hat{\rho} = 0$) indicates that x contributes no information for the prediction of y. See figure 10.7.

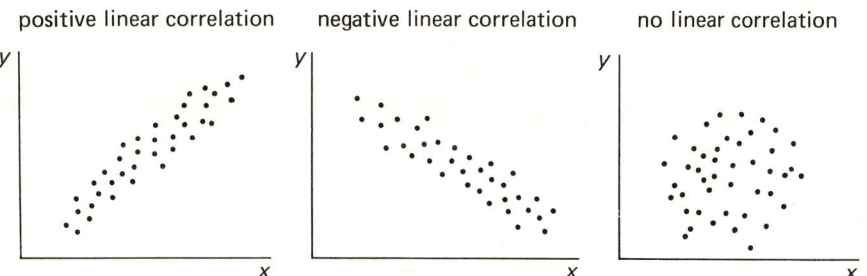

Figure 10.7 Interpreting the Correlation Coefficient

It can be shown that $\hat{\rho}$ must lie in the interval $-1 \leq \hat{\rho} \leq 1$. The implications of various values of $\hat{\rho}$ are indicated in figure 10.8.

Regression and Correlation

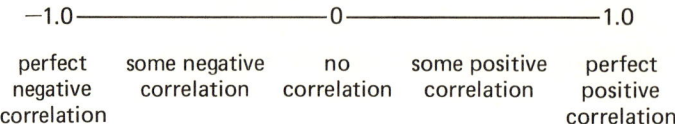

Figure 10.8 Possible Values for $\hat{\rho}$, the Sample Correlation Coefficient

Several misinterpretations of the coefficient of correlation should be noted. First, a coefficient of correlation equal to .5 does not mean that the strength of the relationship between y and x is "halfway" between no correlation and perfect correlation. If we designate $S_{yy} = \Sigma (y - \bar{y})^2$ as the total variability of the y values about their sample mean, it can be shown that an amount equal to $\hat{\rho}^2$ of this total variability can be explained by the variable x. The more closely x and y are linearly related, the more the variability in the y values can be explained by variability in the x values and the closer $\hat{\rho}^2$ will be to 1. If $\hat{\rho} = .5$, the independent variable x is accounting for $\hat{\rho}^2 = .25$ of the total variation in the y values about $\bar{y}$. The quantity $\hat{\rho}^2$ is called a <u>coefficient of determination</u>.

coefficient of determination

Second, y and x could be perfectly related in some way other than in a linear manner when $\hat{\rho} = 0$ or some very small value. See figure 10.9.

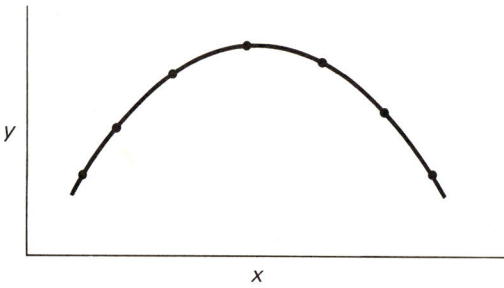

Figure 10.9 Perfect Curvilinear Fit; $\hat{\rho} = 0$

And finally, note that you cannot add correlations. Thus if the simple linear correlations between y and x_1, y and x_2, and y and x_3 are .1, .3, and .2, respectively, it does not follow that x_1, x_2, and x_3 account for $\hat{\rho}_1^2 + \hat{\rho}_2^2 + \hat{\rho}_3^2 = (.1)^2 + (.3)^2 + (.2)^2$ of the variability of the y values about their sample mean. Indeed, x_1, x_2, and x_3 may be highly correlated and contribute the same information for the prediction of y. The relationship between y and several independent variables should not be studied by computing simple correlation coefficients for each of the independent variables. Rather, we should relate y to x_1, x_2, and x_3 by using a single multivariable model. This topic is discussed briefly in section 10.6.

Exercises

Before concluding our discussion of the coefficient of linear correlation, we present a test of significance for ρ. We may be interested in a test of the null hypothesis H_0: $\rho = 0$ (i.e., there is no linear correlation between x and y). Although the results and conclusions drawn from this test are identical to those for H_0: $\beta_1 = 0$, we present the test of ρ for completeness. The assumptions required for this test are identical to those for a test of H_0: $\beta_1 = 0$ (see p. 261).

Test of an Hypothesis Concerning ρ, the Population Correlation Coefficient

Null hypothesis: $\rho = 0$.

Alternative hypothesis:

For a one-tailed test:
1. $\rho > 0$.
2. $\rho < 0$.

For a two-tailed test:
3. $\rho \neq 0$.

Test statistic: $t = \hat{\rho} \sqrt{\dfrac{n-2}{1-\hat{\rho}^2}}$.

Rejection region: For a specified value of α, for df = $(n-2)$, and for a one-tailed test:
1. Reject H_0 if $t > t_a$, where a = α.
2. Reject H_0 if $t < -t_a$, where a = α.

For a specified value of α, for df = $(n-2)$, and for a two-tailed test:
3. Reject H_0 if $|t| > t_a$, where a = $\alpha/2$.

EXERCISES

10.17. Compute the coefficient of linear correlation for the accompanying data.

x	1	2	3	4	5
y	2	4	6	7	9

Regression and Correlation

10.18. Compute the correlation coefficient between the public education expenditure per student for a state and the per capita income of the state (exercise 10.6). Recall that $S_{xx} = 4,881,194$, $S_{xy} = 1,369,845.67$, and $S_{yy} = 449,996.89$.

10.19. Conduct a statistical test of $H_0: \rho = 0$ against the alternative hypothesis $H_a: \rho > 0$ for the data of exercise 10.6. Use $\alpha = .05$. Compare your results to those of exercise 10.13.

10.20. Refer to exercise 10.14. If $S_{xx} = 703.37$ and $S_{xy} = 3087.43$, compute S_{yy} and the sample correlation coefficient for the weight of dental plaque and the total DNA content of plaque.

10.6 MULTIPLE REGRESSION (OPTIONAL)

In the previous sections we have examined methods for obtaining the least squares prediction equation, $\hat{y} = \hat{\beta}_0 + \hat{\beta}_1 x$, based on values from two variables y and x. We turn now to a more complicated situation where we collect data on the dependent variable y and on k ($k > 1$) independent variables x_1, $x_2, \ldots, x_k$. We will use the sample data to obtain the multiple regression prediction equation

$$\hat{y} = \hat{\beta}_0 + \hat{\beta}_1 x_1 + \hat{\beta}_2 x_2 + \cdots + \hat{\beta}_k x_k$$

For example, instead of trying to predict a student's GPA (y) at the end of his college freshman year based on his corresponding high school GPA (x_1), we may wish to use information on several additional variables, such as rank in high school class (x_2), college board verbal score (x_3), and college board math achievement score (x_4). To do this, we would have to examine the records of a sample of n freshmen to obtain information on the variables y, x_1, x_2, x_3, and x_4. The method of least squares can then be used to obtain the least squares multiple regression equation

$$\hat{y} = \hat{\beta}_0 + \hat{\beta}_1 x_1 + \hat{\beta}_2 x_2 + \hat{\beta}_3 x_3 + \hat{\beta}_4 x_4$$

Although there are computational formulas for the least squares estimates of the parameters $\beta_0, \beta_1, \beta_2, \ldots, \beta_k$, the formulas are difficult to work with algebraically beyond the simplest case, $\hat{y} = \hat{\beta}_0 + \hat{\beta}_1 x$, discussed in sections 10.3 and 10.4. Rather than spending time developing expressions for the least squares estimates of the parameters in a multiple regression equation, we will make use of available computer software packages to do the work for us for a particular problem.

Example 10.5

The sales profit (y) of a product in a sales territory is thought to be related to the total population (x_1) of the territory as well as to the advertising expendi-

Multiple Regression (Optional)

Table 10.4 Data for Example 10.5

Sales Territory	Sales Profit per Person (dollars), y	Total Population ($\times 1{,}000{,}000$), x_1	Advertising per Person (dollars), x_2
1	3.6	2.4	.16
2	2.5	1.3	.21
3	4.2	5.1	.12
4	4.1	4.9	.14
5	4.0	3.2	.26
6	5.1	6.7	.10
7	4.3	3.2	.41
8	11.5	.7	.11

ture per person (x_2) in the territory. The data from eight different sales territories are shown in table 10.4. Use these data to find least squares estimates for the coefficients in the multiple regression equation

$$\hat{y} = \hat{\beta}_0 + \hat{\beta}_1 x_1 + \hat{\beta}_2 x_2$$

Predict sales profit for a territory with a population $x_1 = 2.8$ and an advertising expenditure $x_2 = .25$.

Solution

Since it is important for you to be able to locate and identify the least squares estimates from a computer printout, portions of some sample output from several different programs are shown here. The least squares estimates in the printout are shown in color. Note that the least squares prediction equation, $\hat{y} = \hat{\beta}_0 + \hat{\beta}_1 x_1 + \hat{\beta}_2 x_2$, is

$$\hat{y} = 9.04 - .58 x_1 - 11.27 x_2$$

Substituting into this equation, we find that the predicted sales profit per person for a territory with a population $x_1 = 2.8$ and an advertising expenditure $x_2 = .25$ per person is

$$\hat{y} = 9.04 - .58(2.8) - 11.27(.25) = \$4.60$$

Not bad! But before you invest your money, you might also notice that for our data the t test associated with the estimates of β_1 and β_2 have p-values in the neighborhood of .30.

```
                2ND DEGREE REGRESSION EXAMPLE - SAS76.5
                    MODEL : EY= A0 + A1*X1 + A2*X2

                            T FOR H0:        PR > |T|
PARAMETER     ESTIMATE     PARAMETER=0

INTERCEPT     9.04432679      2.90           0.0337
X1           -0.58325688     -1.10           0.3218
X2          -11.26824515     -1.09           0.3237
```

← p value for a two-tailed test

OBSERVATION	OBSERVED VALUE	PREDICTED VALUE
1	3.60000000	5.84159106
2	2.50000000	5.91976137
3	4.20000000	4.71752730
4	4.10000000	4.60881377
5	4.00000000	4.24816104
6	5.10000000	4.00968120
7	4.30000000	2.55792427
8	11.50000000	7.39654001

BMDP 1R

VARIABLE		COEFFICIENT	STD. ERROR	STD. REG COEFF	T	P(2 TAIL)
INTERCEPT		9.044				
X1	2	−0.583	0.531	−0.429	−1.099	0.322
X2	3	−11.268	10.298	−0.427	−1.094	0.324

← p value for two-tailed test

The purpose of example 10.5 is to illustrate that we can rely on available software packages to obtain least squares prediction equations for multiple regression problems. Inferences related to multiple regression prediction equations are beyond the scope of this text. However, as we noted in example 10.5, we should realize that a multiple regression equation is often the first step in examining the relationship between a dependent variable y and several independent variables. And if predictions are made by using the least squares prediction equation, we might ask, "How good is the prediction?"

SUMMARY

This chapter presents an introduction to regression and correlation. We first examined the relationship between a dependent variable y and a single independent variable x. A scatter diagram was used to provide a graphical display of the data, and the method of least squares was used to obtain the regression equation $\hat{y} = \hat{\beta}_0 + \hat{\beta}_1 x$. The sample data can also be used to conduct a test of the null hypothesis that the slope β_1 of the population regression line is equal to zero. This test is summarized in section 10.4. This test tells us whether there is evidence to indicate that x contributes information for the prediction of y. The strength of the linear relationship between y and x can be measured by the sample correlation coefficient $\hat{\rho}$ or by the coefficient of determination $\hat{\rho}^2$.

The fitting of multivariable predictors to experimental data is a very powerful and valuable method of inference. The computer prediction of election-eve outcomes uses this technique. Multivariable predictors are also employed in business forecasting, industrial production, medicine, and many areas of science. In this chapter we discussed, and gave an example to illustrate, how to obtain the least squares regression line $\hat{y} = \hat{\beta}_0 + \hat{\beta}_1 x_1 + \hat{\beta}_2 x_2$ when y is related to only two independent variables. Solutions to multiple regression problems for this situation, as well as for the situation where we are concerned with more than two independent variables, are conveniently obtained by using some of the standard statistical software packages. Students are encouraged to pursue with their professors the computer opportunities available at their institution. Additional details on how to obtain solutions to multiple regression problems can be found in some of the references at the end of this chapter, particularly Mendenhall (1968) and Ott (1977).

REFERENCES

Barr, A. J.; Goodnight, J. H.; Sall, J. P.; and Helwig, J. T. *A User's Guide to SAS 76.* Raleigh, N.C.: SAS Institute, Inc., 1976.

Dixon, W. J. *BMDP: Biomedical Computer Programs.* Berkeley: University of California Press, 1975.

Donkersloot, J. A., and Drummond, J. F. "A Simple Fluorometric Method to Determine Deoxyribonucleic Acid in Dental Plaque." *Journal of Dental Research* 53, 4 (1974): p. 934.

Freund, J. E. *Statistics: A First Course.* 2d ed. Englewood Cliffs, N.J.: Prentice-Hall, 1976.

Huntsberger, D. V., and Billingsley, P. *Elements of Statistical Inference.* 4th ed. Boston: Allyn and Bacon, 1977.

Mendenhall, W. *Introduction to Probability and Statistics.* 5th ed. N. Scituate, Mass.: Duxbury Press, 1979. Chapter 10.

———. *Introduction to Linear Models and the Design and Analysis of Experiments.* Belmont, Calif.: Wadsworth, 1968.

Nie, N.; Hull, C. H.; Jenkins, J. G.; Steinbrenner, K.; and Bent, D. H. *Statistical Package for the Social Sciences.* 2d ed. New York: McGraw-Hill, 1975.

Ott, L. *An Introduction to Statistical Methods and Data Analysis.* N. Scituate, Mass.: Duxbury Press, 1977.

Ryan, T. A.; Joiner, B. L.; and Ryan, B. F. *Minitab Student Handbook.* N. Scituate, Mass.: Duxbury Press, 1976.

Scheaffer, R. L.; Mendenhall, W.; and Ott, L. *Elementary Survey Sampling.* 2d ed. N. Scituate, Mass.: Duxbury Press, 1979.

Walpole, R. E. *Introduction to Statistics.* 2d ed. New York: Macmillan, 1974.

SUPPLEMENTARY EXERCISES

 10.21. An investigator was interested in examining the effect of different doses of a new drug product on the pulse rates of human subjects. Four doses of the drug were to be used in the experiment (1.5, 2.0, 2.5, and 3.0 milligrams per kilogram of body weight). Three persons were randomly assigned to each of the four drug doses. After a prestudy pulse rate was recorded for each individual, subjects were injected with the appropriate drug dose. One hour later, pulse rates were again recorded. The changes in pulse rates are listed in the accompanying table.

Change in Pulse Rate, y	20, 21, 19	16, 17, 17	15, 13, 14	8, 10, 8
Drug Dose, x	1.5	2.0	2.5	3.0

(a) Find the least squares line for these data.
(b) Calculate s^2.
(c) Test the hypothesis that the change in pulse rate y is linearly related to drug dose x. Use $\alpha = .05$.
(d) Predict the change in pulse rate that would accompany a drug dose of 2.3 milligrams per kilogram of body weight.

10.22. Refer to exercise 10.21.
(a) Calculate the correlation coefficient for the data. Interpret your results.
(b) Conduct a test of the alternative hypothesis $H_a: \rho \neq 0$ and compare your conclusions to those of exercise 10.21(c). Use $\alpha = .05$.

10.23. A production foreman was concerned about the quality of the outgoing product from his department. He felt very strongly that the percentage of defective items passing through his assembly line during a 30-minute period increased throughout the day. At nine 30-minute periods throughout the day, the assembly line was closely examined to determine the number of defectives being produced. For each of these 30-minute periods, the number of hours that workers had been working (from 8:00 A.M.) was also recorded. The data are given in the accompanying table.

	Number of Defectives, y	*Number of Hours, x, Workers on Job*
	13	1.0
	14	1.5
	16	2.5
	14	2.0
	15	3.5
	20	4.5
	18	4.0
	18	5.5
	20	6.0
Total	$\Sigma y = 148; \Sigma y^2 = 2490$	$\Sigma x = 30.5; \Sigma x^2 = 128.25; \Sigma xy = 535.5$

(a) Write a linear model relating the number of defectives y to the number of hours on the job for these data.

Supplementary Exercises

(b) Use the method of least squares to fit the model.
(c) Calculate s^2 for the sample data. (Hint: Compute s^2 as in the test for the slope.)
(d) Conduct a test of the hypothesis that there is a positive linear relationship between the number of defectives in a 30-minute interval and the number of hours that the workers have been working that day. Use $\alpha = .05$.
(e) Predict the number of items that would be defective in a 30-minute period if the workers had just completed five hours of work.

10.24. Refer to exercise 10.23.
(a) Compute the sample correlation coefficient to measure the strength of the linear relationship between x and y. Interpret your answer.
(b) Conduct a test of the alternative hypothesis H_a: $\rho > 0$. Compare your results to those in exercise 10.23(d). Use $\alpha = .05$.

10.25. A chain of grocery stores conducted a study to determine the relationship between the amount of money x spent on advertising and the weekly volume y of sales. Six different levels of advertising expenditure were tried in a random order over a six-week period. The accompanying data were observed (in units of $100).

Weekly Sales Volume, y	10.2	11.5	16.1	20.3	25.6	28
Amount Spent on Advertising, x	1.0	1.25	1.5	2.0	2.5	3.0

(a) Plot these data on a scatter diagram.
(b) Use the method of least squares to find the regression equation $\hat{y} = \hat{\beta}_0 + \hat{\beta}_1 x$.
(c) Do the data indicate a significant linear relationship between x and y? Use $\alpha = .05$.
(d) Use the prediction equation of part (b) to estimate sales volume for an expenditure of $220 in advertising.

10.26. Refer to exercise 10.25. Compute the correlation coefficient between the sales volume and advertising volume. Does there appear to be a strong linear relationship between x and y?

10.27. Suppose that the following data were collected on emphysema patients: the number of years x the patient smoked and inhaled and a physician's evaluation y of the patient's lung capacity (measured on a scale of 0 to 100). The results for a sample of ten patients appear in the accompanying table. (Note: $S_{xx} = 876.9$, $S_{yy} = 2510$, and $S_{xy} = 1148$.)
(a) Plot the data on a scatter diagram.
(b) Use the method of least squares to find the regression line $\hat{y} = \hat{\beta}_0 + \hat{\beta}_1 x$.
(c) Is there evidence to indicate a significant positive linear relationship between x and y? Use $\alpha = .05$.

(d) Calculate the correlation coefficient between the variables "lung capacity" and "number of years smoking."
(e) Predict a person's lung capacity after 30 years of smoking.

Patient	Years Smoking, x	Lung Capacity, y
1	25	55
2	36	60
3	22	50
4	15	30
5	48	75
6	39	70
7	42	70
8	31	55
9	28	30
10	33	35

10.28. An experiment was conducted to measure the strength of the linear relationship between two variables, a student's emotional stability (as measured by a guidance counselor's subjective judgment after an encounter session) and the student's score on an achievement test administered to children entering the first grade. The variable of emotional stability was measured on a scale of 0 to 40 (from low to high), and the achievement test was also measured from 0 to 40. Use the accompanying data from a random sample of 15 children to calculate the correlation coefficient. (Note: $S_{xx} = 485.33$, $S_{yy} = 522.93$, and $S_{xy} = 316.67$.)

Student	Emotional Stability, x	Achievement, y	Student	Emotional Stability, x	Achievement, y
1	23	31	9	32	33
2	21	23	10	29	35
3	31	34	11	16	21
4	34	29	12	29	22
5	26	29	13	23	24
6	22	27	14	27	28
7	14	21	15	25	15
8	18	17			

10.29. Refer to exercise 10.28. Rank (from lowest to highest) the 15 children separately on the two variables "emotional stability" and "achievement." For the data presented in this way, calculate the sample correlation coefficient for the two variables.

Experiences with Real Data

10.30. Refer to the data of exercise 10.14.
(a) Construct a 95% confidence interval for β_1, the population slope.
(b) Compare your interval in part (a) to the interval shown in the computer solution to exercise 10.14.
(c) How would the interval change if we wanted a 99% rather than a 95% confidence interval?

10.31. Refer to the data of example 10.4. Use the output shown here to obtain the following:
(a) the least squares prediction equation
(b) 95% confidence limit for the slope
(c) the sample correlation coefficient

LINEAR REGRESSION

	ESTIMATE	95% CONFIDENCE LIMITS	
SLOPE:	0.50500	[0.29981,	0.71019]
INTERCEPT:	9.27500	[8.09626,	10.45374]
		0.165	

CORRELATION COEFFICIENT: 0.97640

X VALUE	OBSERVED Y	PREDICTED Y
1.00000	10.00000	9.78000
3.00000	10.30000	10.79000
5.00000	12.20000	11.80000
7.00000	12.60000	12.81000
9.00000	13.90000	13.82000

EXPERIENCES WITH REAL DATA

Conduct a study to determine whether there is a correlation between a social science major's performance in a math or a statistics course and his or her performance in a course in the social sciences. For example, you may wish to visit the department of sociology to obtain a random sample of 30 senior sociology majors. Contact these students to determine their (numerical) grades in a specific sociology course (such as "Introductory Sociology") and a mathematics (or statistics) course required for graduation. Let x denote a student's sociology grade and y his or her mathematics (statistics) grade.

1. Identify the population from which the sample was drawn.

2. Find the least squares prediction equation, $\hat{y} = \hat{\beta}_0 + \hat{\beta}_1 x$.

3. Calculate the coefficient of linear correlation.

4. Is there evidence to indicate that β_1 is different from zero?

5. Describe the strength of the relationship between the two sets of scores.

11 The Design of an Experiment: Getting More Information for Your Money

CHAPTER OUTLINE

11.1 Introduction

11.2 The Paired-Difference Experiment: An Example of a Designed Experiment

11.3 Choosing the Sample Size to Estimate a Population Mean μ or a Population Proportion p

11.1 INTRODUCTION

design of an experiment

The design of an experiment is essentially a plan for purchasing a specified quantity of information, and, as you might suspect, we hope to make our purchase at the lowest possible cost. The amount of information available in a sample for making an inference about a population parameter can be measured by the bound on the error of estimation, which can be calculated from the sample data. The smaller the bound on the error, the more we know about the population parameter of interest.

The bound on the error of estimation for most commonly used estimators is dependent on the population standard deviation σ and the sample size n. Experimenters can control the quantity of information contained in the sample by determining the number of observations they will include in their sample and by selecting the sample in such a way so as to reduce the variation in the data. Although this is an oversimplification of the problems encountered in designing an experiment, it does summarize the essential points.

Recall from section 6.2 that for n greater than or equal to 30, a bound on the error of estimation for a population mean is $2\sigma/\sqrt{n}$. Thus we obtain more information (a reduction in the bound on error) either by increasing the sample size n or by decreasing σ, the parameter that measures the variability of the sampled data.

A strong similarity exists between audio communication and statistics. Both are concerned with the transmission of a message or signal from one point to another. For example, a stockbroker is responsible for transmitting a verbal message (perhaps the results of a stock sale) from her office in New York City to her client's residence in New Orleans. Or a speaker may wish to communicate with a large audience. If static or background noise is sizable, either in the case of the speaker or stockbroker, the message may arrive in garbled form and hence represent only part of the complete signal. The receiver must use this partial information to infer the nature of the complete message.

Similarly, scientific experimentation is conducted to verify certain theories about natural phenomena. Sometimes we want simply to explore some aspect of nature and to deduce, exactly or with a good approximation, the relationships between certain variables. We can think of experimentation as communication between nature and the scientist. The message consisting of less than complete information about the natural phenomenon is contained in the experimenter's sample data. Imperfections in the measuring instruments, nonhomogeneity of experimental material, and numerous other factors contribute background noise or static, which tend to obscure nature's signal and cause the observed response to vary in a random manner.

For both the communications engineer and the statistician, two factors affect the quantity of information in an experiment: the magnitude of the background noise (the variation) and the volume of the signal (the size of the sample). The greater the noise (the variation), the less information will be contained in the sample. On the other hand, the louder the signal (the larger the sample size), the greater the amplification will be and the greater the chance that the message will penetrate the noise and be received.

The design of experiments is a very broad area of statistics that is primarily concerned with acquiring a fixed amount of information at the lowest possible cost. To this end one can study methods of sampling that will reduce background variation in an experiment and/or amplify nature's signal. Despite the complexity of the subject, some of the important considerations in the design of good experiments can be easily understood and should be presented in an introductory course. We will illustrate the concept of "noise reduction" by using the paired-difference experiment.

11.2 THE PAIRED-DIFFERENCE EXPERIMENT: AN EXAMPLE OF A DESIGNED EXPERIMENT

We observed that two factors affect the quantity of information in an experiment that has been designed for the purpose of making inferences about a population parameter—namely, the variation in the population (as measured by σ) and the size of the sample. Thus experimental designs may be classified

The Paired-Difference Experiment: An Example of a Designed Experiment

as either <u>noise reducers</u> or <u>volume increasers</u>, depending on whether the primary effect on the quantity of information in an experiment is to reduce the variation (σ) in the data or to increase the volume of information.

The <u>paired-difference experiment</u> is an example of a noise-reducing (variation-reducing) design. Suppose that we wish to test the difference between the average wear of two types of automobile tires, using a 20,000-mile road test. If only the rear wheels of each automobile are to be used and five automobiles are to be employed in the experiment, two methods of design might be suggested. First, we could randomly assign five tires of type A and five tires of type B to the 10 rear wheels. Then we might have a random assignment of tires to rear wheels as demonstrated in figure 11.1. The automobiles would then be driven over the 20,000-mile test course and the amount of wear would be recorded for each tire.

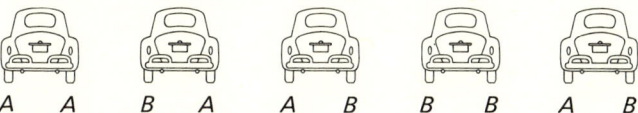

　　A　　A　　B　　A　　A　　B　　B　　B　　A　　B

Figure 11.1 Random Assignment of Five Tires of Type A and Type B

But there is a disadvantage to this design. Since each automobile has a different driver, we would expect the wear measurements to vary greatly from automobile to automobile, depending upon the test driver, his method of accelerating and braking, the balance of the wheels, and the road surface to which the tires were exposed. Thus some wear measurements on tires A or B would be extremely high and others would be extremely low, making the respective sample variances, s_A^2 and s_B^2, large.

Recall that the small-sample test statistic for testing the equality of two population means μ_A and μ_B is

$$t = \frac{\bar{y}_A - \bar{y}_B}{s\sqrt{\dfrac{1}{n_A} + \dfrac{1}{n_B}}}$$

where

$$s^2 = \frac{(n_A - 1)s_A^2 + (n_B - 1)s_B^2}{n_A + n_B - 2}$$

We reject the null hypothesis for large (positive or negative) values of t. If the sample variances s_A^2 and s_B^2 are large, we require large differences in the sample means, $(\bar{y}_A - \bar{y}_B)$, to declare a difference in the mean wear for the two types of tires. Hence the automobile-to-automobile variability, which inflates the sample variances s_A^2 and s_B^2, makes it difficult to detect a difference in the population mean wears, $(\mu_A - \mu_B)$, and thereby reduces the quantity of information in the experiment.

We can improve on the design in figure 11.1 by reducing the variability in the sample data. To do this, we make a comparison of the wear for the two tire types under controlled driver-road conditions. In this way we filter out the variability due to automobiles (drivers) by making comparisons of tire types A and B on each automobile. Thus we randomly assign a type A and a type B tire to the rear wheels of each automobile. Note that this randomization scheme is restricted, since we require both tire types to appear on a car. A typical assignment is shown in figure 11.2. Note that one tire of type A and one of type B are randomly assigned and mounted on the rear wheels of each automobile. The five automobiles could then be driven over the 20,000-mile test course, after which the amount of wear would be recorded for each tire.

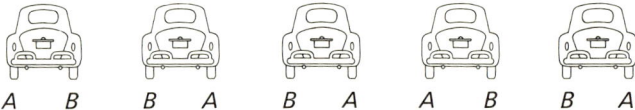

Figure 11.2 Restricted Random Assignment of Tires to Rear Wheels That Eliminates Automobile Variability in the Comparison of Tire Types A and B

The results of a wear test using the restricted random assignment of figure 11.2 are given in table 11.1. The wears for both an A and a B tire, y_A and y_B, are shown in the second and third columns for each automobile. The **difference** between each pair, denoted by the symbol d, is recorded in column four. The averages for y_A, y_B, and d are shown at the bottom of the table. Note that $\bar{d} = (\bar{y}_A - \bar{y}_B)$. We would like to use these data to determine if there is sufficient evidence to indicate a difference in the mean wear for the two types of tires, $(\mu_A - \mu_B)$.

Table 11.1 Results of a Wear Test

	Tire Wear		
Automobile	y_A	y_B	$d = (y_A - y_B)$
1	10.6	10.2	.4
2	9.8	9.4	.4
3	12.3	11.8	.5
4	9.7	9.1	.6
5	8.8	8.3	.5
	$\bar{y}_A = 10.24$	$\bar{y}_B = 9.76$	$\bar{d} = .48$

The Paired-Difference Experiment: An Example of a Designed Experiment

At first glance we may be tempted to employ the methods of comparing two means presented in section 9.4. However, this would not be the proper analysis because one of the assumptions of the t test has been violated—the two samples are not independent because the pairs of observations are linked; they read high or low, depending on the driver-automobile combination to which the pair of tires was assigned. **Hence the data analysis for an experiment is dictated by the design used.** If the design employed was the one obtained by random assignment of the five tires of types A and B to the 10 rear wheels (figure 11.1), we would analyze the data by using the methods of section 9.4. But the experimenter realized that the wear measurements would vary greatly from car to car; thus he could filter or block out this variability if he assigned one tire of type A and one tire of type B to the rear wheels of each car. Hence the restricted random assignment (figure 11.2) dictates that we must perform another analysis to determine whether the mean wear differs for the two types of tires. The appropriate method is to compare the wear between the tire types A and B by using the five difference measurements given in table 11.1.

The proper analysis utilizes the five difference measurements to test the hypothesis that the mean difference between tires A and B is equal to zero. [This hypothesis is equivalent to the hypothesis that the difference in the means for the two types of tires, $(\mu_A - \mu_B)$, is equal to zero.] The testing procedure is identical to the procedure presented in section 9.2 for a small-sample test of an hypothesis concerning a population mean μ. Thus we will use the sample of differences and the one-sample Student's t test to test the null hypothesis $\mu_d = 0$, where μ_d, the mean of the population of differences, is equal to $(\mu_A - \mu_B)$. Elements of the paired-difference test are as follows:

Paired-Difference Test

Null hypothesis: $\mu_d = (\mu_A - \mu_B) = 0$.

Alternative hypothesis:

For a one-tailed test:

1. $\mu_d > 0$.
2. $\mu_d < 0$.

For a two-tailed test:

3. $\mu_d \neq 0$.

Test statistic: $t = \dfrac{\bar{d}}{s_d/\sqrt{n}}$ where $\bar{d} = \dfrac{\Sigma d}{n}$ and

$$s_d = \sqrt{\dfrac{\Sigma d^2 - (\Sigma d)^2/n}{n-1}}$$

Rejection region: For a given value of α, for df $= (n - 1)$, and for a one-tailed test:

1. Reject H_0 if $t > t_a$, with a $= \alpha$.
2. Reject H_0 if $t < -t_a$, with a $= \alpha$.

For a given value of α, for df $= (n - 1)$, and for a two-tailed test:

3. Reject H_0 if $|t| > t_a$, with a $= \alpha/2$.

The notation in the t statistic is slightly different from that used in section 9.2. We use the symbol d_1 to denote the difference between the first pair of observations made on y_A and y_B. Similarly, d_2 is the difference between the second pair, and so on. Then the n differences, $d_1, d_2, \ldots, d_n$, are used to compute

$$\bar{d} = \frac{\Sigma d}{n}$$

the mean of the sample differences.

Keep in mind that for the paired-difference experiment, n refers to the number of sample differences (or number of pairs) rather than to the total number of measurements. And s_d refers to the standard deviation of the sample differences:

$$s_d = \sqrt{\frac{\Sigma d^2 - (\Sigma d)^2/n}{n - 1}}$$

Now consider the data in table 11.1 and test the null hypothesis that $\mu_d = 0$ against the alternative hypothesis that μ_d is not equal to zero. We must first compute $\bar{d}$ and s_d. For these data,

$$\Sigma d = .4 + .4 + .5 + .6 + .5 = 2.4$$
$$\Sigma d^2 = (.4)^2 + (.4)^2 + (.5)^2 + (.6)^2 + (.5)^2 = 1.18$$

Hence for $n = 5$ differences,

$$\bar{d} = \frac{\Sigma d}{n} = \frac{2.4}{5} = .48$$

$$s_d^2 = \frac{\Sigma d^2 - (\Sigma d)^2/n}{n - 1} = \frac{(1.18) - (2.4)^2/5}{4}$$

$$= \frac{1.18 - 1.152}{4} = .007$$

$$s_d = \sqrt{.007} = .0837$$

Substituting into the test statistic t, we have

$$t = \frac{\bar{d}}{s_d/\sqrt{n}} = \frac{.48}{.0837/\sqrt{5}} = \frac{(.48)(2.236)}{.0837} = 12.8$$

The Paired-Difference Experiment: An Example of a Designed Experiment

The rejection region for a two-tailed test can be located by using table 2 in the Appendix. Let's suppose we want $\alpha = .05$. The value of t for a = .025 and $(n - 1) = 4$ degrees of freedom is 2.776. Hence we reject the null hypothesis if the observed value of t is either greater than 2.776 or less than -2.776. Since $t = 12.8$ is greater than 2.776, we reject the null hypothesis that the average difference μ_d is equal to zero. That is, we reject the hypothesis that the difference in mean wear $(\mu_A - \mu_B)$ is equal to zero. We conclude that a difference exists in mean wear for the two types of tires. Indeed, because $\bar{y}_A$ exceeds $\bar{y}_B$, we would venture to state that type A tires show more wear over a comparable test course than type B tires.

A general confidence interval for the mean difference can be obtained by using the general confidence interval for a single population mean, section 9.3.

Confidence Interval for $\mu_d = (\mu_1 - \mu_2)$, Based on a Paired-Difference Experiment

$$\bar{d} \pm t \frac{s_d}{\sqrt{n}}$$

The value of t corresponding to a 90%, a 95%, or a 99% confidence interval is found in table 2 of the Appendix for df = $(n - 1)$ and a = .05, .025, or .005, respectively. For this situation n represents the number of pairs.

Example 11.1

Find a 95% confidence interval for the difference in mean wear for tire types A and B, table 11.1.

Solution

For this example $n = 5$ and a = .025. The value of t corresponding to a = .025 and df = 4 is 2.776. The 95% confidence interval for $\mu_d = (\mu_1 - \mu_2)$ is

$$\bar{d} \pm t \frac{s_d}{\sqrt{n}}$$
$$.48 \pm (2.776)\left(\frac{.0837}{\sqrt{5}}\right)$$
$$.48 \pm .10$$

Thus we estimate the difference in mean tire wear for tire types A and B to be in the interval .38 to .58. The confidence coefficient for this interval estimate

randomized block design

is .95. This means that when we use this estimation procedure, 95% of the time the constructed intervals will enclose μ_d.

The statistical design employed in the tire experiment represents a simple example of a randomized block design. Several points about this design should be emphasized. First, pairing of the measurements occurred when the experiment was planned (actually it occurred in the restricted random assignment of tires to wheels) and not after the data were obtained. Comparisons of wear were then made within each car to eliminate the variability between automobiles (drivers). Second, pairing will not always provide more information for testing the difference between two population means. Recall that the objective of the paired-difference experiment is to reduce the "background noise" (car-to-car variability). If there were absolutely no differences between the cars, we would lose information by pairing because, instead of having 10 sample measurements (one measurement for each of five tires of type A and B), we would have only five differences. This reduces the number of degrees of freedom for the t test statistic and hence makes it more difficult to detect a difference in means when one exists. Thus we recommend pairing only if we can filter out undesirable background noise. The paired-difference experiment is only one of a class of noise-reducing designs referred to as block designs.

Volume-increasing experimental designs are concerned with shifting information in an experiment to display more clearly the parameter(s) of interest. They involve a study of the effect of a set of variables on a response measurement y, and the choice of the combination of variables to be used in the experiment. This study can become quite complex and is beyond the scope of this text. Instead, we will consider the simplest method of controlling the volume of information in an experiment—the selection of the sample size.

EXERCISES

11.1. Consider the accompanying data for a paired-difference experiment.

Pair	y_A	y_B	$d = (y_A - y_B)$
1	21	29	−8
2	28	30	−2
3	17	21	−4
4	24	25	−1
5	27	33	−6
6	18	22	−4
7	20	19	1
8	23	29	−6
9	28	26	2

(a) How many degrees of freedom are associated with the t statistic?
(b) Do the data provide sufficient evidence to indicate a difference between μ_A and μ_B? Test by using $\alpha = .05$.
(c) Construct a 95% confidence interval for $(\mu_A - \mu_B)$. Interpret the interval.

11.2. Psychologists wish to examine the amount of learning exhibited by schizophrenics after taking a specified dose of a tranquilizer. Ten schizophrenics are randomly selected from a patient ward, and before the drug is administered, each is given a standard examination. The amount of time required to complete the exam is recorded for each individual. One hour after the specified drug dose is administered, each patient is given the same standard exam as previously. Again completion times are recorded. If these patients exhibit any learning, we would expect the mean completion time prior to receiving the drug to be more than that needed one hour after receiving the drug. This is our alternative hypothesis.
(a) Use the accompanying data to test the null hypothesis that the mean difference between the before-dose and the after-dose test scores is zero (that is, $H_0: \mu_d = 0$) against the alternative hypothesis $H_a: \mu_d > 0$. Use $\alpha = .05$.

Patient	Before, y_1	After, y_2	Difference, $y_1 - y_2$
1	10	8	2
2	15	13	2
3	30	29	1
4	29	25	4
5	26	21	5
6	28	28	0
7	19	15	4
8	13	10	3
9	14	12	2
10	21	17	4

(b) Construct a 95% confidence interval for $(\mu_1 - \mu_2)$. Interpret the interval.

11.3. Suppose that two independent random samples of $n_1 = n_2 = 10$ observations have been independently selected from two populations to test the null hypothesis $(\mu_1 - \mu_2) = 0$. After the measurements have been recorded in two adjacent columns, side-by-side observations are paired for the $n = 10$ pairs. Will a paired-difference analysis provide more information to test $H_0: \mu_d = 0$ than the unpaired t test of section 9.4? Explain.

11.4. Two analysts, A and B, supposedly of identical abilities, are used to measure the parts per million of a certain type of chemical impurity in drinking water. It is claimed that analyst A tends to give higher readings than analyst B. To test this theory, each of six water samples is divided and then analyzed by A and B. The data are shown in the accompanying table.

Water Sample	Analyst A	Analyst B
1	31.4	28.1
2	37.0	37.1
3	44.0	40.6
4	28.8	27.3
5	59.9	58.4
6	37.6	38.9

(a) Do the data provide sufficient evidence to indicate that the mean reading for analyst A exceeds the mean reading for analyst B? Test by using $\alpha = .05$.
(b) Construct a 90% confidence interval for $(\mu_A - \mu_B)$, the difference in the mean readings for analysts A and B. Interpret the interval.

11.5. An agronomist compared the mean yield of two new varieties of wheat using side-by-side, 10-acre plots at each of eight different farm locations. One 10-acre plot at a location was planted with variety A; the other was planted with variety B. The mean yield per acre for each of the plots is recorded in the accompanying table.

Location	Variety A	Variety B	Location	Variety A	Variety B
1	76	77	5	68	80
2	69	83	6	81	75
3	73	76	7	70	73
4	79	78	8	75	85

(a) Do the data provide sufficient evidence to indicate a difference in the mean yield per acre for the two varieties? Test by using $\alpha = .10$.
(b) Construct a 90% confidence interval for the difference in mean yields for the two varieties of wheat. Interpret the interval.

11.3 CHOOSING THE SAMPLE SIZE TO ESTIMATE A POPULATION MEAN μ OR A POPULATION PROPORTION p

How can we determine the number of observations to include in a sample? The implications of such a question are clear. Observations cost money. If the sample is too large, time and talent are wasted. Conversely, it is also wasteful if the sample is too small because inadequate information has been purchased for the time and effort expended; and it may be impossible to increase the

Choosing the Sample Size to Estimate μ or p

sample size at a later time. **Hence the number of observations the experimenters should buy will depend upon the amount of information they require.**

Suppose we wish to estimate the mean value of accident claims filed against an insurance company. To determine how many claims to examine (or sample), we would have to determine how accurate the company wants us to be. Thus it might specify that the bound on the error of estimation must equal $50. We can then determine the required sample size.

The methods of choosing the sample size for estimating a population mean μ and a population proportion p are identical. We must first specify a desired bound on the error of estimation. We call this bound B. Thus if we use $\bar{y}$ to estimate a population mean μ, we would want $\bar{y}$ to differ from μ by less than some value B. To find the sample size that will produce an estimate with this specified degree of accuracy and with probability approximately equal to .95, we set two standard deviations of the estimator equal to B and solve this expression for the desired sample size n.

Recall that the standard deviation of the sampling distribution for the sample mean $\bar{y}$ is $\sigma_{\bar{y}} = \sigma/\sqrt{n}$. Hence to determine the sample size required to estimate μ, we set two standard deviations of the estimator $\bar{y}$ equal to the bound on the error of estimation B.

$$\frac{2\sigma}{\sqrt{n}} = B$$

Then we solve this equation for n. The sample size required to estimate μ is as shown next.

Sample Size Required to Estimate μ

$$n = \frac{4\sigma^2}{B^2}$$

where B is the bound on the error of estimation. The probability that the error of estimation will be less than B is approximately .95.

You will note that determining a sample size to estimate μ requires knowledge of the population variance σ^2 (or standard deviation σ). We can obtain an approximate sample size by estimating σ^2, using one of two methods:

1. Employ information from a prior experiment to calculate a sample variance s^2. Use this value to approximate σ^2.

2. Use information on the range of the observations (and the Empirical Rule) to obtain an estimate of σ. (This method was employed in section 3.5 as a check on the calculation of s.)

We would then substitute the estimated value of σ^2 in the equation

$$n = \frac{4\sigma^2}{B^2}$$

to determine an approximate sample size n.

We illustrate the procedure for choosing a sample size in the following examples.

Example 11.2

Union officials are concerned about reports of inferior wages paid to employees in an industry under their jurisdiction. It is decided to take a random sample of wage sheets from n employees within the industry to estimate the industry mean hourly wage. It is known that wages within the industry have a range of $10 per hour. Determine the sample size required to estimate μ, the mean hourly wage, with a bound on the error of estimation equal to $.50.

Solution

The desired bound on the error of estimation is $B = \$.50$. Before using the equation $n = 4\sigma^2/B^2$ to determine n, we must estimate the population variance σ^2. We do so using a *range estimate* of σ. Since the range of hourly wages is $10, an estimate of σ (using the method of section 3.5) is

$$s \approx \frac{\text{range}}{4} = \frac{10}{4} = 2.5$$

An approximate sample size can then be found by substituting $(2.5)^2$ for σ^2 in the equation $n = 4\sigma^2/B^2$:

$$n = \frac{4\sigma^2}{B^2} \approx \frac{4(2.5)^2}{(.5)^2} = 100$$

Thus the union officials would have to sample $n = 100$ wage sheets to estimate the mean hourly wage to within a $.50 error.

Example 11.3

A federal agency has decided to investigate the advertised weight displayed on cartons of a certain brand of cereal. The company in question periodically samples cartons of cereal coming off the production line to check their weight. A summary of 1500 of the weights made available to the agency indicates a mean weight of 11.80 ounces per carton and a standard deviation of .75 ounces. Use the company's information to determine the number of cereal

Choosing the Sample Size to Estimate μ or p

cartons the federal agency must examine in order to estimate the average weight of cartons currently produced with a bound on the error of estimation equal to .25 ounces.

Solution

The federal agency has specified that the bound on the error of estimation must be $B = .25$ ounces. To determine the sample size required to achieve this bound, we must obtain an estimate of the population variance σ^2. Assuming the weights made available to the federal agency by the company were randomly selected from its production, we can use the given standard deviation of these weights to form an estimate of σ^2. Thus

$$\sigma^2 \approx (.75)^2 = .5625$$

An approximate sample size can now be found by using the equation,

$$n = \frac{4\sigma^2}{B^2} \approx \frac{4(.75)^2}{.0625} = 36$$

Hence the federal agency must obtain a random sample of $n = 36$ cereal cartons to estimate the mean weight to within .25 ounces.

Selecting the sample size to estimate a binomial parameter p is accomplished in a manner similar to the method employed when estimating μ. You will recall (section 6.5) that $\hat{p} = y/n$ is our point estimator of p and that $\sqrt{pq/n}$ is the standard deviation of the sampling distribution of $\hat{p}$. Then, setting two standard deviations equal to the desired bound on the error of estimation B, we have

$$2\sqrt{\frac{pq}{n}} = B$$

Solving this equation for n, we obtain the sample size as follows:

Sample Size Required to Estimate a Binomial Parameter p

$$n = \frac{4pq}{B^2}$$

where B is the bound on the error of estimation and $q = 1 - p$. The probability that the error of estimation will be less than B is approximately .95.

Note that we must know p to solve for n and that this requirement creates a circular problem, since our final objective is to estimate p. Actually, it is not as complicated as it appears. We often know before the experiment begins that p will lie in a fairly narrow range. For example, the fraction of popular vote for a presidential candidate in a national election is often close to .5. Thus we can substitute for p the value dictated by experience. A second method for finding p is to estimate p by using data collected from a prior experiment. Finally, if there is no prior information, we can use $p = .5$. Substituting .5 for p will yield the largest possible sample size for the bound that you have specified and will thus give a conservative answer to the required sample size. The sample size you calculate this way will likely be larger than required, but you will be on the safe side.

We illustrate the selection of the sample size for estimating a binomial parameter p with an example.

Example 11.4

In a national election poll we wish to estimate the proportion p of voters in favor of candidate A. How many people should be polled so that we can estimate p with a bound on the error of estimation of $B = .02$?

Solution

The pollster has specified that $B = .02$. The sample size n necessary to achieve the desired bound is found by substituting into the equation

$$n = \frac{4pq}{B^2}$$

To solve for n, we must first obtain an approximation for p. If a similar survey has been run recently, we can use the sample proportion from that survey to estimate p. Otherwise, we substitute $p = .5$ to obtain a conservative* sample size (one that is likely to be larger than required). For the case in which no prior survey was run, the sample size is

$$n = \frac{4pq}{B^2} = \frac{4(.5)(.5)}{.0004} = 2500$$

That is, 2500 potential voters must be polled in order to estimate the proportion favoring candidate A, with a bound on the error of estimation of .02.

We have discussed choosing the sample size required to estimate either μ

---

*$p = .5$ gives the largest possible value of pq, that is, $pq = (.5)(.5) = .25$. Since pq appears in the numerator of the formula for n, $p = .5$ will yield the largest value for n.

or p, with a bound on the error of estimation equal to a specified value B. For either case the sample size can be determined by setting two standard deviations of the appropriate sample statistic equal to the desired bound on the error of estimation B and solving for n. When estimating μ, we must supply some estimate of the population variance σ^2. This estimate can usually be obtained by using a value of s^2 from a prior experiment or by using a range estimate to approximate σ. When estimating p, we must supply some value for the binomial parameter appearing in the standard deviation of $\hat{p}$. If data is available from a previous study, we could use this information to estimate p in the formula; otherwise, we substitute $p = .5$ into the formula for n to obtain a conservative sample size (a sample size at least as large as necessary to achieve the specified bound on the error of estimation). The following example shows you how to apply this same procedure to find the sample sizes for a comparison of two population means.

Example 11.5

Suppose we wish to estimate the difference between two population means, correct to within .5 unit, based on independent random samples from the two populations. We know that $\sigma_1^2 \approx \sigma_2^2 \approx 3$, and we plan to select equal sample sizes, $n_1 = n_2 = n$. Find n.

Solution

The sampling distribution of $(\bar{y}_1 - \bar{y}_2)$, the point estimator of $(\mu_1 - \mu_2)$, has a standard deviation of

$$\sigma_{\bar{y}_1 - \bar{y}_2} = \sqrt{\frac{\sigma_1^2}{n_1} + \frac{\sigma_2^2}{n_2}}$$

Then letting $n_1 = n_2 = n$ and $\sigma_1^2 = \sigma_2^2 = 3$, we set

$$2\sigma_{\bar{y}_1 - \bar{y}_2} = B$$

or

$$2\sqrt{\frac{3}{n} + \frac{3}{n}} = .5$$

Solving for n, we obtain

$$2\sqrt{\frac{6}{n}} = .5$$

$$\sqrt{n} = \frac{2\sqrt{6}}{.5}$$

$$n = 96$$

Thus we should select approximately 96 observations in each sample. The resulting error of estimation will be less than $B = .5$ unit with a probability approximately equal to .95.

EXERCISES

11.6. A forester wishes to estimate the mean diameter of pine trees in a 100-acre tree plantation. A preliminary survey of the area suggests that the diameters can be as small as 9 inches and as large as 27 inches. The forester wishes the estimate to be within .5 inch of the trees' mean diameter, with probability approximately equal to .95. How many trees should be included in a random sample?

11.7. Suppose you wish to estimate the mean fair market price of homes, all of similar size and age, in a particular development and you wish your estimate to be within $2000 of the true mean market price, with probability approximately equal to .95. You are fairly certain that no house will sell for less than $40,000 or more than $60,000. If you plan to sample recent sales prices at your local courthouse, approximately how many should be included in your sample? What assumption must you make concerning the sales in order that your estimate be valid?

11.8. A certified public accountant plans to audit the financial records of a company for a random sample of n days to estimate the proportion of account ledger pages that are in error. If the accountant believes that the ledger page error rate is in the neighborhood of 1% and wishes to estimate the true rate correct to within 1%, approximately how many ledger pages should be included in the sample?

11.9. Suppose that you plan to sample the opinions of a particular hospital's nurses to estimate the proportion who are satisfied with working conditions. If you wish to estimate the proportion correct to within .10, how many nurses should be included in your sample? (Hint: Since you have no prior information on the approximate value of p, find the approximate sample size by using $p = .5$.)

SUMMARY

The design of an experiment is actually a plan by which an experimenter can purchase a specified quantity of information about one or more population parameters. The information pertinent to a given parameter is measured by the bound on the error of estimation (that is, by two standard deviations of the point estimator). The factors affecting the quantity of information in an experiment are data variation and sample size or, making an analogy to audio communication, noise and amplification. Good experimental designs are those that filter out unwanted data variation and, at the same time, increase the volume of the signal by increasing the sample size. Proper experimental design can increase the quantity of information per fixed cost or, alternatively, reduce the cost of a specified quantity of information.

The paired-difference experiment, a simple case of a randomized block design, was used to illustrate the concept of noise reduction. Thus one source of unwanted variation (the differences in automobiles and test drivers) was eliminated from the data, thereby unveiling a clear difference in the wearing qualities of tire types A and B. Many more observations would have been required to reveal the difference in mean wear for the two tire types if independent random samples had been employed. Thus filtering out the effect of differences in test drivers and automobiles reduced the cost of a specified amount of information concerning the difference in the wearing qualities of the two types of tires.

A signal is amplified by shifting information in an experiment to (in a sense) increase n and thereby increase the quantity of information about a given parameter. Selecting the sample size necessary to achieve a specified bound on the error of estimation illustrates the effect of the sample size on the bound B. For larger n the bound will be smaller.

The study of experimental design is a separate course in itself, requiring more space than we can allot here. But it is important to note that good designs are those that simultaneously achieve both noise reduction and signal amplification, thus helping the experimenter to acquire a specified amount of information at minimum cost. You will find additional information on this subject in the References.

REFERENCES

Hicks, C. R. *Fundamental Concepts in the Design of Experiments*. 2d ed. New York: Holt, Rinehart and Winston, 1974.

Mendenhall, W. *An Introduction to Linear Models and the Design and Analysis of Experiments*. Belmont, Calif.: Wadsworth, 1968.

——. *Introduction to Probability and Statistics*. 5th ed. N. Scituate, Mass.: Duxbury Press, 1979. Chapter 12.

Ott, L. *An Introduction to Statistical Methods and Data Analysis*. N. Scituate, Mass.: Duxbury Press, 1977. Chapter 9.

SUPPLEMENTARY EXERCISES

11.10. List two factors that affect the quantity of information in an experiment.

11.11. Explain what is meant by noise-reducing experimental designs and by volume-increasing experimental designs.

11.12. Give an example of a noise-reducing experiment.

11.13. Will the paired-difference test always provide more information for testing the difference between two populations than the methods of section 9.3? Explain.

11.14. Consider the accompanying data for a paired-difference experiment.

Pair	y_A	y_B	d
1	4.1	3.9	.2
2	4.0	4.0	0
3	3.8	3.6	.2
4	4.2	4.1	.1
5	3.9	3.8	.1
6	3.9	4.0	−.1

(a) Do the data provide sufficient evidence to indicate a difference between μ_A and μ_B? Test by using $\alpha = .05$.
(b) Construct a 95% confidence interval for $(\mu_A - \mu_B)$. Interpret the interval.
(c) Suppose that the data had resulted from independent random sampling of the two populations—that is, no pairing. Use the method of section 9.5 to construct a 95% confidence interval for $(\mu_A - \mu_B)$. Notice the difference in the interval widths for parts (b) and (c).
(d) If you have conducted a paired-difference experiment, is it valid to analyze the data in an unpaired manner (i.e., by using the method of section 9.5)?
(e) Suppose that you have conducted the experiment by selecting independent samples of equal size from the two populations. In a similar situation some experimenters might pair the observations, one observation from each sample in each pair, and then analyze the data by using the paired-difference test. What is wrong with this procedure?

11.15. An agricultural experiment station was interested in comparing the yields for two new varieties of corn. Because it was felt that there might be a great deal of variability in yield from one farm to another, each variety was randomly assigned to a different 1-acre plot on each of seven farms. The 1-acre plots were planted and the corn was harvested at maturity. The results of the experiment (in bushels of corn) are shown in the accompanying table.
(a) Show that the sample standard deviation of the differences $(y_A - y_B)$ is $s_d = 2.39$.
(b) Use the data to test the null hypothesis that there is no difference in mean yields for the two varieties of corn. Use $\alpha = .05$.

Farm	Variety A	Variety B
1	28.2	21.5
2	24.6	20.1
3	29.7	24.0
4	20.5	21.2
5	34.6	29.8
6	27.1	21.7
7	31.4	26.8

$ **11.16.** Insurance adjusters were concerned about the high repair estimates that they were getting from garage A. To verify their suspicions, each of 10 damaged cars was taken to garage A to obtain a repair estimate and then to another, more reliable garage B. The results are given in the accompanying table (in hundreds of dollars).

Damaged Car	Garage A	Garage B	Damaged Car	Garage A	Garage B
1	2.1	2.0	6	5.4	5.0
2	4.5	3.8	7	7.3	6.5
3	6.3	5.9	8	9.5	8.6
4	3.0	2.8	9	10.1	9.0
5	1.2	1.3	10	7.6	7.2

(a) Show that the sample standard deviation of the differences $(y_A - y_B)$ is $s_d = .38$.

(b) Test the null hypothesis that there is no difference in mean repair estimates for the two garages against the alternative that, on the average, estimates from garage A are higher than those from garage B. Use $\alpha = .01$.

$ **11.17.** Improperly filled orders are a costly problem for mail-order houses. To estimate the mean loss per incorrectly filled order, a large mail-order house plans to randomly sample n incorrectly filled orders and to analyze the added cost associated with each. It is estimated that the added cost of the return and replacement of the incorrect order can never be less than \$40 or more than \$400. If the company wishes to estimate the mean additional cost per incorrect order correct to within \$20, how many incorrect orders must it analyze?

11.18. A manufacturer of dishwashers claims that its machines will operate repair free for 4 years. A quick examination of repair slips on models of this dishwasher indicate that the dates of first repairs have ranged anywhere from 0 to 9 years. Use this information to approximate the number of observations (of repair records) required to estimate the mean time to first repair of the dishwashers correct to within 3 months (.25 years).

11.19. How many repair records (exercise 11.18) must be examined to estimate the mean time to first repair to within 6 months?

11.20. A study is to be conducted to estimate the mean gain in weight for chicks on a specified ration during their first four weeks of growth. Experience has shown that the standard deviation in chick weights during the first four weeks of growth is in the neighborhood of 6 grams. If the experimenter wishes the estimate of mean weight gain to be correct to within 1 gram, with probability approximately equal to .95, how many chicks should be included in the sample?

11.21. Two brands of latex outdoor paints were to be compared for durability at each of six different geographic locations. At each location one-half of a strip of exterior wood was painted with brand A, the other half with brand B. Durability readings were then obtained after a one-year trial period.

Location	Brand A	Brand B
1	3.1	3.2
2	2.9	2.8
3	3.6	3.5
4	4.2	4.0
5	3.8	3.6
6	2.7	2.4

(a) Show that $s_d = .14$.
(b) Use the experimental data to determine if a difference exists between the mean durability for the two brands. Let $\alpha = .05$.

11.22. What advantage(s) does the design proposed in exercise 11.21 have over the corresponding unpaired experiment?

11.23. Random samples of 100 oranges for each of two varieties were collected at each of nine locations and the total sugar content (in pounds) contained in each batch of 100 oranges was recorded. The data are shown in the accompanying table.

Location	Variety A	Variety B	Location	Variety A	Variety B
1	2.1	2.0	6	2.3	2.3
2	1.9	1.7	7	2.0	1.7
3	2.1	1.9	8	2.1	2.2
4	1.9	2.0	9	2.3	2.1
5	2.2	2.1			

(a) Do the data provide sufficient evidence to indicate a difference in the

Supplementary Exercises

mean sugar content per 100 oranges between the two varieties? Test by using $\alpha = .05$.

(b) Construct a 95% confidence interval for $(\mu_1 - \mu_2)$. Interpret the interval.

11.24. A service station manager would like to decrease the price he charges for regular gasoline. Before he does this, he must estimate the average number of gallons sold per week to determine if the decrease would be profitable. Previous records suggest that sales run anywhere from 1000 to 6000 gallons per week. Use this information to determine the sample size required to estimate μ, the average number of gallons sold per week, to within 200 gallons.

11.25. Student government would like to estimate the proportion of students in favor of converting from the semester to the quarter system. Determine the sample size required to estimate p, the proportion of students favoring the change, with a bound on the error of estimation equal to .03. (Hint: Since p is unknown, use $p = .5$ to find the approximate sample size.)

11.26. A large corporation conducted a study and found that approximately 25% of its potential customers were aware of the corporation's product. To increase this percentage, an extensive advertising campaign was carried out. At the end of the campaign, a sample of potential customers was to be contacted to determine the proportion aware of the product. How many potential customers must be contacted to estimate p, the fraction aware of the corporation's product, to within .02 units? (Hint: Since p is unknown, use $p = .5$ to find the approximate sample size.)

11.27. University employees are disturbed about the extremely long time it takes to be reimbursed for out-of-pocket travel expenses. To back up their complaints, they decide to take a sample of travel expense requests filed during the next six months and determine the average amount of time it takes before an individual is reimbursed. Previous experience suggests that these requests take from 7 to 60 days. Determine the sample size needed to estimate μ, the average time for reimbursement, to within 1 day.

11.28. A psychiatrist would like to estimate the average time it takes schizophrenics to react to a specified stimulus. Previous work in this field suggests that the standard deviation for reaction times of schizophrenics is approximately .5 second. Determine the sample size required to estimate μ to within .1 second.

11.29. A small company was concerned about the percentage of its accounts that were overdue on the last day of the previous month. It would be too time-consuming to examine all accounts, since records were not stored in a computer. Hence it was decided to draw a random sample of accounts to estimate the fraction that were overdue on the last day of the previous month. Previous samples have shown that approximately 10% of all accounts were overdue at the end of a month. How large a sample of accounts should be taken in order to estimate p to within .05?

11.30. Refer to exercise 11.14. Suppose you wish to conduct a new paired-difference experiment to estimate the difference between the population means. If you wish to estimate $(\mu_A - \mu_B)$ with an error of less than .03, with probability approximately equal to .95, how many pairs must you sample?

11.31. An experimenter plans to compare the mean number of particles of effluent in water collected at two different points in a water treatment system. The means are expected to be in the neighborhood of 30 and it is known that the variances for the particle counts in samples taken at the two locations are approximately equal to 30. If the experimenter wishes to estimate the difference in the mean counts at the two locations correct to within 1 particle, with probability approximately equal to .95, how many water samples should be analyzed at each location? Assume $n_1 = n_2$.

EXPERIENCES WITH REAL DATA

Scan available newspapers, magazines, and professional journals for the results of a survey where the experimenter estimates the proportion (or percentage) of items in a population possessing a specified attribute. For example, surveys are frequently conducted to estimate the proportion of people in the United States who think that the president is doing an acceptable job. For the survey you choose, perform the following:

1. Identify the sample size.

2. Use the methods of chapter 7 to give a point estimate and place a bound on the error of estimation.

3. Indicate how you might have improved the design of the survey.

4. If, prior to running the survey, you wished to estimate p correct to within .02, how large a sample size would be necessary?

12 Testing the Equality of Population Variances

CHAPTER OUTLINE

12.1 Introduction
12.2 A Test of an Hypothesis Concerning Two Population Variances

12.1 INTRODUCTION

The population parameter that answers an experimenter's practical question will vary from one situation to another, but frequently population variation is the target of interest. For example, a pharmaceutical company can control the *mean potency* of pills by control of the batch mix, but it must also worry about *variation in the potency* from one pill to another. Excessive potency might be as harmful to a patient as an underdose. Thus the company desires the variation in potency, measured by σ^2 or σ, to be as small as possible.

Inferential problems concerning population variation are similar to inferential problems related to the population mean. We could estimate or test hypotheses about a single population variance, compare two variances, or relate a population variance to one or more independent variables. However, due to the introductory nature of this text, we confine our attention to the comparison of two population variances—specifically, to a test of an hypothesis that the two variances are equal. This test is useful in testing the assumption that variances are equal, as they must be if we are to use the Student's t test (introduced in section 9.3). If you wish to make inferences about a single population variance or to relate a population variance to one or more independent variables, we refer you to the text listed in the References at the end of the chapter.

12.2 A TEST OF AN HYPOTHESIS CONCERNING TWO POPULATION VARIANCES

An experimenter is frequently interested in testing an hypothesis concerning the equality of two population variances. For example, we may wish to determine whether a newly designed filament for kitchen ovens produces more variability in oven temperatures than does the filament presently in use. To test equality of variation in the two populations, we hypothesize that the populations of measurements are normally distributed—one population for temperatures run at a given power setting for filament 1 and one for filament 2, the new filament. We label these populations as 1 and 2, respectively. A temperature measurement is acquired by bringing a single oven to a stable temperature for a fixed power setting. Then n_1 and n_2 ovens are randomly selected from the two populations, and the temperature of each oven is observed at the fixed power setting. We are interested in testing whether the variance of population 1, σ_1^2, is equal to the variance of population 2, σ_2^2. Thus we test the null hypothesis that $\sigma_1^2 = \sigma_2^2$.

Sample variances computed from the data in the two samples can be used to construct a test statistic. We will use the ratio s_1^2/s_2^2. If s_1^2/s_2^2 is nearly equal to one, we would have little evidence to indicate that σ_1^2 and σ_2^2 are unequal. However, extremely large or small values would present evidence to contradict the null hypothesis of equality. How large or small must s_1^2/s_2^2 be in order to reject the null hypothesis? The answer to this question depends on the sampling distribution of s_1^2/s_2^2—that is, on the distribution of values of s_1^2/s_2^2 obtained in repeated sampling.

When independent random samples are drawn from two normal populations with equal variances ($\sigma_1^2 = \sigma_2^2$), the quantity s_1^2/s_2^2 possesses a probability distribution in repeated sampling that is known as an *F distribution*. The equation for this probability distribution is omitted, but we note several of its important properties.

F distribution

1. The F distribution, unlike the normal distribution or t distribution, is nonsymmetrical (see figure 12.1).

2. There are many F distributions (and hence many shapes). We can specify a particular one by designating the degrees of freedom associated with s_1^2 and s_2^2. We designate these quantities as ν_1 and ν_2, respectively.

3. Tail-end values for the F distribution are tabulated and appear in tables 4 and 5 of the Appendix.

Table 4 in the Appendix records the upper-tail value of F, call it F_a, that has an area of a equal to .05 to its right (see figure 12.2). A reproduction of part of table 4 is shown in table 12.1. The degrees of freedom for s_1^2, designated by ν_1, are indicated across the top of the table; the degrees of freedom for s_2^2, designated by ν_2, appear in the first column on the left. Thus for $\nu_1 = 8$ and $\nu_2 = 10$, the tabulated value is 3.07. That is, only 5% of the measurements

A Test of an Hypothesis Concerning Two Population Variances

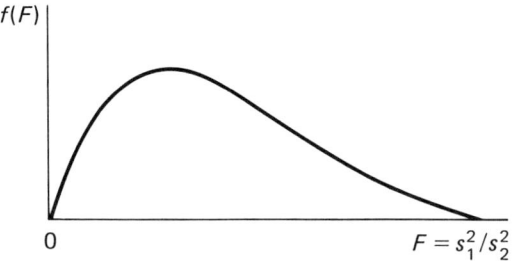

Figure 12.1 Distribution of s_1^2/s_2^2 When Independent Random Samples Are Drawn from Two Normal Populations with Equal Variances

for s_1^2/s_2^2, in repeated sampling from an F distribution with $\nu_1 = 8$ and $\nu_2 = 10$ degrees of freedom, will exceed 3.07. (See figure 12.2.) Similarly, an entry in table 5 in the Appendix is an upper-tail value of the F distribution that has an area equal to .01 to its right. Thus the .01 tail-end value of the F distribution with $\nu_1 = 5$ and $\nu_2 = 3$ degrees of freedom is 28.24.

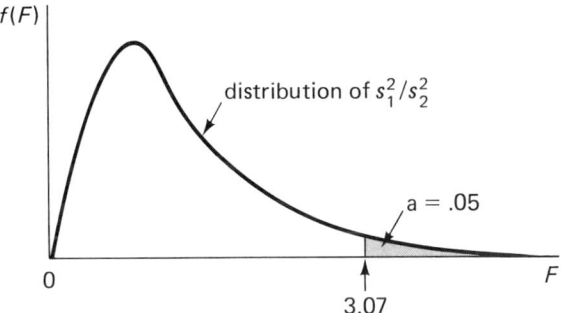

Figure 12.2 Tabulated Value for the F Distribution; $\nu_1 = 8$, $\nu_2 = 10$, a = .05

The statistical test of the null hypothesis $\sigma_1^2 = \sigma_2^2$ utilizes the test statistic

$$F = \frac{s_1^2}{s_2^2}$$

Suppose the alternative hypothesis states that σ_1^2 is greater than σ_2^2, this situation implies a one-tailed test and we may use the tables directly. However, when the alternative hypothesis states that σ_1^2 is not equal to σ_2^2, we need a two-tailed test, with the rejection region divided between the upper and lower tails. The values of F in the lower tail of the F distribution have not been tabulated in the Appendix because the need for them can be circumvented by using the following procedure.

Testing the Equality of Population Variances

Table 12.1 Format of the F Table, Table 4 in the Appendix, a = .05

Degrees of Freedom

$\nu_2 \backslash \nu_1$	1	2	...	6	7	8	9	...	60	120	∞	$\nu_1 \backslash \nu_2$
1	161.4	199.5	...	234.0	236.8	238.9	240.5	...	252.2	253.3	254.3	1
2	18.51	19.00	...	19.33	19.35	19.37	19.38	...	19.48	19.49	19.50	2
3	10.13	9.55	...	8.94	8.89	8.85	8.81	...	8.57	8.55	8.53	3
4	7.71	6.94	...	6.16	6.09	6.04	6.00	...	5.69	5.66	5.63	4
5	6.61	5.79	...	4.95	4.88	4.82	4.77	...	4.43	4.40	4.36	5
6	5.99	5.14	...	4.28	4.21	4.15	4.10	...	3.74	3.70	3.67	6
7	5.59	4.74	...	3.87	3.79	3.73	3.68	...	3.30	3.27	3.23	7
8	5.32	4.46	...	3.58	3.50	3.44	3.39	...	3.01	2.97	2.93	8
9	5.12	4.26	...	3.37	3.29	3.23	3.18	...	2.79	2.75	2.71	9
10	4.96	4.10	...	3.22	3.14	3.07	3.02	...	2.62	2.58	2.54	10
.	.	.		.	.	.	.		.	.	.	.
.	.	.		.	.	.	.		.	.	.	.
.	.	.		.	.	.	.		.	.	.	.
15	4.54	3.68	...	2.79	2.71	2.64	2.59	...	2.16	2.11	2.07	15
16	4.49	3.63	...	2.74	2.66	2.59	2.54	...	2.11	2.06	2.01	16
17	4.45	3.59	...	2.70	2.61	2.55	2.49	...	2.06	2.01	1.96	17

We are at liberty to label either of the two populations as 1. If the population with the larger sample variance is designated as population 2, then $F = s_1^2 / s_2^2$ will be less than one, and we will be concerned about rejection in the lower tail of the F distribution. We can avoid this difficulty by always designating the population with the larger sample variance as population 1. **That is, we will always place the larger sample variance in the numerator of the test statistic**

$$F = \frac{s_1^2}{s_2^2}$$

and identify that population as 1. By using this convention, we will always be concerned with upper-tail rejections. The upper-tail F value for a two-tailed test can then be obtained from tables in the Appendix. The value of α for the two-tailed test will be double the area in the upper tail alone. (See figure 12.3.)

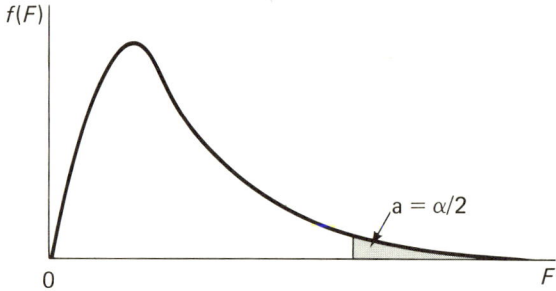

Figure 12.3 Upper-Tail Value of F for a Two-Tailed Test

A Test of an Hypothesis Concerning Two Population Variances

Since tables 4 and 5 have upper-tail tabulated values with a = .05 and .01, respectively, we can conduct a two-tailed test with α = .10 or α = .02 and a one-tailed test with α = .05 or .01. If other values of α are desired, the tabulated values can be found in *Biometrika Tables for Statisticians* (1966).

F Test for Comparing Two Population Variances

Null hypothesis: $\sigma_1^2 = \sigma_2^2$.

Alternative hypothesis:
For a one-tailed test:
1. $\sigma_1^2 > \sigma_2^2$.

For a two-tailed test:
2. $\sigma_1^2 \neq \sigma_2^2$.

Note: Always identify population 1 as the population that produced the larger sample variance.

Test statistic: $F = \dfrac{s_1^2}{s_2^2}$

where s_1^2 is the larger of the two sample variances.

Rejection region: For a given value of α, for $\nu_1 = (n_1 - 1)$, for $\nu_2 = (n_2 - 1)$, and for a one-tailed test:

1. Reject H_0 if F is greater than F_a, the table value for a = α.

For a given value of α, for $\nu_1 = (n_1 - 1)$, for $\nu_2 = (n_2 - 1)$, and for a two-tailed test:

2. Reject H_0 if F is greater than F_a, the table value for a = $\alpha/2$.

Note: See tables 4 and 5 in the Appendix for table values corresponding to a = .05 and .01, respectively.

We illustrate these concepts with some examples.

Example 12.1

The variability in the potency of five-grain aspirin tablets differs from one brand to another. An interested research group would like to compare brand C, a new product recently released, to brand B, the current best-seller. Ran-

Testing the Equality of Population Variances

dom samples of 41 tablets are obtained from bottles of each of the brands; the potency results are given in table 12.2. Use these data to test the hypothesis that the population variances of brands B and C, σ_1^2 and σ_2^2, respectively, are equal. The alternative hypothesis is that the best-seller has more variability in potency; that is σ_1^2 is larger than σ_2^2. Use $\alpha = .01$.

Table 12.2 Data for Example 12.1

	Brand B	Brand C
Sample Size	41	41
Sample Mean	60.2	60.5
Sample Variance	2.20	.98

Solution

We are concerned with a one-tailed statistical test with a = α = .01. The null hypothesis $\sigma_1^2 = \sigma_2^2$ is to be tested against the alternative hypothesis that σ_1^2 is larger than σ_2^2. The test statistic is

$$F = \frac{s_1^2}{s_2^2} = \frac{2.20}{.98} = 2.24$$

To locate the rejection region, we must first designate ν_1 and ν_2, the degrees of freedom associated with s_1^2 and s_2^2, respectively. The degrees of freedom for a sample variance are the same as the degrees of freedom for a small-sample t test of a population mean—namely, one less than sample size. Hence s_1^2 and s_2^2 have $\nu_1 = (n_1 - 1) = 40$ and $\nu_2 = (n_2 - 1) = 40$ degrees of freedom. The upper-tail F value for a = .01, $\nu_1 = 40$, and $\nu_2 = 40$, table 5, is 2.11 (see figure 12.4). Since the observed value of F is larger than 2.11, we reject the null hypothesis of equality of the population variances and conclude that brand B (the best-seller) has more variability in potency than brand C. The chance of our making a wrong decision is $\alpha = .01$ if, in fact, H_0 is true. Note

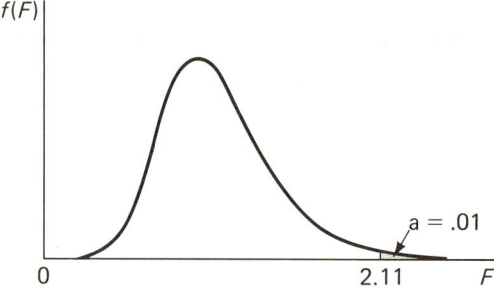

Figure 12.4 Upper-Tail Value of F for a One-Tailed Test: $\nu_1 = \nu_2 = 40$, a = .01

A Test of an Hypothesis Concerning Two Population Variances

that the populations are potencies associated with two production batches of tablets, one each for brands B and C, from which the samples were selected. Large variations in batch quality (potency) could cause brand C to be more variable than brand B at some time in the future.

Example 12.2

Two consumer research groups are vying for a large government contract. Since most subjective evaluations of consumer products will be ratings made by judges, government officials prefer to award the contract to a company that utilizes judges with consistent ratings (of course, other qualifications are also evaluated before reaching a decision). One measure of consistency of ratings is the variability of the judges' scores on the same item.

Before issuing the contract, the federal agency conducts a test in which 10 judges from each company are asked to rate a single item. The sample variances are given in table 12.3. Use these data to test the hypothesis that the variances of the judges' ratings are the same for the two companies. The alternative hypothesis is that they are different. Use $\alpha = .10$.

Table 12.3 Data for Example 12.2

	Company 1	Company 2
Sample Size	10	10
Sample Variance	.50	.15

Solution

We will use a two-tailed test, with $\alpha = .10$. Recall that we have agreed to identify the company with the larger sample variance as population 1. Then the test statistic is

$$F = \frac{s_1^2}{s_2^2} = \frac{.50}{.15} = 3.33$$

To locate the rejection region, we must specify ν_1 and ν_2. Since 10 judges were used from each company, we have

$$\nu_1 = n_1 - 1 = 9 \quad \text{and} \quad \nu_2 = n_2 - 1 = 9$$

So the upper-tail F value for an area of .05 is 3.18 (see table 4). Remember that for a two-tailed test we always place the larger sample variance in the numerator of the F statistic and that this action makes α equal to twice the area a in the upper tail of the F distribution. Therefore, α equals .10 for this two-tailed test.

Since the observed value of F does exceed the tabulated value, we reject the null hypothesis and conclude that the variability of ratings for the judges differs for the two companies. Indeed, it appears that the judges from company 2 have a smaller variability than those from company 1. The probability of making a wrong decision using this F test is $\alpha = .10$ if, in fact, H_0 is true.

As you proceed to solve the exercises at the end of this chapter, you may encounter a difficulty in using the F tables, tables 4 and 5 of the Appendix. To reduce the size of the tables, F values corresponding to some of the larger values of ν_1 and ν_2 have been omitted. But this causes little difficulty because (for large values of ν_1 and ν_2) the F values change very little for small changes in ν_1 and ν_2. For example, see table 12.1, the reproduction of part of table 4. If you let $\nu_1 = 8$ and change ν_2 from $\nu_2 = 15$ to $\nu_2 = 16$, the table value F_a changes only slightly, from 2.64 to 2.59. Consequently, if the value of F_a that you need has been omitted from table 4 (or 5), pick the next largest and smallest values of ν_1 and ν_2, find the corresponding values of F_a, and select a value of F_a between them.

Example 12.3

Find an approximate value of F_a for $a = .05$, $\nu_1 = 11$, and $\nu_2 = 15$.

Solution

Table 4 in the Appendix does not give the value of F_a for $\nu_1 = 11$ and $\nu_2 = 15$, but it gives the values for these:

$$\nu_1 = 10, \quad \nu_2 = 15 \quad F_a = 2.54$$
$$\nu_1 = 12, \quad \nu_2 = 15 \quad F_a = 2.48$$

Since the value F_a for $\nu_1 = 11$ falls between $F_a = 2.54$ and $F_a = 2.48$, we choose the midpoint between these two values—namely, $F_a = 2.51$—as the value corresponding to $\nu_1 = 11$ and $\nu_2 = 15$. (In actuality, the F values corresponding to $\nu_1 = 10$, $\nu_2 = 15$ and $\nu_1 = 12$, $\nu_2 = 15$ are so close together that you could use either one to obtain an adequate approximation to the value of F_a for $\nu_1 = 11$, $\nu_2 = 15$.)

SUMMARY

In this chapter we presented a discussion of one- and two-tailed tests concerning the equality of two population variances. The test statistic s_1^2 / s_2^2 has a sampling distribution known as the F distribution. Although no lower-tail values of the F distributions are tabulated in tables 4 and 5 of the Appendix, we can make use of the upper-tail values for a two-tailed test by labeling the set with the larger sample variance as population 1. For one-tailed tests we always label our populations so that the null hypothesis $\sigma_1^2 = \sigma_2^2$ is tested against the alternative hypothesis that σ_1^2 is greater than σ_2^2. Then we use $F = s_1^2/s_2^2$ and reject H_0 for upper-tail values.

The need to make inferences about two populations' variances occurs when comparing the consistency of psychological tests, the consistency of two machines in manufacturing, or the consistency in the performance of two humans, for example. We have shown how to make an inference about the equality of the measures of consistency σ_1^2 and σ_2^2 for two populations. Thus we have demonstrated another very practical example of statistical inference.

REFERENCES

Mendenhall, W. *Introduction to Probability and Statistics*. 5th ed. N. Scituate, Mass.: Duxbury Press, 1979. Chapter 9.

Ott, L. *An Introduction to Statistical Methods and Data Analysis*. N. Scituate, Mass.: Duxbury Press, 1977. Chapter 12.

Pearson, E. S., and Hartley, H. O. *Biometrika Tables for Statisticians*. Vol. I. 3d ed. Cambridge, England: Cambridge University Press, 1966.

Walpole, R. E. *Introduction to Statistics*. 2d ed. New York: Macmillan, 1974. Chapter 9.

SUPPLEMENTARY EXERCISES

12.1. Find the value of F that locates an area a in the upper tail of the F distribution for these conditions:
(a) $a = .05, \nu_1 = 7, \nu_2 = 12$
(b) $a = .05, \nu_1 = 3, \nu_2 = 10$
(c) $a = .05, \nu_1 = 10, \nu_2 = 20$
(d) $a = .01, \nu_1 = 8, \nu_2 = 15$
(e) $a = .01, \nu_1 = 13, \nu_2 = 25$

12.2. Find approximate values for F_a for these:
(a) $a = .05, \nu_1 = 11, \nu_2 = 24$
(b) $a = .05, \nu_1 = 14, \nu_2 = 14$
(c) $a = .05, \nu_1 = 35, \nu_2 = 22$
(d) $a = .01, \nu_1 = 22, \nu_2 = 24$
(e) $a = .01, \nu_1 = 17, \nu_2 = 25$

(Note: Your answers may not agree with the answers given in the back of the book. As long as your answer is close to the recorded answer, it is satisfactory.)

12.3. Random samples of $n_1 = 8$ and $n_2 = 10$ observations were selected from populations 1 and 2, respectively. The corresponding sample variances were $s_1^2 = 7.4$ and $s_2^2 = 12.7$. Do the data provide sufficient evidence to indicate a difference between σ_1^2 and σ_2^2? Test by using $\alpha = .10$.

12.4. An experiment was conducted to determine whether there was sufficient evidence to indicate that data variation within one population, say population A, exceeded the variation within the second population, population B. Random samples of $n_A = n_B = 8$ measurements were selected from the two populations and the sample variances were calculated to be

$$s_A^2 = 2.87 \qquad s_B^2 = .91$$

Do the data provide sufficient evidence to indicate that σ_A^2 is larger than σ_B^2? Test by using $\alpha = .05$.

12.5. A soft drink firm is debating whether it should invest in a new type of canning machine or whether it should continue operating with the machines presently in use. The company has already determined that it will be able to fill more cans per day for the same cost if the new machines are installed. However, an important factor as yet unsolved is the variability of fills. (The company would, of course, prefer the model with the smaller variance in fills.) Let σ_1^2 and σ_2^2 denote the variances for fills from the new model and the old model, respectively. Obtaining samples of fills from the two models and utilizing the test statistic s_1^2/s_2^2, we can set up either a one-tailed or a two-tailed rejection region, using the F distribution.
(a) What type of rejection region would be most favored by the manager of the soft drink company? Why?
(b) What type of rejection region would be most favored by the salesman for the company manufacturing the model presently in use? Why?

12.6. Refer to exercise 12.5. Suppose random samples of $n_1 = n_2 = 11$ cans from the two machines are examined to determine the amount of fill (in ounces). The means and variances are

$$\bar{y}_1 = 11.70 \qquad \bar{y}_2 = 11.60$$
$$s_1^2 = .06 \qquad s_2^2 = .022$$

Do these data present sufficient evidence to indicate a difference in variability of fills for the two models? Use $\alpha = .10$.

12.7. In a gasoline economy study ten 1-gallon samples of a particular brand of gasoline were used for each of two cars (A and B). Both cars averaged approximately 17 miles per gallon, but the sample standard deviations were .95 and 1.56 for cars A and B, respectively. Use these data to test the hypothesis that the variances in miles per gallon for the two cars are identical. Use $\alpha = .02$.

Supplementary Exercises

12.8. Forty-two undergraduate students were randomly assigned to one of two sections (each of size 21) for the same course. The same instructor was assigned to teach both sections. Unannounced quizzes were used throughout the quarter in section 2 but were not used in section 1. The final section averages were the same; but the variations in final grades for the two sections, as measured by the sample variances, were different:

$$s_1^2 = 9.3 \qquad s_2^2 = 4.1$$

Use these data to test the hypothesis of equality of the population variances for the two methods of teaching (with and without unannounced quizzes) against the alternative hypothesis that unannounced quizzes constitute a teaching method that causes less variability in students' achievement. Use $\alpha = .05$.

12.9. Would your conclusion to exercise 12.8 have changed if you had employed a two-tailed test with $\alpha = .10$? With $\alpha = .02$?

12.10. An important consideration in examining the potency of a pharmaceutical product is the amount of drop in potency for a specific shelf life (time on a pharmacist's shelf). In particular, the variability of these drops in potency is very important. Researchers studied the drops in potency for two different drug products over a six-month period. These data are summarized in the accompanying table. Suppose that drug 1 is an experimental drug product and drug 2 a marketed product. Use a one-tailed test with $\alpha = .01$ to determine whether the data suggest that drug 1 has more variability in potency drop than drug 2.

	Drug 1	Drug 2
Sample Size	10	10
Sample Mean	58	56
Sample Variance	82	23

12.11. Refer to exercise 12.10. Would your result have changed if you had used a two-tailed test with $\alpha = .10$? Why might a two-tailed test be important?

12.12. Blood cholesterol levels for randomly selected patients with similar histories were compared for two diets, one a low-fat-content diet and the other a normal diet. The summary data appear in the accompanying table.

	Low-Fat Content	Normal
Sample Size	19	24
Sample Mean	170	196
Sample Variance	198	435

(a) Do these data present sufficient evidence to indicate a difference in cholesterol level variabilities for the two diets? Use $\alpha = .10$.
(b) What other test might be of interest in comparing the two diets?

12.13. Sales from weight-reducing agents marketed in the United States represent sizable chunks of income for many of the companies that manufacture these products. Many times psychological as well as physical effects contribute to how well a person responds to the recommended therapy. Consider a comparison of two weight-reducing agents, A and B. In particular, we will consider the variabilities in the lengths of times people remain on the therapy. A total of 26 overweight males, matched as closely as possible physically, were randomly divided into two groups. Those in group 1 received preparation A while those assigned to group 2 received preparation B. Use the summary data to compare the variabilities associated with the lengths of time on therapy. Use a two-tailed test with $\alpha = .10$.

	Preparation A	Preparation B
Sample Size	13	13
Sample Mean	25 days	35 days
Sample Variance	50	16

12.14. Refer to exercise 12.13. What might the null and alternative hypotheses have been if preparation A had been a placebo (no active medication) and preparation B a marketed product known to be an effective weight-reducing agent?

12.15. A chemist at an iron ore mine suspects that the variance in the amount (weight, in ounces) of iron oxide per pound of ore tends to increase as the mean amount of iron oxide per pound increases. To test this theory, ten 1-pound specimens of iron ore are selected at each of two locations, one, location 1, containing a much higher mean content of iron oxide than the other, location 2. The amounts of iron oxide contained in the ore specimens are shown in the accompanying table.

Location 1	8.1	7.4	9.3	7.5	7.1	8.7	9.1	7.9	8.4	8.8
Location 2	3.9	4.4	4.7	3.6	4.1	3.9	4.6	3.5	4.0	4.2

(a) Do the data provide sufficient evidence to indicate that the amount of iron oxide per pound of ore is more variable at location 1 than at location 2? Use $\alpha = .05$.
(b) Regardless of the results of part (a), can you give a reason that might support the contention that the variance of the measurements at location 1 should be larger than the variance of the measurements at location 2?

12.16. A personnel officer was planning to use a Student's t test to compare the mean number of monthly absences for two categories of employees, but then she noticed a possible difficulty. The variation in the number of absences per month seemed to differ for the two groups. As a check, the personnel officer randomly selected five months and counted the number of absences for each group. The data are shown in the accompanying table.

Category A	20	14	19	22	25
Category B	37	29	51	40	26

(a) About which assumption necessary for use of the t test was the personnel officer concerned?

(b) Do the data provide sufficient evidence to indicate that the variances differ for the populations of absences for the two employee categories? Use $\alpha = .10$.

EXPERIENCES WITH REAL DATA

Perform an experiment to compare two population variances. For example, since the Arab oil embargo of 1973, everyone has become more energy conscious. Certainly the advertisements for new cars have reflected the importance consumers place on energy conservation. Still, in spite of the encouraging new trends in advertising, very little information is supplied to the consumer. We are told we can expect "23 miles to the gallon under normal driving conditions," but we are left to interpret this statement as we choose. Most people would infer that 23 represents the average number of miles per gallon under normal driving conditions. No mention is made of variability. To give you some notion of variability in miles per gallon, you may wish to compare the variability in miles per gallon for two cars of the same make and model. One way to do this might be as follows. Fill the gasoline tank of your car. Go about your "normal" in-town driving but try to refill the gasoline tank after traveling 100 miles. Compute the miles per gallon for those 100 miles. Repeat this process four more times, refilling the tank and computing the miles per gallon for every 100 miles traveled. Based on the five readings on miles per gallon, compute the sample variance.

Locate a friend with the same make and model of car and ask him or her to join you in this experiment. (If you can't locate a person with the same make or model of car, perhaps you could switch cars with a friend.) Following the same procedure, obtain five readings on the miles per gallon. Compute a sample variance for your friend's data.

1. Using the methods of this chapter, compare the population variances for your miles-per-gallon data and for that of your friend.

Testing the Equality of Population Variances

2. Comment on the possible causes of differences in variabilities for you and your friend.

3. If you and your friend drove the same car in the experiment, what variabilities would you be comparing?

4. If you drove one car and your friend drove another car of the same make and model, what variabilities would you be comparing?

5. How might you design an experiment to measure the variabilities from car to car?

13 Analysis of Variance

CHAPTER OUTLINE

13.1 Introduction

13.2 The Logic Behind an Analysis of Variance

13.3 A Test of an Hypothesis Concerning More Than Two Population Means: An Example of an Analysis of Variance

13.1 INTRODUCTION

Methods for comparing two population means were presented in chapters 8 and 9. Very often the two-sample problem is a simplification of what is encountered in real life. That is, frequently we wish to compare more than two population means.

For example, suppose that we wish to compare the mean incomes of steelworkers for three different ethnic groups, say black, white, and Spanish Americans, in a certain city. Independent random samples of steelworkers would be selected from each of the three ethnic groups (the three populations). We would have to consult the personnel files of the steel companies in the city, list steelworkers in each ethnic group, and select a random sample from each. On the basis of the three sample means, we wish to decide whether the population mean incomes differ and, if so, by how much. Note that the sample means most likely will differ, but this does not necessarily imply a difference in mean income for the three ethnic groups. Even if the population mean incomes were identical, the sample means most probably would differ. Then how do you decide whether the differences among the sample means are large enough to imply a difference among the corresponding population means? We will answer this question by using a technique known as an analysis of variance.

13.2 THE LOGIC BEHIND AN ANALYSIS OF VARIANCE

Why we call the method an analysis of variance can be seen more easily with an example. Assume that we wish to compare three population means based on three samples of five observations each. The data for the three samples are

Table 13.1 A Comparison of Three Population Means (Small Amount of Within-Sample Variation)

Sample 1	Sample 2	Sample 3
29.0	25.1	20.1
29.2	25.0	20.0
29.1	25.0	19.9
28.9	24.9	19.8
28.8	25.0	20.2
$\bar{y}_1 = 29.0$	$\bar{y}_2 = 25.0$	$\bar{y}_3 = 20.0$
$s_1 = .16$	$s_2 = .07$	$s_3 = .16$

shown in table 13.1. Do the data present sufficient evidence to indicate a difference among the three population means? A brief visual analysis of the data in table 13.1 leads us to a rapid, intuitive "yes." A glance at each of the three samples indicates that there is very little variation within each sample—that is, the variation in the measurements within each sample is very small. In contrast, the spread or variation among the sample means is so large in comparison to the within-sample variation that we intuitively conclude that a real difference does exist among the population means.

How does our intuition work when a larger within-sample variation is present? (See table 13.2.) In this case the variation among the sample means is not large relative to the variation within samples. Hence it would be difficult to conclude that the samples were drawn from populations with different means.

The variations in the observations for the two sets of data, tables 13.1 and 13.2, are shown graphically in figure 13.1. The strong evidence to indicate a difference in population means for the data of table 13.1 is apparent in figure

Table 13.2 A Comparison of Three Population Means (Large Amount of Within-Sample Variation)

Sample 1	Sample 2	Sample 3
29.0	33.1	15.2
14.2	7.4	39.3
45.1	17.6	14.8
48.9	44.2	25.5
7.8	22.7	5.2
$\bar{y}_1 = 29.0$	$\bar{y}_2 = 25.0$	$\bar{y}_3 = 20.0$
$s_1 = 18.19$	$s_2 = 14.18$	$s_3 = 12.96$

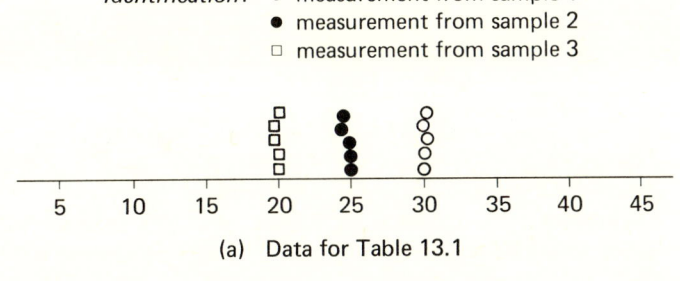

(a) Data for Table 13.1

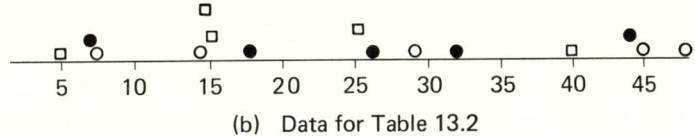

(b) Data for Table 13.2

Figure 13.1 Dot Diagrams for the Data of Tables 13.1 and 13.2

13.1(a). The lack of evidence to indicate a difference in population means for the data of table 13.2 is indicated by the overlapping of data points for the samples in figure 13.1(b).

Figure 13.1 and the data of tables 13.1 and 13.2 indicate very clearly, then, what we mean by an **analysis of variance**. All differences in sample means are judged statistically significant or not significant by comparing them with a measure of the random variation within the population data. Recall that we will measure the variability of the population data by the population standard deviation σ (or, equivalently, the population variance σ^2).

13.3 A TEST OF AN HYPOTHESIS CONCERNING MORE THAN TWO POPULATION MEANS: AN EXAMPLE OF AN ANALYSIS OF VARIANCE

In section 9.4 we presented a method for testing the equality of two population means based on independent random samples from the two populations. We hypothesized two normal populations (1 and 2), with means denoted by μ_1 and μ_2, respectively. To test the null hypothesis that $\mu_1 = \mu_2$, independent random samples of size n_1 and n_2 were drawn from the two populations. The sample data were then used to compute the value of the test statistic

$$t = \frac{(\bar{y}_1 - \bar{y})_2}{s\sqrt{\dfrac{1}{n_1} + \dfrac{1}{n_2}}}$$

where

$$s^2 = \frac{(n_1 - 1)s_1^2 + (n_2 - 1)s_2^2}{(n_1 - 1) + (n_2 - 1)} = \frac{(n_1 - 1)s_1^2 + (n_2 - 1)s_2^2}{n_1 + n_2 - 2}$$

is the pooled estimate of the common population variance σ^2. The rejection region for a specified value of α, the probability of a type I error, was then found by using table 2 in the Appendix.

Suppose now that we wish to extend this method to test the equality of more than two population means. The test procedure described above applies only to two means and therefore is inappropriate. Hence we will employ a more general method of data analysis known as the analysis of variance. We illustrate its use with the following example.

Ecologists are concerned about the level of phosphorus in a large lake in Florida. An environmental action group composed of interested persons in a nearby community has been formed to implement plans for rehabilitating the lake. The ecologists plan to take samples from the lake periodically (say every month) to monitor any changes in the level of phosphorus as the group continues working on the lake. Suppose that nine samples are drawn from various locations of the lake on the first day of each month. At the end of a five-month period the ecologists want to evaluate their progress. In particular, they want to examine the average phosphorus reading for each of the five months.

We label the set of all sample phosphorus readings that could have been obtained from the lake on the first day of the first month as population 1, and we will assume that this population possesses mean μ_1. A random sample of 9 ($n_1 = 9$) measurements (readings) are obtained from this population for the purpose of monitoring the phosphorus level. The set of all readings that could have been obtained on the first day of the second month is labeled population 2, which has mean μ_2. The data from a random sample of 9 readings ($n_2 = 9$) are obtained on the first day of the second month. Similarly, μ_3, μ_4, and μ_5 represent the means of the populations for readings on the first day of the third, fourth, and fifth months, respectively. We also obtain random samples of nine phosphorus readings from these populations.

From each of our five samples we calculate a sample mean and variance. The sample results are summarized in table 13.3.

Table 13.3 Summary of the Sample Results (Phosphorus Levels) for Five Populations

	Population (month)				
	1	2	3	4	5
Sample Mean	$\bar{y}_1$	$\bar{y}_2$	$\bar{y}_3$	$\bar{y}_4$	$\bar{y}_5$
Sample Variance	s_1^2	s_2^2	s_3^2	s_4^2	s_5^2

A Test of an Hypothesis Concerning More Than Two Population Means

If we are interested in testing the equality of the population means (that is, $\mu_1 = \mu_2 = \mu_3 = \mu_4 = \mu_5$), we might be tempted to perform all possible pairwise comparisons of population means. Hence if we assume that the five distributions are approximately normal, with a common variance σ^2, we could conduct the ten t tests comparing the five means, two at a time, as listed in table 13.4 (see section 9.4).

Table 13.4 All Possible Null Hypotheses for Comparing Two Means from Five Populations

$\mu_1 = \mu_2$	$\mu_1 = \mu_4$	$\mu_2 = \mu_3$	$\mu_2 = \mu_5$	$\mu_3 = \mu_5$
$\mu_1 = \mu_3$	$\mu_1 = \mu_5$	$\mu_2 = \mu_4$	$\mu_3 = \mu_4$	$\mu_4 = \mu_5$

One obvious disadvantage to this test procedure is that it is tedious and time-consuming. **But the more important and less apparent disadvantage of running multiple t tests to compare means is that the probability of falsely rejecting at least one of the hypotheses increases as the number of t tests increases.** Thus although we may have the probability of a type I error fixed at $\alpha = .05$ for each individual test, the probability of falsely rejecting H_0 on at least one of these tests is larger than .05. In other words, the combined probability of a type I error for the set of ten hypotheses would be larger than the value .05 set for each individual test. Indeed, it could be as large as .40.

What we need, then, is a single test of the hypothesis "all five population means are equal," which will be less tedious than the individual t tests and which can be performed with a specified probability of a type I error (say $\alpha = .05$). First we assume that the five sets of measurements are normally distributed, with means given by μ_1, μ_2, μ_3, μ_4, and μ_5 and with a common variance σ^2. Consider the quantity

$$s_w^2 = \frac{(n_1 - 1)s_1^2 + (n_2 - 1)s_2^2 + (n_3 - 1)s_3^2 + (n_4 - 1)s_4^2 + (n_5 - 1)s_5^2}{(n_1 - 1) + (n_2 - 1) + (n_3 - 1) + (n_4 - 1) + (n_5 - 1)}$$

$$= \frac{(n_1 - 1)s_1^2 + (n_2 - 1)s_2^2 + (n_3 - 1)s_3^2 + (n_4 - 1)s_4^2 + (n_5 - 1)s_5^2}{n_1 + n_2 + n_3 + n_4 + n_5 - 5}$$

Note that this quantity is merely an extension of

$$s^2 = \frac{(n_1 - 1)s_1^2 + (n_2 - 1)s_2^2}{n_1 + n_2 - 2}$$

which is used as the estimator of the common variance for two populations in a test of the hypothesis $\mu_1 = \mu_2$ (section 9.4). Thus s_w^2 represents a pooled estimate of the common variance σ^2 and measures the <u>variability</u> of the observations <u>within</u> the five <u>populations</u>. (The subscript w refers to the within-population variability.)

variability within populations

Next we consider a quantity that measures the variability between or among the population means. If the null hypothesis $\mu_1 = \mu_2 = \mu_3 = \mu_4 = \mu_5$ is true, the populations are identical, with mean μ and variance σ^2. Drawing single samples from the five populations is then equivalent to drawing five different samples from the same population. What kind of variation might be expected for these sample means? If the variation is too great, we would reject the hypothesis that $\mu_1 = \mu_2 = \mu_3 = \mu_4 = \mu_5$.

To assess the variation from sample mean to sample mean, we need to know the distribution of the mean of a sample of nine observations in repeated sampling. From the Central Limit Theorem (chapter 5) we know that the sampling distribution of $\bar{y}$ based on nine observations will have a mean of $\mu_{\bar{y}} = \mu$ and a variance of $\sigma_{\bar{y}}^2 = \sigma^2/9$. Since we have five different samples of nine observations each, we can estimate the variance $\sigma^2/9$ of the sampling distribution by computing the sample variance for the five sample means:

$$\text{sample variance (of the means)} = \frac{\Sigma \bar{y}^2 - (\Sigma \bar{y})^2/5}{5 - 1}$$

Note that we merely consider the $\bar{y}$'s as a sample of five observations and calculate the sample variance. This quantity is an estimate of $\sigma^2/9$; hence 9 times the sample variance of the means is an estimate of σ^2. We designate this estimate of σ^2 by s_B^2, where the subscript B denotes a measure of the <u>variability among</u> (between) the <u>sample means</u>.

variability among sample means

$$s_B^2 = 9 \times (\text{sample variance of the means})$$

Under the null hypothesis that all five population means are identical, we have two estimates of σ^2, namely s_w^2 and s_B^2. Suppose the ratio s_B^2/s_w^2 is used as a test statistic to test the hypothesis that $\mu_1 = \mu_2 = \mu_3 = \mu_4 = \mu_5$. What would the distribution of this quantity be if we were to repeat the experiment over and over again, each time calculating s_B^2 and s_w^2? As you might surmise from chapter 12, s_B^2/s_w^2 follows an F distribution, with degrees of freedom which can be shown to be $\nu_1 = 4$ for s_B^2 and $\nu_2 = 40$ for s_w^2. The proof of these remarks is beyond the scope of this text. However, we make use of the result for testing the null hypothesis $\mu_1 = \mu_2 = \mu_3 = \mu_4 = \mu_5$.

The test statistic used to test equality of population means is

$$F = \frac{s_B^2}{s_w^2}$$

When the null hypothesis is true, both s_B^2 and s_w^2 are estimates of σ^2, and F would be expected to assume a value near $F = 1$. When the hypothesis of equality is false, s_B^2 will tend to be larger than s_w^2 due to the differences among the population means. Hence we will reject the null hypothesis in the upper tail of the distribution of $F = s_B^2/s_w^2$. For α, the probability of a type I error, equal to .05 or .01, we can locate the rejection region for the one-tailed test by using table 4 or table 5 in the Appendix, with $\nu_1 = 4$ and $\nu_2 = 40$. Thus for

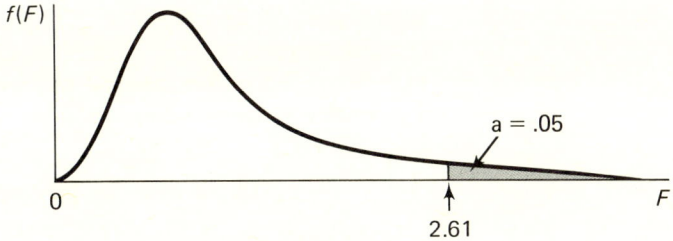

Figure 13.2 Critical Value of F for a $= .05$, $v_1 = 4$, $v_2 = 40$

$\alpha = .05$ the tabulated value of $F = s_B^2/s_w^2$ is 2.61 (see figure 13.2). If the calculated value of F falls in the rejection region, we conclude that at least one of the population means is different from the others.

This procedure can be generalized, with only slight modification in the formulas, to test the equality of k (where k is an integer equal to or greater than two) population means from normal populations, with a common variance σ^2. Random samples of size $n_1, n_2, \ldots, n_k$ are drawn from the respective populations. We then compute the sample means and variances. The null hypothesis $\mu_1 = \mu_2 = \cdots = \mu_k$ is tested against the alternative hypothesis that at least one of the population means is different from the others. The test procedure is given next.

An Analysis of Variance for Testing the Equality of k Population Means

Null hypothesis: $\mu_1 = \mu_2 = \cdots = \mu_k$.

Alternative hypothesis: At least one of the population means is different from the others.

Test statistic: $F = \dfrac{s_B^2}{s_w^2}$

where

$$s_w^2 = \frac{(n_1 - 1)s_1^2 + (n_2 - 1)s_2^2 + \cdots + (n_k - 1)s_k^2}{n_1 + n_2 + \cdots + n_k - k}$$

$$s_B^2 = \frac{\Sigma\, n_i \bar{y}_i^2 - (\Sigma\, n_i \bar{y}_i)^2/n}{k - 1}$$

and $n = n_1 + n_2 + \cdots + n_k$.

Analysis of Variance

Rejection region: Reject the null hypothesis if F is greater than the tabulated value for $a = \alpha$, $\nu_1 = (k - 1)$, and $\nu_2 = (n_1 + n_2 + \cdots + n_k - k)$. (See tables 4 and 5 in the Appendix for F_a values corresponding to $a = .05$ and $a = .01$, respectively.)

Note: When $n_1 = n_2 = \cdots = n_k$, the formula for s_B^2 simplifies to

$$s_B^2 = n' \left[\frac{\Sigma \bar{y}^2 - (\Sigma \bar{y})^2/k}{k - 1} \right]$$ where n' is common sample size

---

We illustrate these ideas in an example.

Example 13.1

A group of psychologists was interested in studying the effect of anxiety on learning as measured by student performance on a series of tests. On the basis of a prestudy test, 27 students were classified into one of three anxiety groups. Group I students were those who scored extremely low on a scale measuring anxiety. Those placed in group III were students who scored extremely high on the anxiety scale. The remaining students were placed in group II. The results of the prestudy anxiety test indicated that 6 students were in group I, 12 were in group II, and 9 were in group III.

Following the prestudy assignment of students to groups, the same battery of tests was given to each of the 27 students. The sample mean and variance of these test scores (based on a total of 100 points) have been summarized in table 13.5 for each group. Use the sample data to test the hypothesis that the average test score for low-, middle-, and high-anxiety students is identical (that is, that anxiety has no effect on a student's performance on this battery of tests).

Table 13.5 Summary of Test Scores

Group I (Low)	Group II (Medium)	Group III (High)
$n_1 = 6$	$n_2 = 12$	$n_3 = 9$
$\bar{y}_1 = 88$	$\bar{y}_2 = 82$	$\bar{y}_3 = 78$
$s_1^2 = 10.1$	$s_2^2 = 14.8$	$s_3^2 = 13.9$

Solution

We first hypothesize a population for each anxiety group that corresponds to all possible test scores for students who could have been included in the study. We will assume the measurements in each population are approximately normally distributed, with the mean of population I equal to μ_1, the mean of II equal to μ_2, and the mean of III equal to μ_3. In addition, we will assume that the populations have a common variance σ^2. From these populations, random samples of size $n_1 = 6$, $n_2 = 12$, and $n_3 = 9$ students were obtained and assigned to the respective groups.

To test the null hypothesis of equality of the population means, $\mu_1 = \mu_2 = \mu_3$, we first compute s_w^2 and s_B^2. We calculate s_w^2 directly from the sample data.

$$s_w^2 = \frac{(n_1 - 1)s_1^2 + (n_2 - 1)s_2^2 + (n_3 - 1)s_3^2}{n_1 + n_2 + n_3 - 3}$$

$$= \frac{5(10.1) + 11(14.8) + 8(13.9)}{6 + 12 + 9 - 3} = \frac{324.5}{24} = 13.52$$

However, before obtaining s_B^2, we must first compute $\Sigma n_i \bar{y}_i$ and $\Sigma n_i \bar{y}_i^2$. From table 13.5 we determine that

$$\Sigma n_i \bar{y}_i = 6(88) + 12(82) + 9(78) = 2{,}214$$
$$\Sigma n_i \bar{y}_i^2 = 6(88)^2 + 12(82)^2 + 9(78)^2 = 181{,}908$$

Hence for $k = 3$ and $n = n_1 + n_2 + n_3 = 27$, we have

$$s_B^2 = \frac{\Sigma n_i \bar{y}_i^2 - (\Sigma n_i \bar{y}_i)^2/n}{k - 1}$$

$$= \frac{181{,}908 - (2{,}214)^2/27}{2} = \frac{181{,}908 - 181{,}548}{2} = 180$$

The test statistic for the null hypothesis $\mu_1 = \mu_2 = \mu_3$ is

$$F = \frac{s_B^2}{s_w^2} = \frac{180.0}{13.52} = 13.31$$

Using the probability of a type I error, $\alpha = .01$, we can locate the upper-tail rejection region for this one-tailed test by using table 5 in the Appendix, with

$$\nu_1 = k - 1 = 2$$
$$\nu_2 = n_1 + n_2 + n_3 - 3 = 24$$

The tabulated value of F is 5.61. Since the observed value of F is greater than 5.61, we reject the hypothesis of equality of the population means (that is, we conclude that at least one of the means is different from the rest). Although the F test does not tell us which of the population means are different, the sample data suggest that anxiety has a detrimental effect on a student's performance in the battery of tests.

SUMMARY

The test procedure described in this chapter is called an analysis of variance because testing the null hypothesis of equality of the means relies on a test statistic that is composed of a measure of variability between populations, s_B^2, and a measure of variability within populations, s_W^2. The rejection region can be fixed so that the probability α of a type I error is some specified value (our tables allow α to be either .05 or .01). The conclusions that are drawn relate to all the means and not to individual ones. Thus if the test statistic falls in the rejection region, we conclude that at least one of the means is different from the others. Note that we cannot specify exactly which means are different from the others. Herein lies the difference between an analysis of variance and the use of individual t tests to compare means.

The analysis of variance is analogous to the use of a floodlight. Large objects (the differences among means) may be recognizable, while smaller objects (differences between individual means) may not. In contrast, the t test is more like a spotlight, which has a smaller field of vision but which provides a more intensified light. Thus we are better able to detect individual differences using the Student's t test.

We conclude from this discussion that we should perform an analysis of variance to indicate overall differences (say among the average phosphorus readings in the lake for the five months). If a significant value of the F statistic is obtained, we can run a few t tests for detecting individual differences. Thus our ecologist might eventually want to compare the mean phosphorus levels between the first and fifth months, third and fifth months, and so forth. However, he should limit himself to only a few comparisons. Otherwise he will increase the risk of detecting a difference that does not exist.

In this chapter we have presented an explanation and an example of an analysis of variance. We needed such a procedure because we were unable to test the equality of more than two population means by using a single test statistic. The analysis of variance is a very general test procedure and can be applied to solve many different problems. However, because of the complexity of the subject, we will go no further. Several useful references are provided at the end of the chapter for those interested in extending their knowledge of analysis of variance.

REFERENCES

Hicks, C. R. *Fundamental Concepts in the Design of Experiments*. 2d ed. New York: Holt, Rinehart and Winston, 1973.

Li, J. C. R. *Introduction to Statistical Inference*. Ann Arbor, Mich.: J. W. Edwards, 1961.

Mendenhall, W. *An Introduction to Linear Models and the Design and Analysis of Experiments*. Belmont, Calif.: Wadsworth, 1968.

Ott, L. *An Introduction to Statistical Methods and Data Analysis*. N. Scituate, Mass.: Duxbury Press, 1977.

SUPPLEMENTARY EXERCISES

13.1. Examine the logic behind an analysis of variance.

13.2. Elasticity readings for random samples of size 6 drawn from three different plastic processes (*A*, *B*, and *C*) are given in the accompanying table.

	Process A	Process B	Process C
	4.2	5.6	3.2
	1.1	5.1	2.5
	3.7	4.4	2.9
	2.6	4.2	3.6
	2.1	4.2	3.2
	3.7	5.1	4.1
Sample Mean	2.90	4.77	3.25
Sample Variance	1.39	.34	.31

Perform an analysis of variance for the experiment to determine if there are differences in the mean elasticity readings among the three plastic processes. Use $\alpha = .05$.

13.3. Refer to exercise 13.2. Use the sample data and the results of the analysis of variance to construct a 95% confidence interval for $(\mu_B - \mu_A)$. (Hint: Use s_w^2 for the pooled sample standard deviation.)

13.4. The length of life of an electronic component was to be studied under five different operating voltages, V_1, V_2, V_3, V_4, and V_5. Ten different components were randomly assigned to each of the five operating voltages. A summary of the resulting lengths of life is recorded for each of the five groups in the accompanying table. Perform an analysis of variance to test the hypothesis that $\mu_1 = \mu_2 = \mu_3 = \mu_4 = \mu_5$. Use $\alpha = .05$.

Voltage V_1	Voltage V_2	Voltage V_3	Voltage V_4	Voltage V_5
$n_1 = 10$	$n_2 = 10$	$n_3 = 10$	$n_4 = 10$	$n_5 = 10$
$\bar{y}_1 = 3.2$	$\bar{y}_2 = 3.8$	$\bar{y}_3 = 4.1$	$\bar{y}_4 = 4.0$	$\bar{y}_5 = 3.7$
$s_1^2 = .46$	$s_2^2 = .51$	$s_3^2 = .39$	$s_4^2 = .20$	$s_5^2 = .28$

330   Analysis of Variance

 13.5. Refer to exercise 13.4. Suppose that two components assigned to voltage V_1 and three assigned to voltage V_3 were found to be damaged prior to experimentation. Assuming that no new components can be found to replace the damaged ones, run an analysis of variance to test equality of the means, using the revised data in the accompanying table. Use $\alpha = .05$.

Voltage V_1	Voltage V_2	Voltage V_3	Voltage V_4	Voltage V_5
$n_1 = 8$	$n_2 = 10$	$n_3 = 7$	$n_4 = 10$	$n_5 = 10$
$s_1^2 = .59$	$s_2^2 = .51$	$s_3^2 = .45$	$s_4^2 = .20$	$s_5^2 = .28$

 13.6. A clinical psychologist wished to compare three methods for reducing hostility levels in university students. A certain psychological test (HLT) was used to measure the degree of hostility. High scores on this test indicate great hostility. Eleven students obtaining high, nearly equal scores were used in the experiment. Five were selected at random from among the 11 problem cases and treated by method A. Three were taken at random from the remaining 6 students and treated by method B. The other 3 students were treated by method C. All treatments continued throughout a semester. Each student was given the HLT test at the end of the semester, with the score results as shown in the accompanying table. Give the level of significance for a test of the null hypothesis "there is no difference in mean HLT test score for the three methods after treatment." (Hint: Use tables 4 and 5 of the Appendix to give an approximate level of significance, such as $p > .05$.)

	Method A	Method B	Method C
	80	70	63
	92	81	76
	87	74	70
	83		
	78		
Sample Mean	84.0	75.0	69.7

13.7. Refer to exercise 13.6. Use the sample data to construct a 99% confidence interval for $(\mu_A - \mu_B)$.

 13.8. Three different methods of instruction in speed-reading were to be compared with respect to the mean level of comprehension. A total of 13 students volunteered for the study; 4 were randomly assigned to instructional procedure 1, 4 to procedure 2, and 5 to procedure 3. After a one-week training period all students were asked to read an identical passage on a film, which was delivered at the rate of 300 words per minute. Students were then asked to answer questions on the film passage. Comprehension grades on these

Supplementary Exercises

questions are listed in the accompanying table. Perform an analysis of variance to test the hypothesis that all three instructional groups have the same average level of comprehension. Use $\alpha = .05$.

	Procedure 1	Procedure 2	Procedure 3
	82	71	91
	80	79	93
	81	78	84
	83	74	90
			88
Sample Mean	81.5	75.5	89.2

13.9. An experiment was conducted to compare the effectiveness of three mouthwashes (A, B, and C) in the treatment of morning halitosis. Although ads were run in the local newspapers at the study site, only 16 people responded and qualified for entrance into the study. Four of these 16 did not complete the study; consequently, only the results for the 12 who completed the study are shown in the computer printout that follows. Scores in the data represent results for individual participants on a pleasurable-nonpleasurable scale. (Higher scores imply greater pleasure.) Use the computer printout to respond to the following statements.
(a) Identify the three sample means and standard deviations.
(b) Compute the standard error of the mean for mouthwash B. Compare this value to the value in the output.
(c) State the results of an F test on the equality of the population mean scores.
(d) Based on the analysis of variance and the magnitudes of the sample means, which mouthwashes appear to be different?

```
ONE WAY ANOVA


MOUTHWASH A                        59.00000
                                   66.00000
                                   58.00000
                                   61.00000

NO. OBSERVATIONS          4
MEAN                      61.00000
STANDARD DEVIATION        3.55903
STANDARD ERROR            1.77951

MOUTHWASH B                        53.00000
                                   56.00000
                                   50.00000

NO. OBSERVATIONS          3
MEAN                      53.00000
```

STANDARD DEVIATION	3.00000	
STANDARD ERROR	1.73205	
MOUTHWASH C		57.00000
		61.00000
		72.00000
		66.00000
		68.00000
NO. OBSERVATIONS	5	
MEAN	64.80000	
STANDARD DEVIATION	5.89067	
STANDARD ERROR	2.63439	

ANOVA TABLE

SOURCE	SS	d.f.	MS	F	P
GROUP	262.11667	2	131.05833 ← s_B^2	6.05506	0.02157 ←p value
ERROR	194.80000	9	21.64444 ← s_w^2		
TOTAL	456.91667	11			

13.10. A horticulturist was investigating the phosphorus content of tree leaves from three different varieties of apple trees, *A, B,* and *C*. Random samples of five leaves from each of the three varieties were analyzed for phosphorus content. (See the table.) Use these data to test the hypothesis of equality of the mean phosphorus levels for the three varieties. Use $\alpha = .05$.

	Variety A	Variety B	Variety C
	.35	.65	.60
	.40	.70	.80
	.58	.90	.75
	.50	.84	.73
	.47	.79	.66
Sample Mean	.46	.78	.71
Sample Variance	.008	.010	.006

13.11. Refer to the data of exercise 13.10. Construct a 95% confidence interval for the difference in mean phosphorus content for varieties *A* and *C*.

13.12. An experiment was conducted to compare the effect of three different paints on the corrosion of pipes. A long pipe was cut into 12 segments, which were randomly assigned to one of the three paints so that each paint would be used on four segments. The segments were painted and allowed to weather for a period of six months. The accompanying corrosion readings were then obtained (table p. 333). Is there evidence to indicate a difference in the mean levels of corrosion for the three paints? Use an analysis of variance with $\alpha = .01$.

13.13. To compare the water-repellent properties of four different chemical coatings, the following tests were performed. Twelve different fabric samples were obtained from the same bolt of material, with three samples randomly

	Paint 1	Paint 2	Paint 3
	10.1	13.4	12.7
	11.4	12.9	11.9
	12.1	13.3	12.5
	10.8	13.1	12.3
Sample Mean	11.10	13.18	12.35
Sample Variance	.727	.049	.117

assigned to each of the four chemical groups *(A, B, C, D)*. Each of the samples was then treated with the assigned chemical coating. Following the chemical treatment, a fixed amount of water was applied to the fabric and the amount of moisture prevention was recorded. These data are recorded in the computer printout that follows. Use the information from the printout to respond to the following statements.

(a) Give the sample mean and standard deviation for the amount of moisture penetration for each chemical coating.

(b) Interpret the results of an analysis of variance to compare the mean moisture penetrations for the four chemicals.

```
ONE WAY ANOVA
GROUP A                         10.10000
                                12.20000
                                11.90000

NO. OBSERVATIONS                3
MEAN                            11.40000
STANDARD DEVIATION              1.13578
STANDARD ERROR                  0.65574

GROUP B                         11.40000
                                12.90000
                                12.70000

NO. OBSERVATIONS                3
MEAN                            12.33333
STANDARD DEVIATION              0.81445
STANDARD ERROR                  0.47022

GROUP C                         9.90000
                                12.30000
                                11.40000

NO. OBSERVATIONS                3
MEAN                            11.20000
STANDARD DEVIATION              1.21244
STANDARD ERROR                  0.70000

GROUP D                         12.10000
                                13.40000
                                12.90000

NO. OBSERVATIONS                3
```

MEAN 12.80000
STANDARD DEVIATION 0.65574
STANDARD ERROR 0.37859

ANOVA TABLE

SOURCE	SS	d.f.	MS	F	P
GROUP	5.20000	3	1.73333 ← s_B^2	1.79931	0.22517 ← p value
ERROR	7.70667	8	0.96333 ← s_w^2		
TOTAL	12.90667	11			

13.14. Use the data of exercise 13.13 to conduct an analysis of variance by using a computer program available to you. Compare your results to those in the output of exercise 13.13.

13.15. Because of the loss in efficiency caused by breakdowns in machinery, production records of each machine in a manufacturing plant must be closely monitored to try to anticipate when equipment is run-down and in need of repair. The data in the accompanying table give the production records for four different machines based on the outputs (in hundreds of pounds) from random samples of five shifts over the past week.

	Machine			
	1	2	3	4
	26.2	20.6	30.7	32.1
	32.0	26.4	35.2	34.7
	34.1	25.1	36.3	35.5
	33.6	24.9	31.9	36.8
	35.6	24.3	30.4	33.3
Sample Mean	32.30	24.26	32.90	34.48
Sample Variance	13.28	4.77	7.24	3.38

(a) Perform an analysis of variance to determine whether the mean outputs differ for the four machines. Use $\alpha = .05$.
(b) Can you recommend that a particular machine is less productive?

13.16. Refer to the data of exercise 13.15.
(a) Construct a 95% confidence interval for $(\mu_4 - \mu_2)$.
(b) Construct a 95% confidence interval for $(\mu_4 - \mu_1)$.

13.17. Sustained-release drug products are now marketed by many pharmaceutical companies. Even though single-dose, nonsustained-release capsules usually get more drug into the bloodstream quicker, sustained-release preparations supposedly achieve and maintain a more even level of release of the drug product over a longer period of time. Consider a study to compare three different drug products, A, B, and C. Equivalent doses of the three preparations were placed in mixtures of fluids similar to the gastric juices of the stomach and observed until 50% of the drug product was released from the capsule formulation. These release times (in minutes) appear in the accompanying table.

	Preparation		
	A	B	C
	15	38	19
	24	33	21
	20	39	27
	16	31	22
	18	26	24
	19	29	18
Sample Mean	18.67	32.67	21.83

(a) Which preparation would you suspect is the sustained-release formulation?

(b) Perform an analysis of variance to test the equality of the population mean release times for the three preparations. Use $\alpha = .05$.

13.18. Students of a class in environmental engineering were assigned the project of comparing the mean dissolved oxygen contents from samples drawn at four different locations of a lake. The four locations were the center of the lake, the north and south edges, and a spot midway between the center and the east side of the lake. At each location five different vial samples were drawn and analyzed for dissolved oxygen content. Use the accompanying data to determine if there is sufficient evidence to indicate a difference in the mean dissolved oxygen contents for the four locations. Use $\alpha = .05$.

	Location			
	1	2	3	4
	4.6	6.7	6.4	5.8
	4.8	6.2	6.3	5.3
	4.3	6.4	6.6	5.7
	4.9	6.5	6.7	5.2
	4.7	6.3	6.5	5.0
Sample Mean	4.66	6.42	6.50	5.40
Sample Variance	.05	.04	.03	.12

13.19. A computer printout for the data of exercise 13.18 is shown here. Compare the computer results to those you obtained in exercise 13.18.

ONE WAY ANOVA

LOCATION 1	
	4.60000
	4.80000
	4.30000
	4.90000
	4.70000
NO. OBSERVATIONS	5
MEAN	4.66000
STANDARD DEVIATION	0.23022
STANDARD ERROR	0.10296

LOCATION 2	
	6.70000
	6.20000
	6.40000
	6.50000
	6.30000
NO. OBSERVATIONS	5
MEAN	6.42000
STANDARD DEVIATION	0.19235
STANDARD ERROR	0.08602

LOCATION 3	
	6.40000
	6.30000
	6.60000
	6.70000
	6.50000
NO. OBSERVATIONS	5
MEAN	6.50000
STANDARD DEVIATION	0.15811
STANDARD ERROR	0.07071

LOCATION 4	
	5.80000
	5.30000
	5.70000
	5.20000
	5.00000
NO. OBSERVATIONS	5
MEAN	5.40000
STANDARD DEVIATION	0.33912
STANDARD ERROR	0.15166

ANOVA TABLE

SOURCE	SS	d.f.	MS		F	P	
GROUP	11.60950	3	3.86983	← s_B^2	67.30145	0.00000	← p value
ERROR	0.92000	16	0.05750	← s_w^2			
TOTAL	12.52950	19					

13.20. A survey was conducted to examine the change in the prices (over the last month) of items included in the typical market basket. Six grocery stores in each of four geographic locations were sampled. The data shown in the accompanying table correspond to the increases in the price of lettuce over the past month for the sampled stores. Use the data to perform an analysis of variance. Draw appropriate conclusions.

(a) Give the sample mean and standard deviation for lettuce price increases at each of the four locations.

(b) Identify all parts of a statistical test concerning the equality of the four population means.

(c) State conclusions, using the results of an analysis of variance.

(d) Which means appear to be different?

	Geographic Location		
1	2	3	4
10.1	15.3	11.8	16.8
11.3	14.8	12.6	9.2
8.2	10.4	14.2	17.5
8.7	9.3	13.9	18.2
12.1	10.7	8.9	10.9
10.4	15.6	7.5	14.5

13.21. Construct a 99% confidence interval for $(\mu_4 - \mu_1)$ for the data of exercise 13.20.

13.22. Use a computer program available to you to perform an analysis of variance for the data of exercise 13.17. Compare the computer results to those you obtained by hand in exercise 13.17.

EXPERIENCES WITH REAL DATA

Every consumer has been faced with the problem of assessing value when either buying or selling goods or services. Frequently, when a major purchase or sale is contemplated, one can enlist the services of a professional appraiser to help assess value. For example, in making arrangements to move furniture and belongings from an apartment or home to another location, one could enlist (purchase) the services of a moving company. A standard procedure is to solicit appraisals from two or three companies before reaching a decision. Similarly, in trying to sell an automobile to a used car dealer, each dealer visited would give an appraisal as to the worth of the car. For those who have had occasion to observe different appraisers, it is interesting to note the difference in the values assigned to the goods or the services. Some appraisers may tend to overvalue worth, others may tend to undervalue worth, and still others may tend to vary up and down with no apparent pattern.

Set up an experiment to compare the mean appraised values of three or more appraisers of a specific item. For example, to illustrate differences in assessing worth, you could choose three bicycle shops in town. With the help of other persons in your class, take the same five bicycles (at different times) to the bicycle shops for an assessment of worth. Try to enlist the services of the same appraiser from each shop for all bicycles. As another example, you may wish to work with three or more real estate agents on multiple listings to obtain appraisals on the same houses. For the experiment you choose, answer the following items.

1. Perform an analysis of variance to compare the mean appraised value. Use $\alpha = .05$.

2. Do any of the appraisers appear to overestimate (underestimate)?

3. Which appraisers appear to agree?

14 Contingency Tables

CHAPTER OUTLINE

14.1 Introduction
14.2 A Test for Determining Whether Two Methods of Classifying Observed Events Are Independent

14.1 INTRODUCTION

enumerative data

Many experiments, particularly in the social sciences, yield enumerative data (data that can be counted). For instance, the classification of people into five income brackets results in an enumeration or count corresponding to the number of persons classified in each of the five income brackets. Or we might be interested in studying the reaction of a mouse to a particular stimulus in a psychological experiment. If a mouse will react in one of three ways when a particular stimulus is applied and if a large number of mice are subjected to the stimulus, the experiment will yield a count for each category, indicating the number of mice that fall into it. Similarly, a traffic study might require a count and a classification of the type of motor vehicles using a particular section of highway. An industrial process manufactures items that fall into one of three quality classes: acceptables, seconds, and rejects. A student of the arts might classify paintings in one of k categories, according to style and period, in order to study trends in style over time. We might wish to classify ideas in a philosophical study or in the field of literature. An advertising campaign would yield classifications of consumer reaction. Indeed, many observations cannot be measured on a continuous scale and hence result in enumerative or classificatory data.

The preceding examples are only a few of the many types of problems that involve count, or enumerative, data. Many such problems are analyzed by means of a statistic developed in the early 1900s by Karl Pearson. We will illustrate the use of this method for an important group of problems involving the investigation of the dependence (or independence) between two methods of data classification. Other applications are discussed in the reference texts listed at the end of this chapter.

14.2 A TEST FOR DETERMINING WHETHER TWO METHODS OF CLASSIFYING OBSERVED EVENTS ARE INDEPENDENT

A problem frequently encountered in the analysis of count data concerns the interdependence of two methods of classification of observed events. For example, a physician in a clinic might wish to classify patients suffering from emphysema according to one of two drug products (either A or B) prescribed for their treatment and then according to the patient's response to the medication (greatly improved, improved, no change). It would be important for the physician to determine if the proportion of patients who show various degrees of improvement when treated by drug A is different from the proportion that show improvement when treated by drug B. Thus she must investigate a contingency (dependence) between the two methods of classification "drug type" and "patient condition."

We illustrate a test of the dependence of two methods of classification with the following example.

A total of 210 emphysema patients ($n = 210$) entering a clinic over a one-year period were treated with one of two drugs (either the standard drug A or an experimental drug B) for a period of one week. After this period of time each patient's condition was rated as either greatly improved, improved, or no change. The number of patients appearing in each category is presented in table 14.1. This two-way classification of the data is called a **contingency table**. The physicians in the clinic wish to determine whether a patient's rating is dependent upon the drug product that was used in his or her treatment. If the two classifications are dependent, this would imply that the proportions of patients classified as greatly improved, improved, or no change are different for the two drugs.

Table 14.1 Results of a One-Year Emphysema Study

| | Patient Condition | | | |
Drug Product	No Change	Improved	Greatly Improved	Total
A (standard)	20	35	45	100
B (experimental)	15	45	50	110
Total	35	80	95	210

The null hypothesis for our test is that the two methods of classification are independent—or, in other words, that the two drug products are equally effective (ineffective) for treating emphysema patients.

A test of the null hypothesis of independence of the drug and the condition

Testing Whether Two Methods Are Independent

observed and expected cell counts

classifications makes use of the <u>observed cell counts</u> and the <u>expected cell counts</u>. The expected cell count is the number of observations we would expect to be classified in a cell if the null hypothesis of independence of the two methods of classification is true. Without proof we state that the expected cell count for a particular cell is computed by multiplying the row total by the column total and dividing by the total number of sample measurements:

$$\text{expected cell count} = \frac{(\text{row total})(\text{column total})}{n}$$

These calculations are given below for the data of table 14.1.

$$\text{expected cell count for drug } A, \text{ no change} = \frac{(100)(35)}{210} = 16.67$$

$$\text{expected cell count for drug } A, \text{ improved} = \frac{(100)(80)}{210} = 38.10$$

$$\text{expected cell count for drug } A, \text{ greatly improved} = \frac{(100)(95)}{210} = 45.23$$

$$\text{expected cell count for drug } B, \text{ no change} = \frac{(110)(35)}{210} = 18.33$$

$$\text{expected cell count for drug } B, \text{ improved} = \frac{(110)(80)}{210} = 41.90$$

$$\text{expected cell count for drug } B, \text{ greatly improved} = \frac{(110)(95)}{210} = 49.77$$

The observed cell counts and the expected cell counts (the numbers in parentheses) are shown in table 14.2 for the data of table 14.1.

Table 14.2 Observed and Expected Cell Counts

	Patient Condition			
Drug	No Change	Improved	Greatly Improved	Total
A	20 (16.67)	35 (38.10)	45 (45.23)	100
B	15 (18.33)	45 (41.90)	50 (49.77)	110
Total	35	80	95	210

Note in table 14.2 that the expected cell counts in a given row may be summed to obtain the corresponding row total. For example, in the first row of table 14.2,

$$16.67 + 38.10 + 45.23 = 100$$

Similarly, for any column the corresponding expected cell counts may be

Contingency Tables

summed to obtain the appropriate column total. **This additive property of the expected cell counts can reduce the computational labor involved in obtaining expected cell counts.** Thus for table 14.2 we would really only need to compute the two entries shown below.

	Patient Condition			
Drug	No Change	Improved	Greatly Improved	Total
A	16.67	38.10		100
B				110
Total	35	80	95	210

The remaining expected cell counts can be obtained by subtracting from the appropriate row or column total.

We would suspect that the null hypothesis is false if the observed cell counts differ appreciably from the expected cell counts. To measure this agreement or disagreement, we compute a quantity known as the <u>chi-square statistic</u>, whose formula is

chi-square statistic

$$\chi^2 = \Sigma \left[\frac{(O - E)^2}{E} \right]$$

That is, we subtract the expected cell count E in each cell from the observed cell count O in that cell. The square of this difference is then divided by E. We do this for all cells and add our results.

If the observed cell counts differ appreciably from the expected cell counts, the quantities $(O - E)^2$ will be large, and hence χ^2 will be large. Thus we will reject our hypothesis of independence of the two methods of classification for large values of χ^2. How large is large? The answer to this question can be found by obtaining the sampling distribution of the quantity χ^2.

It can be shown that when n is large χ^2 has approximately a chi-square probability distribution. **For this approximation to be good, it is desirable for each cell in the contingency table to contain an expected cell count equal to or greater than 5.**

We will not give the formula for the chi-square probability distribution, but we will characterize it with the following properties.

Properties of the Chi-square Distribution

1. The chi-square distribution, like the F distribution, is not a symmetrical distribution (see figure 14.1).

Testing Whether Two Methods Are Independent

2. There are many chi-square probability distributions. We obtain a particular one by specifying the degrees of freedom associated with the chi-square distribution.

Large values of χ^2 are those that fall in the upper (right-hand) tail of the chi-square probability distribution shown in figure 14.1.

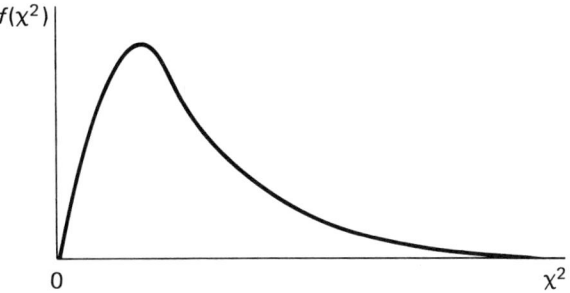

Figure 14.1 The Chi-square Probability Distribution; df = 4

The chi-square test statistic for testing the hypothesis of independence of two methods of classification, one with r categories, the other with c categories, is based on $(r - 1)(c - 1)$ degrees of freedom. Upper-tail values of the test statistic

$$\chi^2 = \sum \frac{(O - E)^2}{E}$$

are shown in table 3 in the Appendix. Entries in the table are chi-square values such that an area of size a lies to the right of χ_a^2 under the curve. We specify the degrees of freedom in the left-hand column of the table. The specified value of a is in the top row of the table. Thus for 14 degrees of freedom and a = .10, the tabulated value of the chi-square distribution is 21.0642 (see figure 14.2).

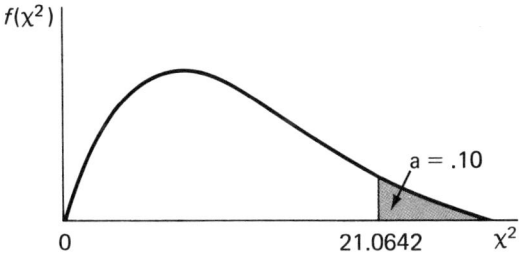

Figure 14.2 Tabulated Value of the Chi-square Distribution with df = 14 and a = .10

Contingency Tables

The rejection region for the one-tailed test of independence can then be determined for a specified probability of a type I error, α, using table 3. If the observed value of χ^2 falls beyond the tabulated value, we reject the null hypothesis of independence of the two classifications. We summarize this test procedure next.

Chi-square Test of Independence

Null hypothesis: The two classifications are independent.

Alternative hypothesis: The two classifications are dependent.

Test statistic: $\chi^2 = \sum \dfrac{(O-E)^2}{E}$.

Rejection region: Reject the null hypothesis if χ^2 exceeds the tabulated value of χ^2 for a = α and df = $(r-1)(c-1)$.

Note: This test will always be a one-tailed, upper-tailed test.

Example 14.1

Use the sample count data in table 14.2 to test the hypothesis of independence of the drug and the condition classifications of emphysema patients. In other words, we wish to test the null hypothesis that the two drug products are equally effective (ineffective) for treating persons suffering from emphysema.

Solution

The observed and expected cell counts are presented in table 14.2. The value of the test statistic χ^2 will be computed and compared with the tabulated value of the chi-square distribution possessing $(r-1)(c-1) = (1)(2) = 2$ degrees of freedom and for a = .05. To find the value of the test statistic, we use the formula

$$\chi^2 = \sum \dfrac{(O-E)^2}{E}$$

Substituting, we obtain

$$\chi^2 = \dfrac{(20-16.67)^2}{16.67} + \dfrac{(35-38.10)^2}{38.10} + \cdots + \dfrac{(50-49.77)^2}{49.77} = 1.75$$

Testing Whether Two Methods Are Independent

The rejection region for this test can be obtained from table 3 of the Appendix for df = 2 and a = .05. This value is 5.99. Since the observed value of χ^2 does not exceed 5.99, we conclude that there is insufficient evidence to reject the null hypothesis of independence. Thus it may not matter which drug is administered; the experimental and the standard drugs may be equally effective.

Example 14.2

A poll of 100 U.S. congressmen was taken to determine their opinions concerning a bill to raise the ceiling on our national debt. Each congressman was then classified according to political party affiliation and opinion on the policy. The survey results are listed in table 14.3. At this point, neither the opponents nor the proponents can claim victory. Test the null hypothesis that these two classifications are independent of one another (that is, congressmen do not hold opinions along party lines) against the alternative hypothesis that congressmen's opinions on the national debt bill are related to their political party affiliations. Use $\alpha = .10$.

Table 14.3 Congressional Opinion Survey Results

Party	Opinion			Total
	Approve of Bill	Do not Approve of Bill	No Opinion Yet	
Republican	28	14	5	47
Democrat	19	28	6	53
Total	47	42	11	100

Solution

Before calculating χ^2, we must first obtain the expected cell counts. Using the row and column totals of table 14.3, and the additivity property of the expected cell counts, we obtain the expected cell counts shown below.

$$\text{Republican, approve} = \frac{(47)(47)}{100} = 22.09$$

$$\text{Republican, disapprove} = \frac{(47)(42)}{100} = 19.74$$

Republican, no opinion = 47 − 22.09 − 19.74 = 5.17
Democrat, approve = 47 − 22.09 = 24.91
Democrat, disapprove = 42 − 19.74 = 22.26
Democrat, no opinion = 11 − 5.17 = 5.83

Contingency Tables

We now compute

$$\chi^2 = \sum \frac{(O - E)^2}{E}$$

and obtain

$$\chi^2 = \frac{(28 - 22.09)^2}{22.09} + \frac{(14 - 19.74)^2}{19.74} + \cdots + \frac{(6 - 5.83)^2}{5.83} = 6.14$$

The rejection region for this test can be located by using table 3, with a = .10 and df = $(r - 1)(c - 1) = (1)(2) = 2$. This tabulated value is 4.61. Since the observed value of χ^2 exceeds the tabulated chi-square value, we reject the null hypothesis of independence of the classifications and conclude that congressmen do seem to hold opinions along party lines. The probability of our making an incorrect decision is $\alpha = .10$ if, in fact, H_0 is true.

The use of the chi-square probability distribution in analyzing enumerative data presented in a two-classification contingency table illustrates the analysis for only one of the many types of classification problems. Many other types of applications are more complicated; hence they will be omitted from this text. However, this short presentation on contingency tables has provided you with an adequate tool for evaluating and making inferences concerning count data that are summarized into two classifications. Frequently, sociological studies are summarized in two-classification contingency tables for publication in magazines or newspapers. Using what you have learned from this section, you will be in a position to determine whether two methods of classifying observed events are independent.

SUMMARY

The material in this chapter has been concerned with a test of the null hypothesis that two methods of classification for enumerative data are independent. When n is large (large enough so that each expected cell count is five or more), χ^2 will be approximately distributed according to a chi-square probability distribution. Values of χ^2 that disagree with the hypothesis of independence are those associated with large deviations between the observed and expected cell counts. Hence the rejection region for the test is always located in the upper tail of the chi-square distribution.

The analysis of a contingency table is just one example of the use of the chi-square test statistic in the analysis of enumerative data. Further reading on this subject can be found in the References.

REFERENCES

Dixon, W. J., *Biomedical Computer Programs*. Berkeley: University of California Press, 1975.

──────, and Brown, M. B., eds. *Biomedical Computer Programs*. Rev. ed. Los Angeles: University of California Press, 1978.

Helwig, J. T. *SAS Supplementary Library User's Guide*. Raleigh, N.C.: SAS Institute, Inc., 1977.

Huntsberger, D. V., and Billingsley, P. *Elements of Statistical Inference*. 4th ed. Boston: Allyn and Bacon, 1977.

Nie, N.; Hull, C. H.; Jenkins, J. G.; Steinbrenner, K.; and Bent, D. H. *Statistical Package for the Social Sciences*. 2d ed. New York: McGraw-Hill, 1975.

Ott, L.; Mendenhall, W.; and Larson, R. *Statistics: A Tool for the Social Sciences*. 2d ed. N. Scituate, Mass.: Duxbury Press, 1978. Chapter 9.

Service, J. *A User's Guide to the Statistical Analysis System*. Raleigh, N.C.: Student Supply Stores, North Carolina State University, 1972.

Walpole, R. E. *Introduction to Statistics*. 2d ed. New York: Macmillan, 1974. Chapter 10.

SUPPLEMENTARY EXERCISES

14.1. State the four parts of a chi-square test of independence.

14.2. In a 2 × 3 contingency table (2 rows, 3 columns), what is the minimum number of expected values that need to be computed if the remaining ones are computed by subtraction? What is the minimum number for a 3 × 3 table?

14.3. A pharmaceutical firm was interested in testing the effectiveness of a new drug product in controlling worms in the small intestine of sheep. A prestudy test was used to select 40 sheep with approximately the same level of infection. These sheep were then randomly divided into two groups of 20. Those in the first group of sheep were given the drug product; those in the second group received no treatment. After a period of two weeks each of the 40 sheep was examined and classified as either "favorable" or "unfavorable," depending on the observed worm count. To be labeled favorable, the

Classification	Group I (Drug-treated)	Group II (Control)
favorable	15 ( )*	7
unfavorable	5	13

*Open parentheses indicate the expected counts that must be computed by using the formula (row total) times (column total) divided by n. The remaining expected counts can be computed by subtraction from the appropriate row or column total.

observed worm count had to be less than 100. The results of the study are presented in the accompanying table (p. 349).

(a) State the four parts of the chi-square test of independence for this situation.

(b) Compute the expected cell counts required for the test of independence.

14.4. Refer to exercise 14.3. Test the hypothesis of independence of the two methods of classification (that is, that the drug product was not effective in controlling worms). Use $\alpha = .10$.

14.5. A preelection survey was conducted in three different districts to compare the fraction of voters favoring the incumbent governor. Random samples of 50 registered voters were polled in each of the districts. The results are presented in the accompanying table. Do these data present sufficient evidence to indicate that the fractions favoring the incumbent governor differ in the three districts? Use $\alpha = .05$.

Opinion	District 1	District 2	District 3
favor incumbent governor	19 ( )	14 ( )	26
do not favor incumbent governor	31	36	24

14.6. A computer program was used to compute the chi-square test statistic for the data of exercise 14.5.

```
OBSERVED VALUES    TITLE: GOV. SURVEY
19.    14.    26.     59.
31.    36.    24.     91.
50.    50.    50.    150.

EXPECTED VALUES
19.7   19.7   19.7
30.3   30.3   30.3

OBS. MINUS EXP.
      -0.7   -5.7    6.3
       0.7    5.7   -6.3

CHI SQUARE BY CELL
0.0    1.6    2.0
0.0    1.1    1.3

CHI SQUARE =   6.091     2 DEGREE(S) OF FREEDOM
```

(a) Compare the computer output to the results you obtained in exercise 14.5.

(b) Give the level of significance for the test. (Hint: Use table 3 of the Appendix to give an approximate p-value.)

14.7. A sociological survey was conducted to compare, for different income categories, the fraction of families with more than two children. A random sample of 150 families was questioned and each one was classified into one of three income categories (under $7,000, $7,000 to $14,000, or over $14,000). The number of children was also recorded for each family. Using the accompanying data, determine if there is sufficient evidence to indicate that the fraction of families having more than two children differs among the three income categories. Use $\alpha = .05$.

Number of Children	Income Category		
	Under $7,000	$7,000–$14,000	Over $14,000
two or fewer	11 ( )	13 ( )	21
more than two	68	28	9

14.8. The marketing research group of a particular firm conducted a survey in three cities to compare the sales potential of a new soft drink. Each person contacted was asked to try the new drink and classify it as either excellent, satisfactory, or unsatisfactory. The results of the survey are summarized in the accompanying table. Use the computer output shown here to conduct a chi-square test of independence. Give the level of significance for your test and draw conclusions.

Classification	City 1	City 2	City 3
excellent	62	51	45
satisfactory	28	30	35
unsatisfactory	10	19	20

```
OBSERVED VALUES    TITLE: CITY SURVEY
  62.    51.    45.    158.
  28.    30.    35.     93.
  10.    19.    20.     49.
 100.   100.   100.    300.
EXPECTED VALUES
 52.7   52.7   52.7
 31.0   31.0   31.0
 16.3   16.3   16.3
OBS. MINUS EXP.
  9.3   -1.7   -7.7
 -3.0   -1.0    4.0
 -6.3    2.7    3.7
CHI SQUARE BY CELL
  1.7    0.1    1.1
  0.3    0.0    0.5
  2.5    0.4    0.8
CHI SQUARE =    7.376    4 DEGREE(S) OF FREEDOM
```

352 Contingency Tables

14.9. A university conducted a self-study to satisfy the requirements for accreditation. One aspect of the self-study concerned faculty evaluations. Through the use of student evaluations of their instructors, each faculty member was classified both by rank and by ability as a teacher. Use the accompanying results to test the null hypothesis of independence of two classifications. Restated, the null hypothesis is that a faculty member's teacher evaluation rating is not related to his or her rank. Use $\alpha = .05$.

Teaching Evaluation	Rank			
	Instructor	Assistant Professor	Associate Professor	Professor
above average	36 ( )	62 ( )	45 ( )	50
average	48 ( )	50 ( )	35 ( )	43
below average	30	13	20	35

14.10. A survey of student opinion concerning a proposed increase in the activities fee was taken to determine if student opinion was independent of sex. The results of 300 student interviews are recorded in the accompanying table. Test, using $\alpha = .05$.

Sex of Student	Opinion		
	Favor Increase	Oppose	Undecided
male	59 ( )	69 ( )	14
female	91	54	13

14.11. The governor of each state was polled to determine his or her opinion concerning a particular domestic policy issue. At the same time party affiliation was also recorded. The data are shown in the accompanying table. Assume that the 50 governors represent a random sample of political leaders throughout the country. Do the data present sufficient evidence to indicate a dependence between party affiliation and the opinion expressed on the domestic policy issue? Use $\alpha = .05$.

Party	Opinion		
	Approve of Policy	Do Not Approve	No Opinion
Republican	18 ( )	5 ( )	5
Democrat	8	8	6

14.12. Patients in a group of 30 suffering from vertigo (dizziness) were randomly assigned to one of two groups. Those in the first group were to be given an antivertigo product, while those in the second group were to receive an identically appearing placebo. Following a specified treatment period, each patient was asked to rate the effectiveness of his or her therapy as either effective, moderately effective, or ineffective in the treatment of vertigo. Use the sample data to test the research hypothesis that higher proportions of patients on the antivertigo compound will have different proportions of patients with ratings of effective or moderately effective as compared to the placebo patients. (Hint: Use chi-square with $\alpha = .05$.)

	Rating		
Treatment	Effective	Moderately Effective	Ineffective
antivertigo	10 ( )	3 ( )	2
placebo	2	6	7

14.13. Two comparable standard metropolitan statistical areas (SMSA) were chosen for trial advertising campaigns using two entirely different promotional plans for a new product. After a two-month trial period in each area, a random sample of 100 shoppers was obtained and each person was asked whether he or she was aware of the new product. These data are summarized in the accompanying table. Assume the two SMSAs are comparable. Is there sufficient evidence to indicate a difference in the proportions of individuals aware of the new product for the two promotional campaigns? Use $\alpha = .05$.

	Aware of Product		
Campaign	Yes	No	Total
1	70 ( )	30	100
2	62	38	100

14.14. Commuter train riders have sometimes been used to answer survey questions because they are a captive audience during the ride. In an experiment with a new schedule of express and local trains along a suburban commuter line, random samples of 50 riders were obtained from the morning rush hour crowd on two local and two express trains. Each rider was asked to classify his or her response as favorable, unfavorable, or undecided about the

	Response			
Train Run	Favorable	Unfavorable	Undecided	Total
local 1	30 ( )	15 ( )	5	50
local 2	32 ( )	16 ( )	2	50
express 1	23 ( )	24 ( )	3	50
express 2	25	21	4	50

new schedule. The data are shown in the accompanying table. Is there evidence to indicate a difference in the proportions of responses falling into the three categories for the four train runs? Use $\alpha = .05$.

14.15. Refer to exercise 14.14. Suppose that the two local trains were two trains along the same run, and, similarly, the two express trains were different trains along the same run. Combine the data to compare the local and express runs for the different categories of response. Use $\alpha = .05$. Does your conclusion differ from that in exercise 14.14?

14.16. A survey of admissions practices at a liberal arts college was conducted to determine whether there appeared to be a difference in the acceptance rates for white and minority (nonwhite) applicants. The results of this survey, which combines information from 4000 applicants, are shown in the accompanying table. Use the accompanying computer printout to conduct a chi-square test of independence. Use table 3 of the Appendix to obtain an approximate level of significance for the test and draw conclusions.

Applicant Accepted?	Applicant Nonwhite	White	Total
yes	38	126	164
no	362	3474	3836
Total	400	3600	4000

```
OBSERVED VALUES    TITLE: RATE SURVEY
   38.    126.    164.
  362.   3474.   3836.
  400.   3600.   4000.

EXPECTED VALUES
  16.4    147.6
 383.6   3452.4

OBS. MINUS EXP.
  21.6    -21.6
 -21.6     21.6

2 × 2
CHI SQUARE BY CELL
  28.4     3.2
   1.2     0.1

CHI SQUARE =  32.96     1 DEGREE(S) OF FREEDOM
```

14.17. A large number of motor vehicle accidents had been occurring along a 10-mile stretch of interstate highway that passes through a large metropolitan area. So the highway patrol safety committee conducted a study to classify these accidents by outcome (whether or not a fatality occurred) and by the single probable cause of the accident. The results of the study are shown in the accompanying table.

	Probable Cause of Accident			
Outcome	Speeding	Driving While Intoxicated	Reckless Driving	Other
fatality	42	61	20	12
no fatality	88	185	100	60

(a) Compute the percentages of accidents with fatalities for each of the cause categories.
(b) Conduct a chi-square test of independence, using $\alpha = .05$.
(c) Draw conclusions.

14.18. Give the level of significance for the test results in exercise 14.17. (Hint: Use table 3 of the Appendix.)

14.19. A carcinogenicity study was conducted to examine the tumor potential of a drug product scheduled for initial testing in humans. A total of 300 rats (150 males and 150 females) were studied for a six-month period. At the beginning of the study 100 rats (50 males, 50 females) were randomly assigned to the control group, 100 to the low-dose group, and the remaining 100 (50 males, 50 females) to the high-dose group. Each day of the six-month period the rats in the control group received an injection of an inert solution while those in the drug groups received an injection of the solution plus drug. The sample data are shown in the accompanying table.

Rat Group	Number of Tumors	
	One or More	None
control	10	90
low-dose	14	86
high-dose	19	81

(a) Give the percentages of rats with one or more tumors for each of the three groups.
(b) Conduct a chi-square test of independence with $\alpha = .05$.
(c) Does there appear to be a drug-related problem regarding tumors for this drug product? That is, as the dose is increased, does there appear to be an increase in the proportion of rats with tumors?

14.20. A computer output for the data of exercise 14.19 is shown here. Compare the output with your results in exercise 14.19.

```
OBSERVED VALUES   TITLE: TUMORS TOTAL
10.    90.    100.
14.    86.    100.
19.    81.    100.
43.   257.    300.
```

Contingency Tables

EXPECTED VALUES
14.3   85.7
14.3   85.7
14.3   85.7

OBS. MINUS EXP.
-4.3   4.3
-0.3   0.3
 4.7 -4.7

CHI SQUARE BY CELL
1.3   0.2
0.0   0.0
1.5   0.3

CHI SQUARE =   3.312   2 DEGREE(S) OF FREEDOM

14.21. Refer to the data of exercise 14.19. Since there were an equal number of male and female rats assigned to each of the treatment groups, it was also decided to examine the sample results by sex. The breakdown by sex is shown in the computer output that follows. Use the computer output to reach some overall conclusions concerning the tumor potential of the drug product.

OBSERVED VALUES   TITLE: TUMORS FEMALE
  4.   46.   50.
 10.   40.   50.
 14.   36.   50.
 28.   122.   150.

EXPECTED VALUES
9.3   40.7
9.3   40.7
9.3   40.7

OBS. MINUS EXP.
-5.3   5.3
 0.7   -0.7
 4.7   -4.7

CHI SQUARE BY CELL
3.0   0.7
0.0   0.0
2.3   0.5

CHI SQUARE =   6.674   2 DEGREE(S) OF FREEDOM

OBSERVED VALUES   TITLE: TUMORS MALES
  6.   44.   50.
  4.   46.   50.
  5.   45.   50.
 15.   135.   150.

EXPECTED VALUES
5.0   45.0
5.0   45.0
5.0   45.0

```
OBS. MINUS EXP.
 1.0   -1.0
-1.0    1.0
 0.0    0.0

CHI SQUARE BY CELL
 0.2    0.0
 0.2    0.0
 0.0    0.0

CHI SQUARE =   0.444      2 DEGREE(S) OF FREEDOM
```

 14.22. The Congress was involved in a heated debate recently as to whether the federal government should continue to make funds available to those who cannot afford to pay for an abortion. The floor vote in the House of Representatives was as shown in the accompanying table. Use the chi-square test of independence to determine whether the variable "voting category" is independent of the variable "political party affiliation." Use $\alpha = .05$.

	Voting Category		
Party	Yes	No	Total
Democrat	181	137	318
Republican	101	111	212
Total	282	248	530

EXPERIENCES WITH REAL DATA

Sociologists and educators periodically point to the changing attitudes of college students, not only within an educational institution over a period of time but also between entering classes. For example, students during the 1950s were labeled complacent and likely to accept the pronouncements of those in authority without question; students of the 1960s were supposed to be the questioning, ever-seeking idealists. (These statements are subject to question!) And some of our colleagues profess now to see a distinct difference between entering and senior classes in their attitude toward study. All these pronouncements are subject to debate, and that leads us to a class problem associated with real data.

A current social question concerns the interpretation that should be given to equal rights and opportunity for minority groups. For example, should equal rights and opportunity imply that each human, regardless of race, religion, or sex, be accorded an equal opportunity for job employment and/or entrance to graduate or professional schools? Or does it mean that society should redress the wrongs of the past and accord members of minority groups

Contingency Tables

a priority status in seeking jobs and/or educational opportunities? On this particular issue the attitudes of entering freshmen may very well differ from those of seniors, who will soon be entering the job market or competing for admission to professional schools.

Conduct a survey to determine whether there is a dependence between college class and the response to the question above. The 12 categories of the study will appear as follows:

	College Class			
Opinion	Freshman	Sophomore	Junior	Senior
favor equal opportunity for each person				
favor priority for minority groups				
no opinion, or neither of the above				

Use the student directory to randomly select 100 students from each college class. Contact each student by telephone to ascertain his or her response to the question. Analyze your data and determine whether the data provide sufficient evidence to indicate a dependence between college class and student attitude to the survey question.

Because seniors face the problems of admission to professional or graduate schools, competition for a limited number of good jobs, and imminent competition for advancement, their attitudes regarding the meaning of "equal opportunity" may differ from the others'. Collapse the 3 × 4 categorization of the data to a 3 × 2 categorization as follows:

Opinion	Seniors	Others
favor equal opportunity for each person		
favor priority for minority groups		
no opinion, or neither of the above		

Experiences with Real Data

Do the data provide sufficient evidence to indicate that the opinions of seniors differ from the opinions of others?

The calculations for this experiment can easily be accomplished on an electronic desk calculator, but you can also use packaged programs and a computer to perform the calculations. Useful packaged programs are available in the Biomed (Biomedical Programs), the SAS (Statistical Analysis Systems), and the SPSS (Statistical Package for the Social Sciences) program libraries (see the References).

15 Nonparametric Statistics

CHAPTER OUTLINE

15.1 Introduction
15.2 A Simple Comparative Test: The Sign Test
15.3 Wilcoxon's Signed-Rank Test
15.4 Wilcoxon's Rank-Sum Test
15.5 Spearman's Rank Correlation Coefficient

15.1 INTRODUCTION

ordinal data

Some studies yield data identified by rank only (ordinal data), either because of the crudeness of the measuring equipment employed or because of the inability of the investigator to further quantify the measurements. For example, in examining the effectiveness of an antihistamine in the treatment of ragweed allergy, we would be interested in measuring the change in symptomatology following treatment with the antihistamine. While it might be desirable to quantify the change in allergic symptoms, it is difficult to find objective measures of nose congestion, stuffiness, and so on. Thus we might be forced to measure improvement on an ordinal scale such as the following: worse, same, mild improvement, moderate improvement. marked improvement. Note that with this scale we cannot measure improvement in absolute terms, but rather we measure it in *relative* terms. For example, someone rated mild would have less improvement than someone rated moderate, but there is no way to measure the difference in improvement using this scale.

Many experiments might result in ordinal data. In the social sciences such variables as prestige, power, and alienation are measured by ordinal scales. In the behavioral sciences variables such as pain-threshold level, emotional stability, and drug reaction might be measured on an ordinal scale.

nonparametric statistical tests

When the variable of interest is measured on an ordinal scale, the test procedures discussed thus far are inappropriate, and we must resort to nonparametric statistical tests to provide a means for analyzing these data. The word "nonparametric" evolves from the type of hypothesis usually tested when dealing with ordinal-level data. Most nonparametric tests do not involve inferences about parameters from the original distribution of measurements. For

Nonparametric Statistics

example, instead of hypothesizing that two populations have the same mean (as in section 9.4), we could hypothesize that the two populations from which the samples were drawn are identical. Note that the practical implications of these two hypotheses are not the same. The first hypothesis is specific to a particular population parameter, while the second hypothesis addresses the question of equality (sameness) of the probability distributions. Two distributions of measurements could be different and still have the same mean.

In sections 15.2 through 15.4 we will discuss several nonparametric statistical tests for comparing two or more populations. Although they have been developed specifically for ordinal-level data, they are also appropriate for quantitative data when one or more of the assumptions underlying a particular parametric statistical test has been violated. For example, in conducting a t test comparing two population means, we make the assumptions that the two populations are normal and have a common variance σ^2. If either of these assumptions is violated, the usual t test for comparing independent samples is inappropriate.

This chapter provides useful alternatives to the parametric test procedures of previous chapters. The nonparametric tests presented here enable us to make statistical inferences even in experimental situations where the usual tests are invalid.

15.2 A SIMPLE COMPARATIVE TEST: THE SIGN TEST

The sign test is a procedure for testing whether two populations have identical probability distributions. Since we will make no assumptions concerning the parameters of the original distributions of measurements, we refer to the sign test as a nonparametric statistical test.

Why use the sign test to make a comparison between two populations? First, some studies yield responses that are hard to quantify. For example, it would be hard to quantify a ranking of the moods of individuals who for the first time have been placed on welfare, or to evaluate the performance of the faculty within a large history department. The sign test works particularly well for these types of data because we do not need to know the exact value of each measurement, only whether one is larger or smaller than the other. Second, it's easy to perform a sign test. Third, we need make no assumptions about the form of the population probability distributions. For example, we need not assume that the populations are normal or mound-shaped.

The sign test is based on the differences in pairs of observations, one observation from each sample. We let y_1 denote an observation from sample 1 and y_2 denote an observation from sample 2, and we note the difference between observations y_1 and y_2 for each pair, showing the sign of the difference. If y_1 is greater than y_2, we show a plus sign; for y_1 less than y_2 we show a minus sign. We omit all pairs for which $y_1 = y_2$. Table 15.1 lists data and plus and minus signs for ten pairs of observations.

A Simple Comparative Test: The Sign Test

Table 15.1 Ten Pairs of Observations

Pair	Sample 1, y_1	Sample 2, y_2	Sign of $(y_1 - y_2)$
1	10.2	10.3	−
2	10.1	10.0	+
3	10.3	10.2	+
4	10.4	10.2	+
5	10.3	10.0	+
6	10.2	10.1	+
7	10.2	10.0	+
8	10.5	10.3	+
9	10.1	10.2	−
10	10.4	10.3	+

The sign test of the null hypothesis "there is no difference in the probability distribution for the two populations" utilizes the number of plus signs, y, for the pairs of observations from the two samples. If the null hypothesis is true, then for each pair the probability that y_1 is greater than y_2 is $p = .5$. That is, there is a 50:50 chance that y_1 will be greater than y_2 or vice versa. Testing the null hypothesis that the distributions are identical is equivalent to testing the hypothesis that a binomial parameter p is equal to .5. This test was discussed in detail for large samples in section 8.4.

We illustrate the application of a binomial test of a proportion for the sign test in example 15.1.

Example 15.1

Each of 20 young mothers was asked to compare two different approaches, I and II, to socializing her young children. After two weeks of employing each approach, the mothers were asked to grade their satisfaction with each approach on an interval scale going from 0 to 5. The results are given in table 15.2. Determine if there is evidence to indicate that the two populations of ratings differ and hence whether approach I is preferred to approach II or vice versa. Use $\alpha = .05$.

Solution

Consider the pair of observations for each mother and let y be the number of times approach I has a higher score (satisfaction) than approach II. The null hypothesis is that the two approaches are equally preferred (and hence the distributions of scores are identical), or, equivalently, the probability that y_1 exceeds y_2 for any pair is $p = .5$. The alternative hypothesis is that the two

Nonparametric Statistics

Table 15.2 Measurements Made on a Random Sample of 20 Mothers

Mother	Approach 1, y_1	Approach 2, y_2	Sign of $(y_1 - y_2)$
1	3	2	+
2	4	2	+
3	3	5	−
4	5	4	+
5	4	3	+
6	3	2	+
7	3	4	−
8	4	3	+
9	3	2	+
10	4	2	+
11	5	4	+
12	3	4	−
13	2	1	+
14	3	2	+
15	5	3	+
16	5	4	+
17	5	3	+
18	2	3	−
19	4	2	+
20	4	3	+

distributions are different; that is, p is greater than or less than .5. Thus we have

$$H_0: p = .5 \quad \text{and} \quad H_a: p \neq .5$$

The test statistic is

$$z = \frac{\hat{p} - p_0}{\sqrt{p_0 q_0 / n}}$$

Substituting $\hat{p} = y/n$ and $p_0 = .5$, and rearranging terms, we can rewrite the formula for z as

$$z = \frac{y - .5n}{\sqrt{.25n}}$$

For $\alpha = .05$ we will reject the null hypothesis if z is greater than 1.96 or less than -1.96. Note that we are using a two-tailed test (as shown in figure 15.1) because we wish to detect values of p that are either greater or smaller than $p = .5$.

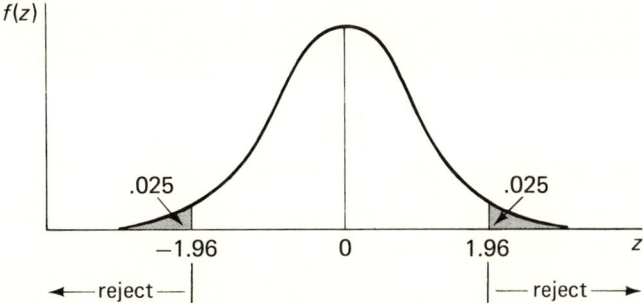

Figure 15.1 Rejection Region for the Sign Test, Example 15.1

From table 15.2 we see that the number of plus signs is $y = 16$, and n is 20. Thus the observed value of z is

$$z = \frac{y - .5n}{\sqrt{.25n}} = \frac{16 - 10}{2.24} = 2.7$$

Since this value is greater than 1.96, z falls in the rejection region. Hence we have evidence to indicate that the degree of satisfaction expressed by mothers differs for approaches I and II. Further examination of the data suggests that mothers prefer approach I.

One problem occasionally encountered when we use the sign test is that ties occur. That is, for one or more pairs $y_1 = y_2$. When this happens we omit the tied pair(s). For example, if you have 20 pairs and one pair results in a tie, you delete the tied pair and work with the remaining $n = 19$ pairs.

Example 15.2

Two psychiatrists were asked to rate (on a scale of 0–12) each of 12 prison inmates concerning their rehabilitative potential. Do the data shown in table 15.3 suggest a difference in the rating scales employed by the two psychiatrists? Use $\alpha = .05$. (Note a low reading on the scale indicates a low rehabilitative potential.)

Solution

The four parts of our statistical test are as follows:

Null hypothesis: $p = .5$ where p is the probability that psychiatrist 1 will rate an inmate higher than psychiatrist 2.

Nonparametric Statistics

Alternative
hypothesis: $p \neq .5$

Test statistic: $z = \dfrac{y - .5n}{\sqrt{.25n}}$

Rejection region: For $\alpha = .05$ reject H_0 if $|z| > 1.96$.

Substituting into the formula for z with $y = 7$ and $n = 11$ (note that there is one tie), we obtain

$$z = \frac{7 - 5.5}{\sqrt{.25(11)}} = .90$$

Since $z = .90$ does not fall in the rejection region, we have insufficient evidence to reject H_0: $p = .5$. That is, according to the sign test there is insufficient evidence to indicate that the population distributions of ratings are different for the two psychiatrists.

Table 15.3 Data for Example 15.2

Inmate	Psychiatrist 1, y_1	Psychiatrist 2, y_2	Sign of $(y_1 - y_2)$
1	6	5	+
2	12	11	+
3	3	4	−
4	9	10	−
5	5	2	+
6	8	6	+
7	1	2	−
8	9	12	−
9	6	5	+
10	7	4	+
11	6	6	tie
12	9	8	+

A summary of the elements of the sign test is presented next.

Sign Test for Comparing Two Populations

Null hypothesis: $p = .5$. (If p denotes the probability of observing a plus sign for a pair, then H_0: $p = .5$ implies that the two population distributions are identical.)

Alternative
hypothesis: For a one-tailed test:

1. $p > .5$.
2. $p < .5$.

For a two-tailed test:

3. $p \neq .5$.

Test statistic: $$z = \frac{y - .5n}{\sqrt{.25n}}$$

where n is the number of pairs, ignoring ties, and y is the number of pairs for which $y_1 > y_2$.

Rejection region: For $\alpha = .05$ (.01) and for a one-tailed test:

1. Reject H_0 if $z > 1.645$ (2.33).
2. Reject H_0 if $z < -1.645$ (-2.33).

For $\alpha = .05$ (.01) and for a two-tailed test:

3. Reject H_0 if $|z| > 1.96$ (2.58).

Note: This test is valid for $n \geq 10$.

To summarize, the advantages of the sign test are that it makes no assumptions concerning the nature of the original distributions of measurements, it can be applied to small samples ($n \geq 10$ pairs), and it is easy to apply.

Conversely, the disadvantage of the sign test is also clear. Since we use only the signs of the differences between y_1 and y_2, not the actual values of the measurements, we sacrifice some of the information available in the experiment. Consequently, a sign test based on small samples may not detect a difference between two populations that other tests, which utilize information about the magnitudes of y_1 and y_2, may detect. Although this is a serious disadvantage, it does not detract from the value of the sign test as a rapid method of detecting a difference between two populations.

The sign test is only one of a number of useful nonparametric statistical tests. Others are presented in Mendenhall (1979), Ott (1977), and Siegel (1956), which are listed in the References at the end of this chapter.

EXERCISES

 15.1. Fifty people were asked to rate each of two products, A and B. Use the sign test with $n = 50$ and $\alpha = .05$ to determine whether the distributions of ratings (for products A and B) are the same if $y = 29$ people rated product A higher than product B.

15.2. Refer to exercise 15.1. State the level of significance for your test.

15.3. Two judges were asked to rate each of 22 male inmates on his rehabilitative potential. The results are shown in the accompanying table. Use a sign test with $\alpha = .05$ to determine whether the distributions of scores are different for the two judges.

Inmate	Judge 1 Rating	Judge 2 Rating	Inmate	Judge 1 Rating	Judge 2 Rating
1	6	5	12	9	8
2	12	11	13	10	8
3	3	4	14	6	7
4	9	10	15	12	9
5	5	2	16	4	3
6	8	6	17	5	5
7	1	2	18	6	4
8	12	9	19	11	8
9	6	5	20	5	3
10	7	4	21	10	9
11	6	6	22	10	11

15.4. A test of manual dexterity was administered to a large kindergarten class. Twelve scores were randomly selected from the boys' scores and twelve from the girls'. The results are recorded in the accompanying table.

Boys	78	57	59	63	65	79	58	75	63	81	72	65
Girls	84	62	58	76	82	72	61	74	69	70	80	73

Use the sign test to decide if the two groups have different distributions of scores. Use $\alpha = .05$.

15.5. Typical diets of 12 families subsisting on a poverty-level income were rated for their nutritional value. The ratings used were poor, fair, good, and excellent. The mothers of the families then participated in a series of classes on nutrition and the preparation of balanced meals. A month later the diet of each family was rated again. The results are shown in the accompanying table.

Family	Before	After	Family	Before	After
1	poor	fair	7	fair	good
2	poor	fair	8	good	fair
3	fair	fair	9	poor	fair
4	good	excellent	10	poor	good
5	fair	good	11	poor	good
6	poor	good	12	good	excellent

Use the sign test with $\alpha = .05$ to decide whether the classes improved the quality of nutrition in the families' diets.

15.6. In a before-and-after study a social psychologist reported the accompanying data for 10 small groups. Use the sign test to determine whether scores changed after the instruction. Set up all parts of the statistical test, using $\alpha = .05$.

Group	Before Instruction	After Instruction	Group	Before Instruction	After Instruction
1	8	9	6	3	5
2	10	13	7	5	8
3	7	9	8	10	9
4	4	7	9	12	13
5	6	8	10	11	15

15.7. The effect of Benzedrine on the heart rate of dogs (in beats per minute) was to be examined in an experiment. Fourteen dogs were chosen for the study and each dog served as its own control. Half of the dogs were assigned to receive Benzedrine during the first study period while the other seven were given a placebo (saline solution). All dogs were examined to determine the heart rates after two hours on the medication. After two weeks in which no medication was given, the regimens for the dogs were switched for the second study period; that is, the dogs on Benzedrine were given the placebo while the others received Benzedrine. Again heart rates were measured two hours after administration of the medication. The accompanying sample data are not arranged in the order in which they were taken but are summarized by regimen. Use these data to test the alternative hypothesis $H_0: p > .5$ (i.e., that measurements for Benzedrine tend to be higher than those for the placebo). Use a one-tailed test with $\alpha = .05$.

Dog	Placebo	Benzedrine	Dog	Placebo	Benzedrine
1	250	258	8	296	305
2	271	285	9	301	319
3	243	245	10	298	308
4	252	250	11	310	320
5	266	268	12	286	293
6	272	278	13	306	305
7	293	280	14	309	313

15.3 WILCOXON'S SIGNED-RANK TEST

The Wilcoxon signed-rank test, which makes use of the sign and magnitude of the rank of the differences between pairs of measurements, provides a way to compare two populations when the variable of interest is measured on an ordinal scale. As with the sign test of the previous section, Wilcoxon's signed-

Nonparametric Statistics

rank test also provides an alternative to the paired t test. Utilizing the pairs of measurements with a nonzero difference, we rank the differences from lowest to highest, ignoring their signs. If two or more measurements have the same nonzero difference (ignoring sign), we assign each difference a rank equal to the average of the occupied ranks. The appropriate sign is then attached to the rank of each difference.

Before summarizing the Wilcoxon signed-rank test, we define the following notation:

n = number of pairs of observations with a nonzero difference
T_+ = sum of the positive ranks
T_- = sum of the negative ranks
T = the smaller of T_+ and T_-, ignoring their signs
$$\mu_T = \frac{n(n+1)}{4}$$
$$\sigma_T = \sqrt{\frac{n(n+1)(2n+1)}{24}}$$

The Wilcoxon signed-rank test is presented next.

Wilcoxon Signed-Rank Test ($n \leq 50$)

Null hypothesis: The distributions are identical.

Alternative hypothesis: The distributions are different (for a two-tailed test).

Test statistic: T, the smaller of T_+ and T_-, ignoring their signs.

Rejection region: For a specified value of α (either .05, .02, or .01) and sample size n, obtain the entry from table 8 of the Appendix. If T is less than or equal to the tabulated value, reject H_0.

Note: The Wilcoxon signed-rank test can be modified for a one-tailed test. Details of this modification are beyond the scope of this text but are presented in Ott (1977).

Example 15.3

Consider the following experimental situation. Each of 10 rats was weighed and then given a fixed diet supplemented with a dose of 10 milligrams per kilogram of a new drug product. After three months the rats were weighed individually again. Use the data in table 15.4 and the Wilcoxon signed-rank

Table 15.4 Data for Example 15.3

Rat	Predrug Weight	Postdrug Weight	Difference, (Postdrug − Predrug)
1	66.3	122.0	55.7
2	75.9	128.4	52.5
3	84.8	142.7	57.9
4	79.8	137.8	58.0
5	81.6	136.8	55.2
6	87.2	143.0	55.8
7	66.3	110.4	44.1
8	80.2	125.9	45.7
9	79.6	133.6	54.0
10	78.2	74.6	−3.6

test to decide whether there has been a change in weight for the three-month period. (Note: One reason for conducting such an experiment is to examine the drug product for potential toxicity. Toxicity of compounds is sometimes manifested through a weight loss over a three-month period.)

Solution

The null hypothesis is that the population distributions of weights for predrug and postdrug weights are identical. The alternative hypothesis is that these two distributions of weights are different. To determine the test statistic T, we must first rank the ten (postdrug − predrug) differences from lowest to highest. These are shown in table 15.5. The appropriate sign is then attached to each rank. The sums for the positive and for the negative ranks are then

$$T_- = -1$$
$$T_+ = 2 + 3 + \cdots + 10 = 54$$

and T, the smaller of T_+ and T_-, ignoring the sign, is 1.

Table 15.5 Ranks for the Data of Table 15.4

Rat	Rank of Difference	Rat	Rank of Difference
1	7	6	8
2	4	7	2
3	9	8	3
4	10	9	5
5	6	10	−1

Nonparametric Statistics

The rejection region for a two-tailed test with $n = 10$ and $\alpha = .05$ is found by using table 8 in the Appendix: we will reject H_0 if T is less than or equal to 8. Since the observed value of T is 1, we reject H_0 and conclude that the distributions of predrug and postdrug weights are different. Practically, since all but one rat had a postdrug weight higher than the predrug weight, we can say that there has been an increase in weight for rats over the three months.

EXERCISES

15.8. Two different brands of fertilizer (A and B) were compared on each of 10 different two-acre plots. Each plot was subdivided into one-acre subplots with brand A randomly assigned to one subplot and brand B to the other. Fertilizers were then applied to subplots at the rate of 60 pounds per acre. Barley yields (in bushels per acre) are listed by fertilizer and plot in the accompanying table. Use the Wilcoxon signed-rank test to test the hypothesis that the distributions of barley yields for the two brands of fertilizer are identical against the alternative that they are different. Use $\alpha = .05$.

Plot	Fertilizer A, y_1	Fertilizer B, y_2	Difference, $(y_1 - y_2)$
1	312	346	−34
2	333	372	−39
3	356	392	−36
4	316	351	−35
5	310	330	−20
6	352	364	−12
7	389	375	14
8	317	315	2
9	316	327	−11
10	346	378	−32

15.9. Refer to the data of example 15.1. Use Wilcoxon's signed-rank test to compare the two different approaches. Use $\alpha = .05$. Are your conclusions similar to those obtained using the sign test?

15.10. A single leaf was taken from each of 11 different tobacco plants. Each was divided in half: one half was chosen at random and treated with preparation I; the other half received preparation II. The object of the experiment was to compare the effects of the two preparations for the control of mosaic virus as measured by the number of lesions on the half leaves after a fixed period of time. The fewer the number of lesions, the better the preparation is. These data are recorded in the accompanying table. For $\alpha = .05$, use Wilcoxon's signed-rank test to examine the alternative hypothesis that the distributions of lesions are different for the two preparations.

Tobacco Plant	Number of Lesions on the Half Leaf	
	Preparation I	Preparation II
1	18	14
2	20	15
3	9	6
4	14	12
5	38	32
6	26	30
7	15	9
8	10	2
9	25	18
10	7	3
11	13	6

15.11. Refer to exercise 15.10. Conduct a sign test and compare your results to those of exercise 15.10. Use $\alpha = .05$.

15.12. Refer to exercise 15.7. Use Wilcoxon's signed-rank test to compare the distributions of heart rates for dogs on Benzedrine and on the placebo. Use a two-tailed test with $\alpha = .05$.

15.4 WILCOXON'S RANK-SUM TEST

The Wilcoxon rank-sum test (not to be confused with the Wilcoxon signed-rank test of section 15.3) provides a procedure for testing whether two populations are identical when *independent* random samples are selected from the two populations. Under the null hypothesis the two populations are assumed to be identical. Hence independent random samples from the respective populations should be similar. One way to measure the similarity between the samples is to jointly rank (from lowest to highest) the measurements from the combined samples and examine the sum of the ranks for measurements in sample 1 (or, equivalently, sample 2). Let T denote the sum of the ranks for sample 1. Intuitively, we would have evidence to reject the null hypothesis that the two populations are identical if T is extremely small (or large).

The procedure is as follows. Independent random samples are obtained from the two populations. The combined sample data from the two populations are then jointly ranked. If there are ties among any measurements in the combined sample data, we assign each tied measurement a rank equal to the average of the occupied ranks. Then when the sample sizes n_1 and n_2 are both larger than 10, the sum of the ranks T for sample 1 will be approximately normally distributed, with mean and variance given by

$$\mu_T = \frac{n_1(n_1 + n_2 + 1)}{2}$$

$$\sigma_T^2 = \frac{n_1 n_2 (n_1 + n_2 + 1)}{12}$$

The details of the test procedure are summarized next.

Wilcoxon's Rank-Sum Test ($n_1 > 10$ and $n_2 > 10$)

Null hypothesis: The two populations are identical.

Alternative hypothesis: For a one-tailed test:

1. The distribution of measurements for population 1 is above (to the right of) the distribution of measurements for population 2.
2. The distribution of measurements for population 1 is below (to the left of) the distribution of measurements for population 2.

For a two-tailed test:

3. The two populations are different.

Test statistic: $z = \dfrac{T - \mu_T}{\sigma_T}$

where T is the sum of the ranks in sample 1,

$\mu_T = n_1(n_1 + n_2 + 1)/2$

and

$$\sigma_T = \sqrt{\frac{n_1 n_2 (n_1 + n_2 + 1)}{12}}$$

Rejection region: For $\alpha = .05\,(.01)$ and for a one-tailed test:

1. Reject H_0 if $z > 1.645\,(2.33)$.
2. Reject H_0 if $z < -1.645\,(-2.33)$.

For $\alpha = .05\,(.01)$ and for a two-tailed test:

3. Reject H_0 if $|z| > 1.96\,(2.58)$.

Note: This is equivalent to the Mann-Whitney U test (Conover, 1971).

Wilcoxon's Rank-Sum Test

Example 15.4

Environmental engineers were interested in determining whether a cleanup project on a nearby lake was effective. Prior to initiation of the project, 12 samples of water had been obtained at random from the lake and analyzed for the amount of dissolved oxygen (in ppm). Due to diurnal fluctuations in the dissolved oxygen, all measurements were obtained at the 2 P.M. peak period. The data are presented in table 15.6; for convenience, the data are arranged in ascending order. Use $\alpha = .05$ to test the following hypotheses:

H_0: The distributions of measurements for before and six months after the cleanup project began are identical.

H_a: The distribution of dissolved oxygen measurements after the cleanup project is above (to the right of) the corresponding distribution of measurements before initiating the cleanup project. (It should be noted that a cleanup project has been effective in one sense if the dissolved oxygen increases over a period of time.)

Table 15.6 Dissolved Oxygen Measurements (ppm)

Before Cleanup	After Cleanup	Before Cleanup	After Cleanup
10.2	11.0	10.8	11.6
10.3	11.2	10.8	11.7
10.4	11.2	10.9	11.8
10.6	11.2	11.1	11.9
10.6	11.4	11.1	11.9
10.7	11.5	11.3	12.1

Solution

In order to compute the value of the test statistic z, we must first jointly rank the combined sample of 24 observations by assigning the rank of 1 to the smallest observation, the rank of 2 to the next smallest, and so on. When two or more measurements are the same, we assign all of them a rank equal to the average of the ranks they occupy. For example, there are two measurements equal to 10.6. Since these two measurements occupy the ranks 4 and 5, they are assigned a rank of 4.5, the average of the occupied ranks. The sample measurements and associated ranks (in parentheses) are listed in table 15.7. Summing the ranks, we find that the "before cleanup" ranks have the smaller sum, which is $T = 84$.

If we are trying to detect a shift to the right in the distribution for the measurements after the cleanup, we would expect the sum of the ranks for observations in sample 1 (before cleanup) to be small. Thus we will reject H_0 for $z < -1.645$. Substituting $n_1 = n_2 = 12$, we have

Nonparametric Statistics

$$\mu_T = \frac{12(25)}{2} = 150$$
$$\sigma_T^2 = 12(25) = 300$$
$$\sigma_T = \sqrt{300} = 17.32$$
$$z = \frac{84 - 150}{17.32} = -3.81$$

Since the computed value of z is less than -1.645, the table value for $\alpha = .05$ and a one-tailed test, we reject H_0 and conclude that the population distribution of measurements after cleanup is shifted to the right of the corresponding population distribution of measurements before cleanup.

Table 15.7 Dissolved Oxygen Measurements and Ranks

Before Cleanup	After Cleanup	Before Cleanup	After Cleanup
10.2 (1)	11.0 (10)	10.8 (7.5)	11.6 (19)
10.3 (2)	11.2 (14)	10.8 (7.5)	11.7 (20)
10.4 (3)	11.2 (14)	10.9 (9)	11.8 (21)
10.6 (4.5)	11.2 (14)	11.1 (11.5)	11.9 (22.5)
10.6 (4.5)	11.4 (17)	11.1 (11.5)	11.9 (22.5)
10.7 (6)	11.5 (18)	11.3 (16)	12.1 (24)

EXERCISES

15.13. Refer to example 15.3. Ten rats were weighed and then placed on a fixed diet supplemented with 10 milligrams per kilogram of a new drug product for a period of three months. As part of the same experiment, an additional 10 rats were placed in a control group. The same procedure was followed except that no drug product was mixed with the feed. The prestudy and poststudy

	Drug-treated Rats			Control Rats	
Prestudy Weight	Poststudy Weight	Difference, (Poststudy − Prestudy)	Prestudy Weight	Poststudy Weight	Difference (Poststudy − Prestudy)
66.3	122.0	55.7	65.0	127.6	62.6
75.9	128.4	52.5	68.3	124.3	56.0
84.8	142.7	57.9	77.4	133.8	56.4
79.8	137.8	58.0	76.5	140.1	63.6
81.6	136.8	55.2	82.0	153.9	71.9
87.2	143.0	55.8	79.4	145.6	66.2
66.3	110.4	44.1	69.8	139.3	69.5
80.2	125.9	45.7	70.1	130.4	60.3
79.6	133.6	54.0	66.7	132.7	66.0
78.2	74.6	−3.6	71.3	144.2	72.9

weights for the 10 drug-treated rats of example 15.3 and the 10 control rats are shown in the accompanying table. Use the Wilcoxon rank-sum test to compare the weight gains for the two groups. Use $\alpha = .05$. Draw some conclusions.

15.14. Refer to the data of exercise 15.13. Note the importance of using a control group in the experiment. Even though both groups of animals showed statistically significant weight gain [you can verify this with a sign test on the (poststudy − prestudy) differences in each group], the weight gain for the drug-treated rats was less than that for the corresponding control group. What conclusions could you draw if the control group was not included in the experiment? Can historical information on control animals substitute for the use of a concurrent control group?

15.15. Refer to exercise 15.13. Are the two treatment groups comparable with regard to the pretreatment weights? Use Wilcoxon's rank-sum test to compare the two groups. What problem might you encounter if the pretreatment weights were significantly different?

15.16. The accompanying data resulted from an experiment conducted to compare the heart rates (in beats per minute) for a group of 12 control animals and another group of 12 rats treated with an antihypertensive product. The response variable measured was the change in heart rate for each rat. Use the methods of section 9.4 to conduct a two-sample t test for comparing the mean change in heart rate for the two groups. Give the level of significance for your test.

Control	Treated	Control	Treated
9.0	59.0	−17.0	51.0
12.0	44.0	36.0	−9.0
36.0	63.0	42.0	75.5
77.5	87.5	65.0	28.0
−7.5	30.5	30.5	82.0
32.5	57.5	45.5	65.0

15.17. Wilcoxon's rank-sum test was performed for the data of exercise 15.16. Use the computer output shown here to compare the results with those you obtained in exercise 15.16. (Note: You have to compute z and draw a conclusion.)

WILCOXON RANK SUM TEST

CONTROL	TREATED
9.00000	59.00000
12.00000	44.00000
36.00000	63.00000
77.50000	87.50000
−7.50000	30.50000

	32.50000	57.50000
	−17.00000	51.00000
	36.00000	−9.00000
	42.00000	75.50000
	65.00000	28.00000
	30.50000	82.00000
	45.50000	65.00000
NO. OBSERVATIONS:	12	12
MEDIAN:	34.25000	58.25000
SUM OF RANKS:	118.0000	182.0000
AVERAGE RANK:	9.8333	15.1667

T = 118.0000

15.5 SPEARMAN'S RANK CORRELATION COEFFICIENT

The rank correlation coefficient provides a nonparametric procedure for measuring the strength of the relationship between two variables. It is used when we are working with the rankings of individual values for the two variables. In addition, it can be used to provide a general measure of the tendency for one variable to increase with another when the variables are not linearly related (and hence when the correlation coefficient of section 10.5 is inappropriate). To see how it is used, we will look at an example.

Each of ten judges was asked to rate an experimental mattress according to both firmness and comfort, with scores to be assigned in the range from 0 to 7.0. Higher scores indicate greater firmness or comfort. The results of the study are presented in the second and fourth columns of table 15.8.

Table 15.8 Results of the Experimental Mattress Study

Judge	Firmness	Rank, x	Comfort	Rank, y
1	2.5	2	5.0	7
2	3.0	4	4.7	6
3	5.0	8	3.0	1.5
4	4.0	7	4.2	4
5	3.5	6	4.5	5
6	2.0	1	3.0	1.5
7	3.3	5	5.9	9
8	5.2	9	5.5	8
9	2.8	3	3.2	3
10	5.8	10	6.1	10

The rank correlation coefficient is computed as follows. First, rank the measurements on each of the variables from smallest to largest. The score 2.0 is the smallest score for firmness; it receives rank 1. The measurement 2.5,

Spearman's Rank Correlation Coefficient

which is the next smallest, receives rank 2, and so on. These ranks are denoted by x and are listed in the third column of table 15.8. The smallest measurement for comfort is 3.0, but two judges gave this same rating. The procedure used for all tied scores is to give each of them a rank equal to the average of their occupied ranks. In this case the tied scores occupy ranks 1 and 2. Hence they both receive a rank of 1.5, the average of 1 and 2. Ranks assigned to comfort scores, denoted by y, are listed in the fifth column of table 15.8.

Spearman rank correlation coefficient

The Spearman rank correlation coefficient $\hat{\rho}_s$ for the sample data can be calculated in the same way as the linear correlation coefficient of section 10.5, with the exception that x and y now denote ranks rather than actual measurements. Thus

$$\hat{\rho}_s = \frac{S_{xy}}{\sqrt{S_{xx}S_{yy}}}$$

When no ties are present, the preceding complicated formula for $\hat{\rho}_s$ reduces to this simpler form:

$$\hat{\rho}_s = 1 - \frac{6(\Sigma d^2)}{n(n^2 - 1)}$$

where n is the number of pairs of observations and d represents the difference $(x - y)$ between a pair of ranks x and y. Even when ties occur, little error will result when using this formula if the number of ties is small in relation to the number of data points.

Example 15.5

Calculate the Spearman rank correlation coefficient $\hat{\rho}_s$ to measure the strength of the relationship between firmness and comfort for the sample data on experimental mattresses. (See table 15.8.)

Solution

It is convenient to construct a table of ranks, squares, and cross products to aid in our calculations; see table 15.9.

Using the data in table 15.9, we can determine the values of S_{xx}, S_{yy}, and S_{xy}, and hence we can obtain the value of $\hat{\rho}_s$ by using these values.

$$S_{xx} = \Sigma x^2 - \frac{(\Sigma x)^2}{n} = 385 - \frac{(55)^2}{10} = 82.5$$

$$S_{yy} = \Sigma y^2 - \frac{(\Sigma y)^2}{n} = 384.5 - \frac{(55)^2}{10} = 82.0$$

$$S_{xy} = \Sigma xy - \frac{(\Sigma x)(\Sigma y)}{n} = 335.5 - \frac{(55)(55)}{10} = 33.0$$

Nonparametric Statistics

Table 15.9 Computations for Example 15.5

Judge	Firmness	Rank, x	Comfort	Rank, y	x^2	y^2	xy
1	2.5	2	5.0	7	4	49	14
2	3.0	4	4.7	6	16	36	24
3	5.0	8	3.0	1.5	64	2.25	12
4	4.0	7	4.2	4	49	16	28
5	3.5	6	4.5	5	36	25	30
6	2.0	1	3.0	1.5	1	2.25	1.5
7	3.3	5	5.9	9	25	81	45
8	5.2	9	5.5	8	81	64	72
9	2.8	3	3.2	3	9	9	9
10	5.8	10	6.1	10	100	100	100
Sum		55		55	385	384.5	335.5

Thus for

$$\hat{\rho}_s = \frac{S_{xy}}{\sqrt{S_{xx}S_{yy}}}$$

we have

$$\hat{\rho}_s = \frac{33}{\sqrt{(82.5)(82)}} = .40$$

Other than observing the sign of $\hat{\rho}_s$ (which indicates a positive or a negative association) and the relative magnitude of $\hat{\rho}_s$, the only way for us to interpret $\hat{\rho}_s$ is to conduct a statistical test to see if the corresponding population rank correlation coefficient ρ_s is different from zero. But before introducing the statistical test, we will show the equivalence of the shortcut formula for $\hat{\rho}_s$ with an example.

Example 15.6

As indicated previously, the formula $\hat{\rho}_s = S_{xy}/\sqrt{S_{xx}S_{yy}}$ reduces to $\hat{\rho}_s = 1 - [6 \Sigma d^2/n(n^2 - 1)]$ when no ties occur. Even when a few ties occur, very little error will be introduced if we use the simpler formula. Compute $\hat{\rho}_s$ for the ranked data of example 15.5 by using the simpler computational formula.

Solution

The data shown in table 15.10 are useful for computing $\hat{\rho}_s$. Substituting into the formula, with $n = 10$, we have

Spearman's Rank Correlation Coefficient

$$\hat{\rho}_s = 1 - \frac{6(98.50)}{10(100-1)} = 1 - .60 = .40$$

Note that this result is identical to that obtained in example 15.5.

Table 15.10 Computations for Calculating $\hat{\rho}_s$

x	y	d = x − y	d²
2	7	−5	25
4	6	−2	4
8	1.5	6.5	42.25
7	4	3	9
6	5	1	1
1	1.5	−.5	.25
5	9	−4	16
9	8	1	1
3	3	0	0
10	10	0	0
Sum		0	98.50

If the rank correlation coefficient $\hat{\rho}_s$ is computed from ranks of two variables, we can test whether or not there is a relationship (positive or negative) between the two variables by using the following procedure.

Large-Sample Rank Correlation Test for ρ_s

Null hypothesis: $\rho_s = 0$ (i.e., no association between the ranks x and y).

Alternative hypothesis: For a one-tailed test:
1. $\rho_s > 0$.
2. $\rho_s < 0$.

For a two-tailed test:
3. $\rho_s \neq 0$.

Test statistic: $z = \hat{\rho}_s \sqrt{n-1}$.

Rejection region: For $\alpha = .05$ (.01) and for a one-tailed test:
1. Reject H_0 if $z > 1.645$ (2.33).
2. Reject H_0 if $z < -1.645$ (−2.33).

For $\alpha = .05$ (.01) and for a two-tailed test:

3. Reject H_0 if $|z| > 1.96$ (2.58).

Note: The number n of pairs of ranks must be 10 or more.

Example 15.7

Test the hypothesis that there is a positive association between pairs of comfort and firmness scores for the experimental mattress data in table 15.8. Use the rank order correlation coefficient.

Solution

The four parts of the statistical test are as follows:

Null hypothesis: $\rho_s = 0$.

Alternative hypothesis: $\rho_s > 0$.

Test statistic: $z = \hat{\rho}_s \sqrt{n - 1}$.

Rejection region: For $\alpha = .05$, reject H_0 if $z > 1.645$.

Recall that the computed value of $\hat{\rho}_s$ for this problem was found to be .40. The test statistic z is then

$$z = (.40)\sqrt{10 - 1} = (.40)(3) = 1.20$$

Since $z = 1.20$, z does not fall in the rejection region, and we have insufficient evidence to conclude that there is a positive association between the ranks of firmness and comfort. Consequently, we have insufficient information to suggest that the actual firmness and comfort scores are related.

In this section we presented a nonparametric measure of the association between two variables, which is useful for data that can be ranked. No assumption is made concerning the distribution of the populations. This test is also useful when the two variables are related but not in a linear sense. The test statistic for a test of the hypothesis "no association" involves the rank correlation coefficient, which is merely the correlation coefficient for the ranked observations. A test, based on the Student's t test, that is more exact for small samples is described in the text by Bradley (see the References at the end of the chapter).

SUMMARY

We have presented in this chapter a few of the many nonparametric statistical procedures that are available to users of statistics. It is important to be aware of these techniques as potential alternatives to standard parametric procedures when one or more of the underlying assumptions of those procedures is violated.

The sign test, perhaps the simplest of all nonparametric statistical tests, can be used to compare two populations based on paired data. The null hypothesis tested is that the two populations have identical probability distributions. The test is based on the number of pairs for which the measurement in sample 1 is greater than the corresponding measurement in sample 2.

The Wilcoxon signed-rank test provides an alternative to the two-sample paired t test of section 11.2. In contrast, the Wilcoxon rank-sum test is used as an alternative to the two-sample unpaired t test of section 9.4 when one or more of the underlying assumptions for the parametric test is violated. It, too, is easy to use and is based on the ranks of the measurements in the two samples.

Spearman's rank correlation coefficient $\hat{\rho}_s$ is an alternative to the correlation coefficient (of section 10.5) and can be used to measure the strength of the linear relation between two variables based on the ranks of individual values. A test of the null hypothesis $H_0: \rho_s = 0$ can also be performed.

REFERENCES

Bradley, J. V. *Distribution-Free Statistical Tests*. Englewood Cliffs, N.J.: Prentice-Hall, 1968.

Conover, W. J. *Practical Nonparametric Statistics*. New York: Wiley, 1971.

Hollander, M., and Wolfe, D. A. *Nonparametric Statistical Methods*. New York: Wiley, 1973.

Mendenhall, W. *Introduction to Probability and Statistics*. 5th ed. N. Scituate, Mass.: Duxbury Press, 1979.

Omstead, P. S. and Tukey, J. W. "A Corner Test for Association." *Annals of Mathematical Statistics* 18:495–513.

Ott, L. *An Introduction to Statistical Methods and Data Analysis*. N. Scituate, Mass.: Duxbury Press, 1977.

Siegel, S. *Nonparametric Statistics for the Behavioral Sciences*. New York: McGraw-Hill, 1956.

SUPPLEMENTARY EXERCISES

 15.18. Ten sets of identical twins were administered drug products, and the increase in their pulse rates was observed. One twin of each set received drug

A; the other received drug B. The purpose of the experiment was to determine if differences exist in the mean increases in pulse rate for the population of pulse rates for those administered drug A and the population corresponding to persons on drug B. The accompanying data were observed. Test the hypothesis "no difference in mean increase," using the sign test. (Note: We have no reason to believe one mean increase is greater than the other prior to conducting the test.) Use $\alpha = .05$.

Identical Twin Pair	Drug A	Drug B	Identical Twin Pair	Drug A	Drug B
1	12	19	6	12	15
2	14	13	7	13	10
3	8	6	8	15	18
4	11	24	9	18	21
5	14	12	10	17	22

15.19. A panel of 12 taste testers were asked to compare a domestic and a foreign wine (denoted by A and B, respectively). Judges were asked to rate each wine on a seven-point scale (1, 2, ..., 7), where 7 denotes the highest score. The results of the experiment are listed in the accompanying table. Use the sign test with $\alpha = .05$ to determine if the data indicate a difference in ratings for the domestic and the foreign wine.

Judge	Wine A	Wine B	Judge	Wine A	Wine B
1	7	4	7	4	4
2	5	4	8	6	4
3	4	3	9	5	3
4	7	6	10	4	5
5	6	7	11	7	3
6	5	2	12	6	4

15.20. Two judges rated each of 12 beauty pageant contestants on their poise, using a ten-point scale (1, 2, ..., 10). The results of the judging are listed in the accompanying table.

	Contestant											
	1	2	3	4	5	6	7	8	9	10	11	12
Judge 1	5	4	3	10	3	9	10	1	8	6	3	4
Judge 2	7	8	4	6	5	8	10	3	7	5	8	4

(a) Calculate the rank correlation coefficient to measure the strength of the linear relationship between the scores given by the two judges.

(b) Conduct a test of the hypothesis that there is no linear relationship between the two judges' scores. Use $\alpha = .05$.

15.21. Ten pairs of identical twins participated in an experiment to investigate two methods of teaching children to read music. One child from each pair was taught to read music by method A; the other received instruction according to method B. The results of an examination given at the end of a six-week training period are presented in the accompanying table.

	\multicolumn{10}{c}{Pair}
	1	2	3	4	5	6	7	8	9	10
Method A	75	80	67	73	93	88	70	95	84	92
Method B	73	76	65	70	95	82	65	85	83	95

(a) Replace the scores for the children taught by method A with ranks. Do the same for the scores from method B. Calculate the rank correlation coefficient to measure the strength of the linear relationship between methods A and B.

(b) Conduct a test of the alternative hypothesis $H_a: \rho_s \neq 0$. Use $\alpha = .05$.

15.22. Use a sign test on the original data of exercise 15.21 to determine if there is a difference in the levels of achievement for the two methods of teaching children to read music ($\alpha = .05$). Are your results the same as those found in exercise 15.21?

15.23. Two deans rated each of 15 fellowship applicants in terms of their expected academic success. The ratings, based on a scale of 1 to 5, are presented in the accompanying table. Do the data present sufficient evidence to indicate a difference in rating scales for the two deans? Base your conclusions on the sign test.

Applicant	Dean A	Dean B	Applicant	Dean A	Dean B
1	2	4	9	5	4
2	3	5	10	2	3
3	2	1	11	5	5
4	1	3	12	1	3
5	2	5	13	4	5
6	4	1	14	3	3
7	4	4	15	4	5
8	1	5			

15.24. Examine the accompanying computer output for the data of example 15.4.

(a) Locate T in the output.

(b) Set up all parts of a test of the null hypothesis that the two populations are identical.
(c) Draw conclusions, using $\alpha = .05$.

WILCOXON RANK SUM TEST

	X	Y
OBSERVATIONS:	Before Cleanup	After Cleanup
	10.20	11.00
	10.30	11.20
	10.40	11.20
	10.60	11.20
	10.60	11.40
	10.70	11.50
	10.80	11.60
	10.80	11.70
	10.90	11.80
	11.10	11.90
	11.10	11.90
	11.30	12.10
NO. OBSERVATIONS	12.00	12.00
MEDIANS	10.75	11.55
SUM OF RANKS	84.00	216.00

EXPERIENCES WITH REAL DATA

Conduct a class experiment involving some aspect of marketing research or consumer preference sampling. For example, it might be of interest to conduct a consumer survey to determine whether there is a preference for one of two marketed products in a given category (e.g., brands of potato chips, beers, etc.). Alternatively, the class experiment could be involved with an election or local bond issue. For the problem the class should do the following:

1. Develop a questionnaire (this could include a rating scale).

2. Develop a procedure for identifying persons to be included in the survey.

3. Collect the sample data.

4. Analyze the data, using one or more of the nonparametric methods of this chapter.

5. Draw conclusions and critique the procedure.

16 Lying with Statistics

CHAPTER OUTLINE

16.1 Lying in General
16.2 Graphical Distortions
16.3 Biased Samples (Loading the Dice)
16.4 The Honest Inference (What is the Sample Size?)
16.5 Distorting the Truth (and Being Honest)
16.6 Just Plain Lying

16.1 LYING IN GENERAL

Being statisticians, we find it difficult to talk about "lying" and to associate that ancient evil with our profession. However, true or false, the statistician is frequently regarded as a distorter of the truth, and, unfortunately, the idea is not new! The famous British statesman Disraeli is quoted as saying, "There are three kinds of lies: lies, damned lies, and statistics."* A book has been written on the subject of lying with statistics; furthermore, most statisticians can personally attest to receiving friendly barbs that are invariably a distortion of Disraeli's statement: "There are three kinds of liars: liars, damned liars, and statisticians." One person even had the audacity to state, "Statistics can be made to support almost anything—particularly statisticians!" The foregoing remarks may lead you to believe that our field is in the grips of a crisis in confidence. Do some, most, or all statisticians merit a measure of distrust and disbelief? The answer to this question is "Yes and no." We will explain.

First of all, note that lying—the purposeful distortion of truth—can occur only when we communicate. And since communication can be accomplished with graphs, pictures, sound, aroma, taste, words, numbers, or any other means devised to reach our senses, lying can be achieved by using any one or any combination of these methods of communication. We have all been misled by magazine or billboard advertisements about the quality of a product and by persuasive television or radio advertisements that quote statistics to show product A as superior "to all other brands." Why pick on numbers—that is,

*Darrell Huff, *How to Lie with Statistics* (New York: Norton, 1954).

statistics—as being the major method for distorting the truth? If a person is dishonest, he or she can lie in any number of ways, and we wager that more intentional lies are expressed by picture or by word than were ever perpetrated by statistics. To counter Disraeli, we believe that the public is much more concerned with the honesty and veracity of politicians than it is with the honesty and veracity of statisticians. Now if these arguments have successfully shifted you away from a blanket accusation of all statisticians to a position of objectivity, we are ready to answer the question, "Do statisticians merit a measure of distrust and disbelief?"

Some trained statisticians undoubtedly lie just like other dishonest humans, but we suspect that their lying intentionally with their statistics occurs very infrequently. Most consulting statisticians, like other professionals, are too concerned with their reputations to distort the truth purposely, although they do sometimes err. In this respect, misleading statements by trained statisticians are often errors of omission rather than errors of commission. A statistician may unintentionally fail to explain clearly the meaning of a numerical statement, or the statistician may omit some background information that is necessary for a clear interpretation of the results. A correct statement by the statistician may also appear to be distorted because the reader lacks knowledge of elementary statistics. Thus the statistician may present a very clear expression of an inference by using a 95% confidence interval, which is meaningless to a person who has not been exposed to the introductory concepts of statistics. And, on occasion, a statistician may make an outright error—he or she may employ the wrong technique or make a mistake in the computations. Unfortunately, trained statisticians are no less subject to error than physicians, engineers, administrators, or others.

But what can be said of *un*trained statisticians—people who do *not* have training equivalent to a professional degree in statistics? In all fairness, many people working in applied fields do an excellent job with the statistical analysis and reporting of their data. Unfortunately, some do not. These so-called statisticians have little training or experience to justify that professional classification. As a consequence, some statistical analyses are conducted with the wrong statistical methods and the results of many analyses are incorrectly interpreted.

The results of a statistical analysis are often distorted by the persons responsible for their publication. For example, when a statistician gives the results of statistical analyses to business managers, advertising departments, public relations people, and editorial departments, the accidental distortions or misinterpretation of statistical results will usually make a statistician turn prematurely gray.

Let us say no more. Our message to the public is a generalization of the age-old warning, "Let the buyer beware," since it applies to statistics that relate to all topics. If you want to avoid being "taken," you need to be knowledgeable and to be aware of possible pitfalls.

Graphical Distortions

We have come full circle, then, back to H. G. Wells's comment: "Statistical thinking will one day be as necessary for efficient citizenship as the ability to read and write." This course, which has covered some of the elementary concepts of statistical inference, should assist you in identifying some of the obvious ways of misrepresenting the truth with statistics and should help you in evaluating the worth of published statistical results.

16.2 GRAPHICAL DISTORTIONS

Pictures provide an excellent way to distort the truth. You have seen tobacco advertisements featuring beautiful people who are able to create an almost uncontrollable urge in us to dash out and buy a pack of cigarettes. Mail-order catalog sketches of products are frequently more attractive than the real thing, but we usually take this type of "lying" for granted. We submit to these mini-frauds with much less distress than we should. Statistical pictures are the histograms, frequency polygons, pie charts, and bar graphs of chapter 2. These drawings or displays of numerical results are difficult to combine with sketches of lovely women or handsome men and, hence, are secure from the most common form of graphic distortion. But other distortions are possible. One can shrink or stretch the axes to imply the desired results, based on the idea that shallow and steep slopes are associated with small and large increases, respectively.

For example, suppose that the values of a leading consumer price index over the first six months of the year were 160, 165, 178, 189, 196, and 210. If you are an economic advisor to the administration in power, you might show the upward movement of this consumer price index by using the frequency polygon of figure 16.1. In this graph the increase in the index is apparent, but it does not appear to be very great. On the other hand, as an economic advisor to the minority party, you might see these figures in a much different light and

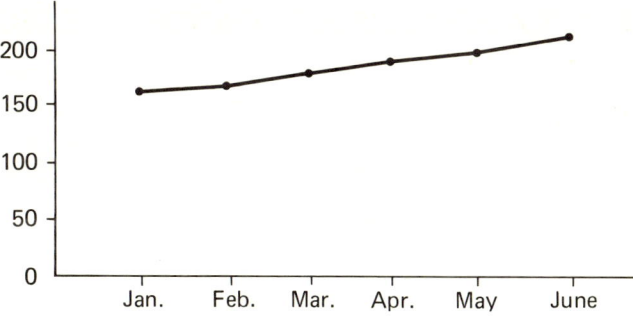

Figure 16.1 Changes in a Consumer Price Index

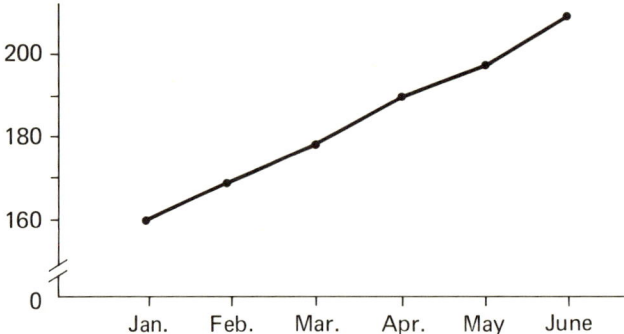

Figure 16.2 Changes in a Consumer Price Index

display the data as shown in figure 16.2. For this graph the vertical axis is stretched and does not include zero. Note the impression of a substantial rise that is indicated by the steeper slope.

Another way to achieve the same effect—to decrease or increase a slope—is to stretch or shrink the horizontal axis. Of course, you are sometimes limited in the amount of shrinking or stretching you can apply and still achieve a picture that appears reasonable to the viewer. For example, you could not greatly shrink or stretch the horizontal axes of figures 16.1 and 16.2 because of the limited number of data points ($n = 6$).

Shrinking or stretching axes to increase the slopes in bar graphs, histograms, frequency polygons, or other figures usually catches the hasty reader off guard; the distortions are apparent only if you look closely at the axes. **The important point, however, is that increases or decreases in responses are judged large or small depending on the arbitrary importance to the observer of the change, not on the slopes shown in graphic representations.**

16.3 BIASED SAMPLES (LOADING THE DICE)

One of the most common statistical distortions of the truth occurs because the experimenter, unwittingly or sometimes knowingly, samples the wrong population. That is he or she draws the sample from a set of measurements that is not the proper population of interest.

For example, suppose that you wish to assess the reaction of taxpayers to a proposed park and recreation center for children. A random sample of households is selected, and interviewers are sent to those households in the sample. Unfortunately, no one is at home in 40% of the sample households, so you randomly select and substitute other households in the city to make up the deficit. The resulting sample is selected from the *wrong* population and the sample is said to be *biased*. Why?

Biased Samples (Loading the Dice)

The specified population of interest in the household survey is the collection of opinions that would be obtained from the complete set of all households in the city. In contrast, the sample was drawn from a much smaller population or subset of this group—the set of opinions from householders who were at home when the sample was taken. It is possible that the fractions of householders favoring the park in these two populations are equal and no damage was done by confining the sampling to those at home. But it is much more likely that those at home had small children and that this group would yield a higher fraction in favor of the park than would the city as a whole. Thus we have a biased sample because it is loaded in favor of families with small children. Or perhaps a better way to see the difficulty is to note that the experimenter unwittingly selected the sample only from a special subset of the population of interest.

Biased samples frequently result from surveys that utilize mailed questionnaires. In a sense the investigator lets the selection and number of the sampling units depend upon the interests, available time, and various other personal characteristics of the individuals who receive the questionnaires. Extremely busy and energetic people may drop the questionnaires in the nearest wastebasket; you rarely hear from those low-energy folk who are uninterested or who are engrossed with other activities. Most often, the respondents are activists—those who are highly in favor, those who are opposed, or those who have something to gain from a certain outcome of the survey.

Although numerous well-known newscasters and analysts utilize election results as an expression of public opinion on major issues, it is a well-known fact that voting results represent a biased sample of public opinion. Those who vote represent much less than half of the eligible voters; they are individuals who desire to exercise their rights and responsibilities as citizens or are individuals who have been specially motivated to participate. The resultant subset of voters is not representative of the interests and opinions of all eligible voters in the country.

Sampling the wrong population also occurs when people attempt to extrapolate experimental results from one population to another. Numerous experimental results have been published about the effect of various products (e.g., saccharin) in inducing cancer in moles, rats, the breasts of beagles, and so forth; these results are often used to imply that humans have a high risk of developing cancer after frequent or extended exposure to the product. These inferences are not always justified, because the experimental results were not obtained on humans. It is quite possible that humans are capable of resisting much higher doses than rats, or perhaps humans may be completely resistant for some reason. Drug induction of cancer in small mammals *does* indicate a need for concern and caution by humans, but it does not prove that the drug is definitely harmful to humans. Note that we are not criticizing experimentation in various species of animals, because it is frequently the only way we can obtain any information about potential toxicity in human beings; researchers certainly cannot apply deadly doses of drugs to humans. We simply point out that the experimenter is knowingly sampling a population that is only similar (and quite likely not too similar) to the one of interest.

Engineers frequently test "rats" instead of "humans," "rats" in this context being miniature models or pilot plants of a new engineering system. Experiments on the models occasionally yield results that differ substantially from the results of the larger, real systems. So again we see a sampling from the wrong population, but it is the best that the engineer can do because of the economics of the situation. Funds are not usually available to test a number of full-scale models prior to production.

Many other examples could be given of biased samples or of sampling from the wrong populations. But, in particular, you should be inquisitive about how the sample was drawn and whether it was *randomly* selected from the population of interest. If this information is not given in the published results of a survey or experiment, you are advised to take the inferences with a grain of salt. Fragmentary reporting most frequently occurs in the news media, and you will occasionally find that unscrupulous opinion pollsters will purposely load the results. Have you ever noticed how the polls conducted by competing political candidates frequently disagree and almost always favor the man who pays the bill for the poll? So much for loading the dice!

16.4   THE HONEST INFERENCE (WHAT IS THE SAMPLE SIZE?)

Many who like to distort the truth with numbers are given to hiding or not revealing the sample size. Thus they possess a sample that contains insufficient information to prove or disprove the point at issue. If the sample *appears* to favor their point, it can still be used to good advantage by keeping the sample size a secret.

For example, suppose you read that a survey indicates that approximately 75% of a sample favor a new high-rise building complex. Further investigation might reveal that the investigator sampled only four people. When three out of the four favored his project, he decided to stop the survey. Of course, we exaggerate with this example; but we could also have revealed inconclusive results based on a sample of 25, even though many buyers would consider this sample size to be large enough. As you well know, *very large samples* are required to achieve adequate information in sampling binomial populations.

Now think back about some recent newspaper and newsmagazine reports of public opinion surveys. How many times did they include the sample size? It has only been in recent years that the major public opinion polls have published the survey sample size upon which their results are based.

Again, the moral of the story is to beware and not to let observed sample differences imply a difference in the corresponding populations. Whether a difference actually exists depends on the sample sizes. The value of published statistical results that omit sample sizes is difficult to ascertain. Consequently, the results cannot be taken as reliable.

16.5 DISTORTING THE TRUTH (AND BEING HONEST)

The most ethical and subtle way to distort the truth with numbers is to present data exactly "as it is" (of course, you only do this when it appears to favor your position). That is, tell the truth and leave the rest to the ignorance of the reader. Most readers have never been exposed to the elementary concepts of statistical inference and do not distinguish the difference between a *sample* and a *population*.

For example, suppose that an article reports 7 of 15 recovered from a serious illness when treated with drug 1 and 9 of 14 recovered when treated with drug 2. What would you as a reader conclude? Although the data appear to favor 2, we hope you recall that a difference in sample proportions does not imply a corresponding difference in the population parameters (chapters 5 and 8). The truth is that, considering the size of the samples ($n_1 = 15$, $n_2 = 14$), the difference between the two fractions of survival, $\hat{p}_1 - \hat{p}_2$, is too small to conclude that a difference exists between p_1 and p_2. (Recall that the standard deviation of the estimated difference between two binomial parameters is

$$\sigma_{\hat{p}_1 - \hat{p}_2} = \sqrt{\frac{p_1 q_1}{n_1} + \frac{p_2 q_2}{n_2}}$$

and that fairly large samples were required in chapter 8 to show a difference between binomial parameters.) Thus the presentation of data described above appears to indicate a difference in population fractions of survival when actually insufficient evidence exists to reach this conclusion. Purposely or accidentally, this type of presentation does mislead readers. Physicians (who receive a large number of advertisements on the efficacy of drugs) should be cautious in evaluating this type of "honest" information.

16.6 JUST PLAIN LYING

By "just plain lying" we mean the distortion of statistical results by the purposeful omission of background information on how the sample was collected, by the omission of important descriptive characteristics of the experimental conditions, or by a falsification of the statistical inference. There is no way that the public or the honest statistician can be protected against "just plain lying." Distortions of the type described above may be observed in some advertisements, and you should not be too surprised if they occur occasionally in published scientific articles. Unfortunately, the reading public seems to have complete faith in anything that appears in print. Accordingly, we conclude by stressing the importance of closely evaluating all published material and the importance of obtaining factual information concerning data upon which statistical inferences are based.

REFERENCE

Huff, D. *How to Lie with Statistics*. New York: Norton, 1954.

17 A Summary

The preceding 16 chapters construct a picture of statistics that is centered about the objective of statistics, statistical inference, which runs as a thread through the entire book, from the phrasing or describing of the inference, through the discussion of the probabilistic mechanism and the presentation of the reasoning involved in making the inference, to the first formal elementary discussion of the theory of statistical inference. What is statistics? What is its purpose? How does it accomplish its objective? If we have answered these questions to your satisfaction, if each chapter and section seems to fulfill a purpose and to complete a portion of the picture, we have in some measure accomplished our objective.

Chapter 1 used examples to present statistics as a scientific tool for making inferences, predictions, or decisions concerning a population of measurements based upon information contained in a sample. Thus statistics is part of the evolutionary process known as the scientific method—which, in essence, is the observation of nature—in order that we may form inferences or theories concerning the structure of nature and test these theories against repeated observations. Sampling and experimentation are used to obtain a quantity of information that, it is hoped, will be employed to provide the best inferences concerning the population from which the sample was drawn.

The ways in which we describe a set of measurements were presented in chapters 2 and 3 in terms of a frequency distribution and the associated numerical descriptive measures. In particular, we noted that the frequency distribution is subject to a probabilistic interpretation and that numerical descriptive measures are more suitable for inferential purposes because we can more easily associate a measure of goodness with them. Finally, a secondary but extremely important result of our study of numerical descriptive measures involved the notion of variation, its measurement in terms of a standard deviation, and its interpretation for mound-shaped distributions of data using the Empirical Rule. Thus, while concerned with describing a set of measurements—namely, the population—we provided the basis for a description of the sampling distributions of estimators and the two-standard-deviation bound on an error of estimation encountered in chapter 6. This basis also helped to locate regions of contradiction, called rejection regions, for the statistical tests of hypotheses in chapter 7.

Chapter 4 introduced some elementary concepts of probability that provide the foundation for making statistical inferences. In addition, continuous and discrete random variables were presented with an important example of each type: the binomial random variable (discrete) and the normal random variable (continuous). Most of the inferential statistics problems discussed in this text involve one of these two random variables.

In chapter 5 the notion of random sampling was presented as a lead-in to a discussion of the Central Limit Theorem, which provides a basis for describing the sampling distribution of sample means and sums. As a consequence, we use the Central Limit Theorem to justify the use of the Empirical Rule and the normal probability distribution as an approximation to the binomial probability distribution when the number of trials n is large. Through examples we attempted to reinforce the probabilistic concept of statistical inference introduced in preceding chapters and to encourage you, as a matter of intuition, to employ statistical reasoning in making inferences.

Chapters 6 and 7 discussed statistical inferences—estimation and statistical tests—and the methods of measuring the goodness of the inference. Because most of the point estimators and test statistics discussed possess normal distributions, the estimation and test procedures are quite similar for the problems in this text. The notions of statistical inference were extended to a discussion of comparisons in chapter 8 and of small-sample tests and estimation in chapter 9. Inferences concerning the linear relationship and correlation of two variables y and x were presented in chapter 10.

Chapter 11 presented an elementary discussion of experimental design—controlling the factors that affect the quantity of information in a sample. We saw how the sample size affects the quantity of information pertinent to a given parameter—information that is measured by the standard deviation of the parameter estimator. Similarly, we saw how making comparisons within relatively homogeneous blocks of experimental material reduces variation that tends to obscure differences in treatments.

With the exception of chapter 16, the remaining chapters of the text introduce topics that are optional for an introductory course. Chapter 12 considers the problem of comparing two variances, a topic that is a prerequisite to a discussion of the analysis of variance, chapter 13. The analysis of variance extends the comparison of means to more than two populations. Simultaneously, it introduces another way to view the analysis of data through the analysis (or partitioning) of variance. The analysis of contingency tables, chapter 14, yields a test of the independence of two directions of classification based on the cell counts for a two-way classification table. Thus, in some respects, the analysis of a contingency table is equivalent to a test of no correlation for the quantitative data of chapter 10. Chapter 15 pulls together a number of different nonparametric procedures that can be used as alternatives to the parametric tests of previous chapters.

Special problems of inference are traded for a lighter but more basic topic in chapter 16, "Lying with Statistics." This chapter points to some of the distortions that you may encounter in popular literature and is consequently an important topic for inclusion in any introductory course.

A Summary

Now that you have waded through part or all of the preceding 16 chapters, we might ask you, "What is statistics? How does it work? What are some simple but practical applications to problems that occur in real life? And, most importantly, of what value is statistics to you?" A synopsis of the text shows statistics to be a theory of information with inference as its goal. A sample is selected from a population, and an inference is made based on information contained in the sample. How good is the inference? Can we trust a particular numerical claim? Or is there doubt concerning its validity? This is your reason for studying statistics: You need to know. Requoting H. G. Wells, "Statistical thinking will one day be as necessary for efficient citizenship as the ability to read."*

*From *How to Lie with Statistics,* by Darrell Huff.

Appendix I
Useful Statistical Tests and Confidence Intervals

I. Inferences concerning the mean of a population.
   A. Sample size n is large ($n \geq 30$).
   1. Test:

   Null hypothesis: $\mu = \mu_0$ (μ_0 is specified).

   Alternative hypothesis: For a one-tailed test:
   1. $\mu > \mu_0$.
   2. $\mu < \mu_0$.

   For a two-tailed test:
   3. $\mu \neq \mu_0$.

   Test statistic: $z = \dfrac{\bar{y} - \mu_0}{\sigma_{\bar{y}}}$ where $\sigma_{\bar{y}} = \dfrac{\sigma}{\sqrt{n}}$

   Rejection region: For $\alpha = .05$ (or .01) and for a one-tailed test:
   1. Reject H_0 if $z > 1.645$ (or 2.33).
   2. Reject H_0 if $z < -1.645$ (or -2.33).

   For $\alpha = .05$ (or .01) and for a two-tailed test:
   3. Reject H_0 if $|z| > 1.96$ (or 2.58).

   Note: When $n \geq 30$ you may substitute s for σ in the formula for $\sigma_{\bar{y}}$. You must use the methods of chapter 9 when $n < 30$.

   2. Large-sample confidence interval:

   $$\bar{y} \pm z\sigma_{\bar{y}}$$

   where $\sigma_{\bar{y}} = \sigma/\sqrt{n}$ Note: The values of z for a 90%, a 95%, or a 99% confidence interval for μ are 1.645, 1.96, or 2.58, respectively. When $n \geq 30$, you may substitute s for σ in the formula for $\sigma_{\bar{y}}$. Refer to chapter 9 when $n < 30$.

Appendix I

B. Small samples, $n < 30$, and the observations are nearly normally distributed.

1. Test:

   Null hypothesis: $\mu = \mu_0$ (μ_0 is specified).

   Alternative hypothesis: For a one-tailed test:

   1. $\mu > \mu_0$.
   2. $\mu < \mu_0$.

   For a two-tailed test:

   3. $\mu \neq \mu_0$.

   Test statistic: $t = \dfrac{\bar{y} - \mu_0}{s/\sqrt{n}}$.

   Rejection region: For a specified value of α, df = $(n - 1)$, and for a one-tailed test:

   1. Reject H_0 if $t > t_a$, where a = α.
   2. Reject H_0 if $t < -t_a$, where a = α.

   For a specified value of α, df = $(n - 1)$, and for a two-tailed test:

   3. Reject H_0 if $|t| > t_a$, where a = $\alpha/2$.

2. Small-sample confidence interval:

$$\bar{y} \pm \frac{ts}{\sqrt{n}}$$

The value of t corresponding to a 90%, a 95%, or a 99% confidence interval is found in table 2 of the Appendix for df = $(n - 1)$ and a = .05, .025, or .005, respectively.

II. Inferences concerning the difference between the means of two populations.

A. Large samples.

1. Assumptions:

   (a) Population 1 has mean equal to μ_1 and variance equal to σ_1^2.

   (b) Population 2 has mean equal to μ_2 and variance equal to σ_2^2.

2. Some results:

   (a) The sampling distribution of $(\bar{y}_1 - \bar{y}_2)$ is approximately normal for large samples.

   (b) The mean of the sampling distribution, $\mu_{\bar{y}_1 - \bar{y}_2}$, is equal to the difference between the populations means, $(\mu_1 - \mu_2)$.

   (c) The standard deviation of the sampling distribution is

$$\sigma_{\bar{y}_1 - \bar{y}_2} = \sqrt{\frac{\sigma_1^2}{n_1} + \frac{\sigma_2^2}{n_2}}$$

3. Test:

   Null hypothesis: $\mu_1 - \mu_2 = 0$ (i.e., $\mu_1 = \mu_2$).

   Alternative hypothesis: For a one-tailed test:

   1. $\mu_1 - \mu_2 > 0$.
   2. $\mu_1 - \mu_2 < 0$.

   For a two-tailed test:

   3. $\mu_1 - \mu_2 \neq 0$.

   Test statistic: $z = \dfrac{\bar{y}_1 - \bar{y}_2}{\sigma_{\bar{y}_1 - \bar{y}_2}}$ where $\sigma_{\bar{y}_1 - \bar{y}_2} = \sqrt{\dfrac{\sigma_1^2}{n_1} + \dfrac{\sigma_2^2}{n_2}}$

   Rejection region: For $\alpha = .05$ (or .01) and for a one-tailed test:

   1. Reject H_0 if $z > 1.645$ (or 2.33).
   2. Reject H_0 if $z < -1.645$ (or -2.33).

   For $\alpha = .05$ (or .01) and for a two-tailed test:

   3. Reject H_0 if $|z| > 1.96$ (or 2.58).

   Note: For $n_1 \geq 30$ and $n_2 \geq 30$, s_1^2 and s_2^2 can be substituted for σ_1^2 and σ_2^2.

4. Large-sample estimation:

$$\bar{y}_1 - \bar{y}_2 \pm z\sigma_{\bar{y}_1 - \bar{y}_2}$$

where

$$\sigma_{\bar{y}_1 - \bar{y}_2} = \sqrt{\dfrac{\sigma_1^2}{n_1} + \dfrac{\sigma_2^2}{n_2}}$$

and $z = 1.645$, 1.96, or 2.58 for a 90%, a 95%, or a 99% confidence interval, respectively. Note: For $n_1 \geq 30$ and $n_2 \geq 30$, s_1^2 and s_2^2 can be substituted for σ_1^2 and σ_2^2.

B. Small samples.

1. Assumptions: Both populations are approximately normally distributed and $\sigma_1^2 = \sigma_2^2$.

2. Test:

   Null hypothesis: $\mu_1 - \mu_2 = 0$.

   Alternative hypothesis: For a one-tailed test:

   1. $\mu_1 - \mu_2 > 0$.
   2. $\mu_1 - \mu_2 < 0$.

   For a two-tailed test:

   3. $\mu_1 - \mu_2 \neq 0$.

Test statistic: $$t = \frac{\bar{y}_1 - \bar{y}_2}{s\sqrt{\dfrac{1}{n_1} + \dfrac{1}{n_2}}}$$

where

$$s = \sqrt{\frac{(n_1 - 1)s_1^2 + (n_2 - 1)s_2^2}{n_1 + n_2 - 2}}$$

Rejection region: For a specified value of α, for df = $(n_1 + n_2 - 2)$, and for a one-tailed test:

1. Reject H_0 if $t > t_a$, where a = α.
2. Reject H_0 if $t < -t_a$, where a = α.

For a specified value of α, for df = $(n_1 + n_2 - 2)$, and for a two-tailed test:

3. Reject H_0 if $|t| > t_a$, where a = $\alpha/2$.

3. Small-sample confidence interval:

$$\bar{y}_1 - \bar{y}_2 \pm ts\sqrt{\frac{1}{n_1} + \frac{1}{n_2}}$$

where

$$s = \sqrt{\frac{(n_1 - 1)s_1^2 + (n_2 - 1)s_2^2}{n_1 + n_2 - 2}}$$

The value of t corresponding to a 90%, a 95%, or a 99% confidence interval is found in table 2 of the Appendix for df = $(n_1 + n_2 - 2)$ and a = .05, .025, or .005, respectively.

III. Inferences concerning a population proportion p.

A. Assumptions for a binomial experiment:

1. Experiment consists of n identical trials, each resulting in one of two outcomes, say, success and failure.
2. The probability of success is equal to p and remains the same from trial to trial.
3. The trials are independent of each other.
4. The variable measured is y, the number of successes observed during the n trials.

B. Results:

1. The estimator of p is $\hat{p} = y/n$.
2. The mean of $\hat{p}$ is p.
3. The variance of $\hat{p}$ is pq/n.

C. Large-sample test:

Statistical Tests and Confidence Intervals

Null hypothesis: $p = p_0$ (p_0 is specified).

Alternative hypothesis: For a one-tailed test:

1. $p > p_0$.
2. $p < p_0$.

For a two-tailed test:

3. $p \neq p_0$.

Test statistic: $z = \dfrac{\hat{p} - p_0}{\sigma_{\hat{p}}}$ where $\sigma_{\hat{p}} = \sqrt{\dfrac{p_0 q_0}{n}}$

Rejection region: For $\alpha = .05$ (or .01) and for a one-tailed test:

1. Reject H_0 if $z > 1.645$ (or 2.33).
2. Reject H_0 if $z < -1.645$ (or -2.33).

For $\alpha = .05$ (or .01) and for a two-tailed test:

3. Reject H_0 if $|z| > 1.96$ (or 2.58).

Note: $q_0 = 1 - p_0$. This test is valid when np_0 and nq_0 are both 10 or more.

D. Large-sample confidence interval:

$$\hat{p} \pm z\sigma_{\hat{p}}$$

where

$$\sigma_{\hat{p}} = \sqrt{\dfrac{pq}{n}}$$

Note: The z values corresponding to a 90%, a 95%, or a 99% confidence interval are, respectively, 1.645, 1.96, or 2.58. If $n\hat{p}$ and $n\hat{q}$ are both 10 or greater, we may substitute $\hat{p}$ and $\hat{q}$ for p and q in the formula for $\sigma_{\hat{p}}$.

IV. Inferences comparing two population proportions p_1 and p_2.

A. Assumption: Independent random samples are drawn from each of two binomial populations.

	Pop. 1	Pop. 2
Probability of Success:	p_1	p_2
Sample Size:	n_1	n_2
Observed Successes:	y_1	y_2

B. Results:

1. The estimated difference between p_1 and p_2 is

$$\hat{p}_1 - \hat{p}_2 = \dfrac{y_1}{n_1} - \dfrac{y_2}{n_2}$$

2. The mean of $(\hat{p}_1 - \hat{p}_2)$ is $(p_1 - p_2)$.

Appendix I

3. The variance of $(\hat{p}_1 - \hat{p}_2)$ is
$$\frac{p_1 q_1}{n_1} + \frac{p_2 q_2}{n_2}$$

C. Large-sample test:

Null hypothesis: $p_1 - p_2 = 0$.

Alternative hypothesis: For a one-tailed test:
1. $p_1 - p_2 > 0$.
2. $p_1 - p_2 < 0$.

For a two-tailed test:
3. $p_1 - p_2 \neq 0$.

Test statistic: $z = \dfrac{\hat{p}_1 - \hat{p}_2}{\sigma_{\hat{p}_1 - \hat{p}_2}}$

where
$$\sigma_{\hat{p}_1 - \hat{p}_2} = \sqrt{pq\left(\frac{1}{n_1} + \frac{1}{n_2}\right)}$$

and p is approximated by
$$\hat{p} = \frac{y_1 + y_2}{n_1 + n_2} \quad \text{with} \quad \hat{q} = 1 - \hat{p}$$

Rejection region: For $\alpha = .05$ (or .01) and for a one-tailed test:
1. Reject H_0 if $z > 1.645$ (or 2.33).
2. Reject H_0 if $z < -1.645$ (or -2.33).

For $\alpha = .05$ (or .01) and for a two-tailed test:
3. Reject H_0 if $|z| > 1.96$ (or 2.58).

Note: $n\hat{p}$ and $n\hat{q}$ must be greater than or equal to 10 for both populations.

D. Large-sample confidence interval:
$$\hat{p}_1 - \hat{p}_2 \pm z\sigma_{\hat{p}_1 - \hat{p}_2}$$

where
$$\sigma_{\hat{p}_1 - \hat{p}_2} = \sqrt{\frac{p_1 q_1}{n_1} + \frac{p_2 q_2}{n_2}}$$

with $q_1 = (1 - p_1)$ and $q_2 = (1 - p_2)$. Substitute $z = 1.645$, 1.96, or 2.58 for a 90%, a 95%, or a 99% confidence interval, respectively. Note: Use $\hat{p}_1$ and $\hat{p}_2$ for the unknown parameters p_1 and p_2 in the formula for $\sigma_{\hat{p}_1 - \hat{p}_2}$. Very little error will result provided the sample sizes are large.

V. Tests for comparing the equality of two variances.

A. Assumptions:

1. Population 1 has a normal distribution, with mean μ_1 and variance σ_1^2.

Statistical Tests and Confidence Intervals

2. Population 2 has a normal distribution, with mean μ_2 and variance σ_2^2.

3. Two independent random samples are drawn, n_1 measurements from population 1, n_2 from population 2.

B. Test for comparing two population variances:

Null hypothesis: $\sigma_1^2 = \sigma_2^2$.

Alternative hypothesis: For a one-tailed test:

1. $\sigma_1^2 > \sigma_2^2$.

For a two-tailed test:

2. $\sigma_1^2 \neq \sigma_2^2$

Note: Always identify population 1 as the population that produced the larger sample variance.

Test statistic: $F = \dfrac{s_1^2}{s_2^2}$

where s_1^2 is the larger of the two sample variances.

Rejection region: For a given value of α, for $\nu_1 = (n_1 - 1)$, for $\nu_2 = (n_2 - 1)$, and for a one-tailed test:

1. Reject H_0 if F is greater than F_a, the table value for $a = \alpha$.

For a given value of α, for $\nu_1 = (n_1 - 1)$, for $\nu_2 = (n_2 - 1)$, and for a two-tailed test:

2. Reject H_0 if F is greater than F_a, the table value for $a = \alpha/2$.

Note: See tables 4 and 5 in the Appendix for table values corresponding to $a = .05$ and $.01$, respectively.

Appendix II
Statistical Tables

Table 1 Normal Curve Areas

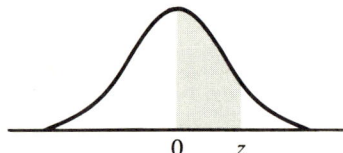

z	.00	.01	.02	.03	.04	.05	.06	.07	.08	.09
0.0	.0000	.0040	.0080	.0120	.0160	.0199	.0239	.0279	.0319	.0359
0.1	.0398	.0438	.0478	.0517	.0557	.0596	.0636	.0675	.0714	.0753
0.2	.0793	.0832	.0871	.0910	.0948	.0987	.1026	.1064	.1103	.1141
0.3	.1179	.1217	.1255	.1293	.1331	.1368	.1406	.1443	.1480	.1517
0.4	.1554	.1591	.1628	.1664	.1700	.1736	.1772	.1808	.1844	.1879
0.5	.1915	.1950	.1985	.2019	.2054	.2088	.2123	.2157	.2190	.2224
0.6	.2257	.2291	.2324	.2357	.2389	.2422	.2454	.2486	.2517	.2549
0.7	.2580	.2611	.2642	.2673	.2704	.2734	.2764	.2794	.2823	.2852
0.8	.2881	.2910	.2939	.2967	.2995	.3023	.3051	.3078	.3106	.3133
0.9	.3159	.3186	.3212	.3238	.3264	.3289	.3315	.3340	.3365	.3389
1.0	.3413	.3438	.3461	.3485	.3508	.3531	.3554	.3577	.3599	.3621
1.1	.3643	.3665	.3686	.3708	.3729	.3749	.3770	.3790	.3810	.3830
1.2	.3849	.3869	.3888	.3907	.3925	.3944	.3962	.3980	.3997	.4015
1.3	.4032	.4049	.4066	.4082	.4099	.4115	.4131	.4147	.4162	.4177
1.4	.4192	.4207	.4222	.4236	.4251	.4265	.4279	.4292	.4306	.4319
1.5	.4332	.4345	.4357	.4370	.4382	.4394	.4406	.4418	.4429	.4441
1.6	.4452	.4463	.4474	.4484	.4495	.4505	.4515	.4525	.4535	.4545
1.7	.4554	.4564	.4573	.4582	.4591	.4599	.4608	.4616	.4625	.4633
1.8	.4641	.4649	.4656	.4664	.4671	.4678	.4686	.4693	.4699	.4706
1.9	.4713	.4719	.4726	.4732	.4738	.4744	.4750	.4756	.4761	.4767
2.0	.4772	.4778	.4783	.4788	.4793	.4798	.4803	.4808	.4812	.4817
2.1	.4821	.4826	.4830	.4834	.4838	.4842	.4846	.4850	.4854	.4857
2.2	.4861	.4864	.4868	.4871	.4875	.4878	.4881	.4884	.4887	.4890
2.3	.4893	.4896	.4898	.4901	.4904	.4906	.4909	.4911	.4913	.4916
2.4	.4918	.4920	.4922	.4925	.4927	.4929	.4931	.4932	.4934	.4936
2.5	.4938	.4940	.4941	.4943	.4945	.4946	.4948	.4949	.4951	.4952
2.6	.4953	.4955	.4956	.4957	.4959	.4960	.4961	.4962	.4963	.4964
2.7	.4965	.4966	.4967	.4968	.4969	.4970	.4971	.4972	.4973	.4974
2.8	.4974	.4975	.4976	.4977	.4977	.4978	.4979	.4979	.4980	.4981
2.9	.4981	.4982	.4982	.4983	.4984	.4984	.4985	.4985	.4986	.4986
3.0	.4987	.4987	.4987	.4988	.4988	.4989	.4989	.4989	.4990	.4990

This table is abridged from Table I of *Statistical Tables and Formulas,* by A. Hald (New York: John Wiley & Sons, 1952). Reproduced by permission of A. Hald and the publishers, John Wiley & Sons.

Statistical Tables

Table 2 Percentage Points of the t Distribution

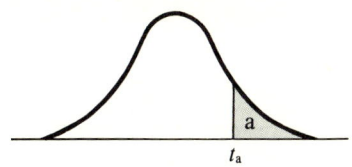

df	a = .10	a = .05	a = .025	a = .010	a = .005
1	3.078	6.314	12.706	31.821	63.657
2	1.886	2.920	4.303	6.965	9.925
3	1.638	2.353	3.182	4.541	5.841
4	1.533	2.132	2.776	3.747	4.604
5	1.476	2.015	2.571	3.365	4.032
6	1.440	1.943	2.447	3.143	3.707
7	1.415	1.895	2.365	2.998	3.499
8	1.397	1.860	2.306	2.896	3.355
9	1.383	1.833	2.262	2.821	3.250
10	1.372	1.812	2.228	2.764	3.169
11	1.363	1.796	2.201	2.718	3.106
12	1.356	1.782	2.179	2.681	3.055
13	1.350	1.771	2.160	2.650	3.012
14	1.345	1.761	2.145	2.624	2.977
15	1.341	1.753	2.131	2.602	2.947
16	1.337	1.746	2.120	2.583	2.921
17	1.333	1.740	2.110	2.567	2.898
18	1.330	1.734	2.101	2.552	2.878
19	1.328	1.729	2.093	2.539	2.861
20	1.325	1.725	2.086	2.528	2.845
21	1.323	1.721	2.080	2.518	2.831
22	1.321	1.717	2.074	2.508	2.819
23	1.319	1.714	2.069	2.500	2.807
24	1.318	1.711	2.064	2.492	2.797
25	1.316	1.708	2.060	2.485	2.787
26	1.315	1.706	2.056	2.479	2.779
27	1.314	1.703	2.052	2.473	2.771
28	1.313	1.701	2.048	2.467	2.763
29	1.311	1.699	2.045	2.462	2.756
inf.	1.282	1.645	1.960	2.326	2.576

From "Table of Percentage Points of the t-distribution." Computed by Maxine Merrington, *Biometrika*, Vol. 32 (1941), p. 300. Reproduced by permission of the *Biometrika* Trustees.

Table 3 Percentage Points of the Chi-square Distribution

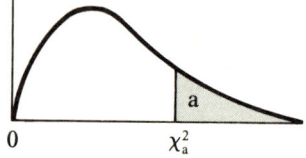

df	a = .995	a = .990	a = .975	a = .950	a = .900
1	0.0000393	0.0001571	0.0009821	0.0039321	0.0157908
2	0.0100251	0.0201007	0.0506356	0.102587	0.210720
3	0.0717212	0.114832	0.215795	0.351846	0.584375
4	0.206990	0.297110	0.484419	0.710721	1.063623
5	0.411740	0.554300	0.831211	1.145476	1.61031
6	0.675727	0.872085	1.237347	1.63539	2.20413
7	0.989265	1.239043	1.68987	2.16735	2.83311
8	1.344419	1.646482	2.17973	2.73264	3.48954
9	1.734926	2.087912	2.70039	3.32511	4.16816
10	2.15585	2.55821	3.24697	3.94030	4.86518
11	2.60321	3.05347	3.81575	4.57481	5.57779
12	3.07382	3.57056	4.40379	5.22603	6.30380
13	3.56503	4.10691	5.00874	5.89186	7.04150
14	4.07468	4.66043	5.62872	6.57063	7.78953
15	4.60094	5.22935	6.26214	7.26094	8.54675
16	5.14224	5.81221	6.90766	7.96164	9.31223
17	5.69724	6.40776	7.56418	8.67176	10.0852
18	6.26481	7.01491	8.23075	9.39046	10.8649
19	6.84398	7.63273	8.90655	10.1170	11.6509
20	7.43386	8.26040	9.59083	10.8508	12.4426
21	8.03366	8.89720	10.28293	11.5913	13.2396
22	8.64272	9.54249	10.9823	12.3380	14.0415
23	9.26042	10.19567	11.6885	13.0905	14.8479
24	9.88623	10.8564	12.4011	13.8484	15.6587
25	10.5197	11.5240	13.1197	14.6114	16.4734
26	11.1603	12.1981	13.8439	15.3791	17.2919
27	11.8076	12.8786	14.5733	16.1513	18.1138
28	12.4613	13.5648	15.3079	16.9279	18.9392
29	13.1211	14.2565	16.0471	17.7083	19.7677
30	13.7867	14.9535	16.7908	18.4926	20.5992
40	20.7065	22.1643	24.4331	26.5093	29.0505
50	27.9907	29.7067	32.3574	34.7642	37.6886
60	35.5346	37.4848	40.4817	43.1879	46.4589
70	43.2752	45.4418	48.7576	51.7393	55.3290
80	51.1720	53.5400	57.1532	60.3915	64.2778
90	59.1963	61.7541	65.6466	69.1260	73.2912
100	67.3276	70.0648	74.2219	77.9295	82.3581

From "Tables of the Percentage Points of the χ^2-Distribution." *Biometrika*, Vol. 32 (1941), pp. 188–189, by Catherine M. Thompson. Reproduced by permission of the *Biometrika* Trustees.

Table 3 (Continued)

a = .10	a = .05	a = .025	a = .010	a = .005	df
2.70554	3.84146	5.02389	6.63490	7.87944	1
4.60517	5.99147	7.37776	9.21034	10.5966	2
6.25139	7.81473	9.34840	11.3449	12.8381	3
7.77944	9.48773	11.1433	13.2767	14.8602	4
9.23635	11.0705	12.8325	15.0863	16.7496	5
10.6446	12.5916	14.4494	16.8119	18.5476	6
12.0170	14.0671	16.0128	18.4753	20.2777	7
13.3616	15.5073	17.5346	20.0902	21.9550	8
14.6837	16.9190	19.0228	21.6660	23.5893	9
15.9871	18.3070	20.4831	23.2093	25.1882	10
17.2750	19.6751	21.9200	24.7250	26.7569	11
18.5494	21.0261	23.3367	26.2170	28.2995	12
19.8119	22.3621	24.7356	27.6883	29.8194	13
21.0642	23.6848	26.1190	29.1413	31.3193	14
22.3072	24.9958	27.4884	30.5779	32.8013	15
23.5418	26.2962	28.8454	31.9999	34.2672	16
24.7690	27.5871	30.1910	33.4087	35.7185	17
25.9894	28.8693	31.5264	34.8053	37.1564	18
27.2036	30.1435	32.8523	36.1908	38.5822	19
28.4120	31.4104	34.1696	37.5662	39.9968	20
29.6151	32.6705	35.4789	38.9321	41.4010	21
30.8133	33.9244	36.7807	40.2894	42.7956	22
32.0069	35.1725	38.0757	41.6384	44.1813	23
33.1963	36.4151	39.3641	42.9798	45.5585	24
34.3816	37.6525	40.6465	44.3141	46.9278	25
35.5631	38.8852	41.9232	45.6417	48.2899	26
36.7412	40.1133	43.1944	46.9630	49.6449	27
37.9159	41.3372	44.4607	48.2782	50.9933	28
39.0875	42.5569	45.7222	49.5879	52.3356	29
40.2560	43.7729	46.9792	50.8922	53.6720	30
51.8050	55.7585	59.3417	63.6907	66.7659	40
63.1671	67.5048	71.4202	76.1539	79.4900	50
74.3970	79.0819	83.2976	88.3794	91.9517	60
85.5271	90.5312	95.0231	100.425	104.215	70
96.5782	101.879	106.629	112.329	116.321	80
107.565	113.145	118.136	124.116	128.299	90
118.498	124.342	129.561	135.807	140.169	100

Table 4 Percentage Points of the F Distribution

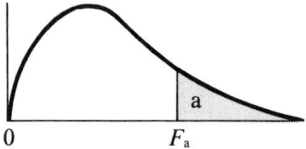

Degrees of Freedom (a = .05)

v_1 \ v_2	1	2	3	4	5	6	7	8	9
1	161.4	199.5	215.7	224.6	230.2	234.0	236.8	238.9	240.5
2	18.51	19.00	19.16	19.25	19.30	19.33	19.35	19.37	19.38
3	10.13	9.55	9.28	9.12	9.01	8.94	8.89	8.85	8.81
4	7.71	6.94	6.59	6.39	6.26	6.16	6.09	6.04	6.00
5	6.61	5.79	5.41	5.19	5.05	4.95	4.88	4.82	4.77
6	5.99	5.14	4.76	4.53	4.39	4.28	4.21	4.15	4.10
7	5.59	4.74	4.35	4.12	3.97	3.87	3.79	3.73	3.68
8	5.32	4.46	4.07	3.84	3.69	3.58	3.50	3.44	3.39
9	5.12	4.26	3.86	3.63	3.48	3.37	3.29	3.23	3.18
10	4.96	4.10	3.71	3.48	3.33	3.22	3.14	3.07	3.02
11	4.84	3.98	3.59	3.36	3.20	3.09	3.01	2.95	2.90
12	4.75	3.89	3.49	3.26	3.11	3.00	2.91	2.85	2.80
13	4.67	3.81	3.41	3.18	3.03	2.92	2.83	2.77	2.71
14	4.60	3.74	3.34	3.11	2.96	2.85	2.76	2.70	2.65
15	4.54	3.68	3.29	3.06	2.90	2.79	2.71	2.64	2.59
16	4.49	3.63	3.24	3.01	2.85	2.74	2.66	2.59	2.54
17	4.45	3.59	3.20	2.96	2.81	2.70	2.61	2.55	2.49
18	4.41	3.55	3.16	2.93	2.77	2.66	2.58	2.51	2.46
19	4.38	3.52	3.13	2.90	2.74	2.63	2.54	2.48	2.42
20	4.35	3.49	3.10	2.87	2.71	2.60	2.51	2.45	2.39
21	4.32	3.47	3.07	2.84	2.68	2.57	2.49	2.42	2.37
22	4.30	3.44	3.05	2.82	2.66	2.55	2.46	2.40	2.34
23	4.28	3.42	3.03	2.80	2.64	2.53	2.44	2.37	2.32
24	4.26	3.40	3.01	2.78	2.62	2.51	2.42	2.36	2.30
25	4.24	3.39	2.99	2.76	2.60	2.49	2.40	2.34	2.28
26	4.23	3.37	2.98	2.74	2.59	2.47	2.39	2.32	2.27
27	4.21	3.35	2.96	2.73	2.57	2.46	2.37	2.31	2.25
28	4.20	3.34	2.95	2.71	2.56	2.45	2.36	2.29	2.24
29	4.18	3.33	2.93	2.70	2.55	2.43	2.35	2.28	2.22
30	4.17	3.32	2.92	2.69	2.53	2.42	2.33	2.27	2.21
40	4.08	3.23	2.84	2.61	2.45	2.34	2.25	2.18	2.12
60	4.00	3.15	2.76	2.53	2.37	2.25	2.17	2.10	2.04
120	3.92	3.07	2.68	2.45	2.29	2.17	2.09	2.02	1.96
∞	3.84	3.00	2.60	2.37	2.21	2.10	2.01	1.94	1.88

From "Tables of Percentage Points of the Inverted Beta (F)-Distribution," *Biometrika*, Vol. 33 (1943), pp. 73–88, by Maxine Merrington and Catherine M. Thompson. Reproduced by permission of the *Biometrika* Trustees.

Table 4 (Continued)

10	12	15	20	24	30	40	60	120	∞	ν_1 / ν_2
241.9	243.9	245.9	248.0	249.1	250.1	251.1	252.2	253.3	254.3	1
19.40	19.41	19.43	19.45	19.45	19.46	19.47	19.48	19.49	19.50	2
8.79	8.74	8.70	8.66	8.64	8.62	8.59	8.57	8.55	8.53	3
5.96	5.91	5.86	5.80	5.77	5.75	5.72	5.69	5.66	5.63	4
4.74	4.68	4.62	4.56	4.53	4.50	4.46	4.43	4.40	4.36	5
4.06	4.00	3.94	3.87	3.84	3.81	3.77	3.74	3.70	3.67	6
3.64	3.57	3.51	3.44	3.41	3.38	3.34	3.30	3.27	3.23	7
3.35	3.28	3.22	3.15	3.12	3.08	3.04	3.01	2.97	2.93	8
3.14	3.07	3.01	2.94	2.90	2.86	2.83	2.79	2.75	2.71	9
2.98	2.91	2.85	2.77	2.74	2.70	2.66	2.62	2.58	2.54	10
2.85	2.79	2.72	2.65	2.61	2.57	2.53	2.49	2.45	2.40	11
2.75	2.69	2.62	2.54	2.51	2.47	2.43	2.38	2.34	2.30	12
2.67	2.60	2.53	2.46	2.42	2.38	2.34	2.30	2.25	2.21	13
2.60	2.53	2.46	2.39	2.35	2.31	2.27	2.22	2.18	2.13	14
2.54	2.48	2.40	2.33	2.29	2.25	2.20	2.16	2.11	2.07	15
2.49	2.42	2.35	2.28	2.24	2.19	2.15	2.11	2.06	2.01	16
2.45	2.38	2.31	2.23	2.19	2.15	2.10	2.06	2.01	1.96	17
2.41	2.34	2.27	2.19	2.15	2.11	2.06	2.02	1.97	1.92	18
2.38	2.31	2.23	2.16	2.11	2.07	2.03	1.98	1.93	1.88	19
2.35	2.28	2.20	2.12	2.08	2.04	1.99	1.95	1.90	1.84	20
2.32	2.25	2.18	2.10	2.05	2.01	1.96	1.92	1.87	1.81	21
2.30	2.23	2.15	2.07	2.03	1.98	1.94	1.89	1.84	1.78	22
2.27	2.20	2.13	2.05	2.01	1.96	1.91	1.86	1.81	1.76	23
2.25	2.18	2.11	2.03	1.98	1.94	1.89	1.84	1.79	1.73	24
2.24	2.16	2.09	2.01	1.96	1.92	1.87	1.82	1.77	1.71	25
2.22	2.15	2.07	1.99	1.95	1.90	1.85	1.80	1.75	1.69	26
2.20	2.13	2.06	1.97	1.93	1.88	1.84	1.79	1.73	1.67	27
2.19	2.12	2.04	1.96	1.91	1.87	1.82	1.77	1.71	1.65	28
2.18	2.10	2.03	1.94	1.90	1.85	1.81	1.75	1.70	1.64	29
2.16	2.09	2.01	1.93	1.89	1.84	1.79	1.74	1.68	1.62	30
2.08	2.00	1.92	1.84	1.79	1.74	1.69	1.64	1.58	1.51	40
1.99	1.92	1.84	1.75	1.70	1.65	1.59	1.53	1.47	1.39	60
1.91	1.83	1.75	1.66	1.61	1.55	1.50	1.43	1.35	1.25	120
1.83	1.75	1.67	1.57	1.52	1.46	1.39	1.32	1.22	1.00	∞

Table 5 Percentage Points of the F Distribution

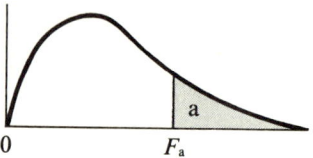

Degrees of Freedom ($a = .01$)

v_2 \ v_1	1	2	3	4	5	6	7	8	9
1	4052	4999.5	5403	5625	5764	5859	5928	5982	6022
2	98.50	99.00	99.17	99.25	99.30	99.33	99.36	99.37	99.39
3	34.12	30.82	29.46	28.71	28.24	27.91	27.67	27.49	27.35
4	21.20	18.00	16.69	15.98	15.52	15.21	14.98	14.80	14.66
5	16.26	13.27	12.06	11.39	10.97	10.67	10.46	10.29	10.16
6	13.75	10.92	9.78	9.15	8.75	8.47	8.26	8.10	7.98
7	12.25	9.55	8.45	7.85	7.46	7.19	6.99	6.84	6.72
8	11.26	8.65	7.59	7.01	6.63	6.37	6.18	6.03	5.91
9	10.56	8.02	6.99	6.42	6.06	5.80	5.61	5.47	5.35
10	10.04	7.56	6.55	5.99	5.64	5.39	5.20	5.06	4.94
11	9.65	7.21	6.22	5.67	5.32	5.07	4.89	4.74	4.63
12	9.33	6.93	5.95	5.41	5.06	4.82	4.64	4.50	4.39
13	9.07	6.70	5.74	5.21	4.86	4.62	4.44	4.30	4.19
14	8.86	6.51	5.56	5.04	4.69	4.46	4.28	4.14	4.03
15	8.68	6.36	5.42	4.89	4.56	4.32	4.14	4.00	3.89
16	8.53	6.23	5.29	4.77	4.44	4.20	4.03	3.89	3.78
17	8.40	6.11	5.18	4.67	4.34	4.10	3.93	3.79	3.68
18	8.29	6.01	5.09	4.58	4.25	4.01	3.84	3.71	3.60
19	8.18	5.93	5.01	4.50	4.17	3.94	3.77	3.63	3.52
20	8.10	5.85	4.94	4.43	4.10	3.87	3.70	3.56	3.46
21	8.02	5.78	4.87	4.37	4.04	3.81	3.64	3.51	3.40
22	7.95	5.72	4.82	4.31	3.99	3.76	3.59	3.45	3.35
23	7.88	5.66	4.76	4.26	3.94	3.71	3.54	3.41	3.30
24	7.82	5.61	4.72	4.22	3.90	3.67	3.50	3.36	3.26
25	7.77	5.57	4.68	4.18	3.85	3.63	3.46	3.32	3.22
26	7.72	5.53	4.64	4.14	3.82	3.59	3.42	3.29	3.18
27	7.68	5.49	4.60	4.11	3.78	3.56	3.39	3.26	3.15
28	7.64	5.45	4.57	4.07	3.75	3.53	3.36	3.23	3.12
29	7.60	5.42	4.54	4.04	3.73	3.50	3.33	3.20	3.09
30	7.56	5.39	4.51	4.02	3.70	3.47	3.30	3.17	3.07
40	7.31	5.18	4.31	3.83	3.51	3.29	3.12	2.99	2.89
60	7.08	4.98	4.13	3.65	3.34	3.12	2.95	2.82	2.72
120	6.85	4.79	3.95	3.48	3.17	2.96	2.79	2.66	2.56
∞	6.63	4.61	3.78	3.32	3.02	2.80	2.64	2.51	2.41

From "Tables of Percentage Points of the Inverted Beta (F)-Distribution," *Biometrika*, Vol. 33 (1943), pp. 73–88, by Maxine Merrington and Catherine M. Thompson. Reproduced by permission of the *Biometrika* Trustees.

Table 5 (Continued)

10	12	15	20	24	30	40	60	120	∞	v_1 / v_2
6056	6106	6157	6209	6235	6261	6287	6313	6339	6366	1
99.40	99.42	99.43	99.45	99.46	99.47	99.47	99.48	99.49	99.50	2
27.23	27.05	26.87	26.69	26.60	26.50	26.41	26.32	26.22	26.13	3
14.55	14.37	14.20	14.02	13.93	13.84	13.75	13.65	13.56	13.46	4
10.05	9.89	9.72	9.55	9.47	9.38	9.29	9.20	9.11	9.02	5
7.87	7.72	7.56	7.40	7.31	7.23	7.14	7.06	6.97	6.88	6
6.62	6.47	6.31	6.16	6.07	5.99	5.91	5.82	5.74	5.65	7
5.81	5.67	5.52	5.36	5.28	5.20	5.12	5.03	4.95	4.86	8
5.26	5.11	4.96	4.81	4.73	4.65	4.57	4.48	4.40	4.31	9
4.85	4.71	4.56	4.41	4.33	4.25	4.17	4.08	4.00	3.91	10
4.54	4.40	4.25	4.10	4.02	3.94	3.86	3.78	3.69	3.60	11
4.30	4.16	4.01	3.86	3.78	3.70	3.62	3.54	3.45	3.36	12
4.10	3.96	3.82	3.66	3.59	3.51	3.43	3.34	3.25	3.17	13
3.94	3.80	3.66	3.51	3.43	3.35	3.27	3.18	3.09	3.00	14
3.80	3.67	3.52	3.37	3.29	3.21	3.13	3.05	2.96	2.87	15
3.69	3.55	3.41	3.26	3.18	3.10	3.02	2.93	2.84	2.75	16
3.59	3.46	3.31	3.16	3.08	3.00	2.92	2.83	2.75	2.65	17
3.51	3.37	3.23	3.08	3.00	2.92	2.84	2.75	2.66	2.57	18
3.43	3.30	3.15	3.00	2.92	2.84	2.76	2.67	2.58	2.49	19
3.37	3.23	3.09	2.94	2.86	2.78	2.69	2.61	2.52	2.42	20
3.31	3.17	3.03	2.88	2.80	2.72	2.64	2.55	2.46	2.36	21
3.26	3.12	2.98	2.83	2.75	2.67	2.58	2.50	2.40	2.31	22
3.21	3.07	2.93	2.78	2.70	2.62	2.54	2.45	2.35	2.26	23
3.17	3.03	2.89	2.74	2.66	2.58	2.49	2.40	2.31	2.21	24
3.13	2.99	2.85	2.70	2.62	2.54	2.45	2.36	2.27	2.17	25
3.09	2.96	2.81	2.66	2.58	2.50	2.42	2.33	2.23	2.13	26
3.06	2.93	2.78	2.63	2.55	2.47	2.38	2.29	2.20	2.10	27
3.03	2.90	2.75	2.60	2.52	2.44	2.35	2.26	2.17	2.06	28
3.00	2.87	2.73	2.57	2.49	2.41	2.33	2.23	2.14	2.03	29
2.98	2.84	2.70	2.55	2.47	2.39	2.30	2.21	2.11	2.01	30
2.80	2.66	2.52	2.37	2.29	2.20	2.11	2.02	1.92	1.80	40
2.63	2.50	2.35	2.20	2.12	2.03	1.94	1.84	1.73	1.60	60
2.47	2.34	2.19	2.03	1.95	1.86	1.76	1.66	1.53	1.38	120
2.32	2.18	2.04	1.88	1.79	1.70	1.59	1.47	1.32	1.00	∞

Appendix II

Table 6 Squares and Square Roots

n	n^2	$\sqrt{n}$	$\sqrt{10n}$	n	n^2	$\sqrt{n}$	$\sqrt{10n}$
				35	1 225	5.916 08	18.70829
1	1	1.000 00	3.162 28	36	1 296	6.000 00	18.97367
2	4	1.414 21	4.472 14	37	1 369	6.082 76	19.23538
3	9	1.732 05	5.477 23	38	1 444	6.164 41	19.49359
4	16	2.000 00	6.324 56	39	1 521	6.245 00	19.74842
5	25	2.236 07	7.071 07	40	1 600	6.324 56	20.00000
6	36	2.449 49	7.745 97	41	1 681	6.403 12	20.24846
7	49	2.645 75	8.366 60	42	1 764	6.480 74	20.49390
8	64	2.828 43	8.944 27	43	1 849	6.557 44	20.73644
9	81	3.000 00	9.486 83	44	1 936	6.633 25	20.97618
10	100	3.162 28	10.00000	45	2 025	6.708 20	21.21320
11	121	3.316 63	10.48809	46	2 116	6.782 33	21.44761
12	144	3.464 10	10.95445	47	2 209	6.855 66	21.67948
13	169	3.605 55	11.40175	48	2 304	6.928 20	21.90890
14	196	3.741 66	11.83216	49	2 401	7.000 00	22.13594
15	225	3.872 98	12.24745	50	2 500	7.071 07	22.36068
16	256	4.000 00	12.64911	51	2 601	7.141 43	22.58318
17	289	4.123 11	13.03840	52	2 704	7.211 10	22.80351
18	324	4.242 64	13.41641	53	2 809	7.280 11	23.02173
19	361	4.358 90	13.78405	54	2 916	7.348 47	23.23790
20	400	4.472 14	14.14214	55	3 025	7.416 20	23.45208
21	441	4.582 58	14.49138	56	3 136	7.483 32	23.66432
22	484	4.690 42	14.83240	57	3 249	7.549 83	23.87467
23	529	4.795 83	15.16575	58	3 364	7.615 77	24.08319
24	576	4.898 98	15.49193	59	3 481	7.618 15	24.28992
25	625	5.000 00	15.81139	60	3 600	7.745 97	24.49490
26	676	5.099 02	16.12452	61	3 721	7.810 25	24.69818
27	729	5.196 15	16.43168	62	3 844	7.874 01	24.89980
28	784	5.291 50	16.73320	63	3 969	7.937 25	25.09980
29	841	5.385 17	17.02939	64	4 096	8.000 00	25.29822
30	900	5.477 23	17.32051	65	4 225	8.062 26	25.49510
31	961	5.567 76	17.60682	66	4 356	8.124 04	25.69047
32	1 024	5.656 85	17.88854	67	4 489	8.185 35	25.88436
33	1 089	5.744 56	18.16590	68	4 624	8.246 21	26.07681
34	1 156	5.830 95	18.43909	69	4 761	8.306 62	26.26785

Statistical Tables

Table 6 (Continued)

n	n^2	$\sqrt{n}$	$\sqrt{10n}$	n	n^2	$\sqrt{n}$	$\sqrt{10n}$
70	4 900	8.366 60	26.45751	105	11 025	10.24695	32.40370
71	5 041	8.426 15	26.64583	106	11 236	10.29563	32.55764
72	5 184	8.485 28	26.83282	107	11 449	10.34408	32.71085
73	5 329	8.544 00	27.01851	108	11 664	10.39230	32.86335
74	5 476	8.602 33	27.20294	109	11 881	10.44031	33.01515
75	5 625	8.660 25	27.38613	110	12 100	10.48809	33.16625
76	5 776	8.717 80	27.56810	111	12 321	10.53565	33.31666
77	5 929	8.774 96	27.74887	112	12 544	10.58301	33.46640
78	6 084	8.831 76	27.92848	113	12 769	10.63015	33.61547
79	6 241	8.888 19	28.10694	114	12 996	10.67708	33.76389
80	6 400	8.944 27	28.28427	115	13 225	10.72381	33.91165
81	6 561	9.000 00	28.46050	116	13 456	10.77033	34.05877
82	6 724	9.055 39	28.63564	117	13 689	10.81665	34.20526
83	6 889	9.110 43	28.80972	118	13 924	10.86278	34.35113
84	7 056	9.165 15	28.98275	119	14 161	10.90871	34.49638
85	7 225	9.219 54	29.15476	120	14 400	10.95445	34.64102
86	7 396	9.273 62	29.32576	121	14 641	11.00000	34.78505
87	7 569	9.327 38	29.49576	122	14 884	11.04536	34.92850
88	7 744	9.380 83	29.66479	123	15 129	11.09054	35.07136
89	7 921	9.433 98	29.83287	124	15 376	11.13553	35.21363
90	8 100	9.486 83	30.00000	125	15 625	11.18034	35.35534
91	8 281	9.539 39	30.16621	126	15 876	11.22497	35.49648
92	8 464	9.591 66	30.33150	127	16 129	11.26943	35.63706
93	8 649	9.643 65	30.49590	128	16 384	11.31371	35.77709
94	8 836	9.695 36	30.65942	129	16 641	11.35782	35.91657
95	9 025	9.746 79	30.82207	130	16 900	11.40175	36.05551
96	9 216	9.797 96	30.98387	131	17 161	11.44552	36.19392
97	9 409	9.848 86	31.14482	132	17 424	11.48913	36.33180
98	9 604	9.899 50	31.30495	133	17 689	11.53256	36.46917
99	9 801	9.949 87	31.46427	134	17 956	11.57584	36.60601
100	10 000	10.00000	31.62278	135	18 225	11.61895	36.74235
101	10 201	10.04998	31.78050	136	18 496	11.66190	36.87818
102	10 404	10.09950	31.93744	137	18 769	11.70470	37.01351
103	10 609	10.14889	32.09361	138	19 044	11.74734	37.14835
104	10 816	10.19804	32.24903	139	19 321	11.78983	37.28270

Appendix II

Table 6 (Continued)

n	n^2	$\sqrt{n}$	$\sqrt{10n}$	n	n^2	$\sqrt{n}$	$\sqrt{10n}$
140	19 600	11.83216	37.41657	175	30 625	13.22876	41.83300
141	19 881	11.87434	37.54997	176	30 976	13.26650	41.95235
142	20 164	11.91638	37.68289	177	31 329	13.30413	42.07137
143	20 449	11.95826	37.81534	178	31 684	13.34166	42.19005
144	20 736	12.00000	37.94733	179	32 041	13.37909	42.30829
145	21 025	12.04159	38.07887	180	32 400	13.41641	42.42641
146	21 316	12.08305	38.20995	181	32 761	13.45362	42.54409
147	21 609	12.12436	38.34058	182	33 124	13.49074	42.66146
148	21 904	12.16553	38.47077	183	33 489	13.52775	42.77850
149	22 201	12.20656	38.60052	184	33 856	13.56466	42.89522
150	22 500	12.24745	38.72983	185	34 225	13.60147	43.01163
151	22 801	12.28821	38.85872	186	34 596	13.63818	43.12772
152	23 104	12.32883	38.98718	187	34 969	13.67479	43.24350
153	23 409	12.36932	39.11521	188	35 344	13.71131	43.35897
154	23 716	12.40967	39.24283	189	35 721	13.74773	43.47413
155	24 025	12.44990	39.37004	190	36 100	13.78405	43.58899
156	24 336	12.49000	39.49684	191	36 481	13.82027	43.70355
157	24 649	12.52996	39.62323	192	36 864	13.85641	43.81780
158	24 964	12.56981	39.74921	193	37 249	13.89244	43.93177
159	25 281	12.60952	39.87480	194	37 636	13.92839	44.04543
160	25 600	12.64911	40.00000	195	38 025	13.96424	44.15880
161	25 921	12.68858	40.12481	196	38 416	14.00000	44.27189
162	26 244	12.72792	40.24922	197	38 809	14.03567	44.38468
163	26 569	12.76715	40.37326	198	39 204	14.07125	44.49719
164	26 806	12.80625	40.49691	199	39 601	14.10674	44.60942
165	27 225	12.84523	40.62019	200	40 000	14.14214	44.72136
166	27 556	12.88410	40.74310	201	40 401	14.17745	44.83302
167	27 889	12.92285	40.86563	202	40 804	14.21267	44.94441
168	28 224	12.96148	40.98780	203	41 209	14.24781	45.05552
169	28 561	13.00000	41.10961	204	41 616	14.28286	45.16636
170	28 900	13.03840	41.23106	205	42 025	14.31782	45.27693
171	29 241	13.07670	41.35215	206	42 436	14.35270	45.38722
172	29 584	13.11488	41.47288	207	42 849	14.38749	45.49725
173	29 929	13.15295	41.59327	208	43 264	14.42221	45.60702
174	30 276	13.19091	41.71331	209	43 681	14.45683	45.71652

Table 6 (Continued)

n	n^2	$\sqrt{n}$	$\sqrt{10n}$	n	n^2	$\sqrt{n}$	$\sqrt{10n}$
210	44 100	14.49138	45.82576	245	60 025	15.65248	49.49747
211	44 521	14.52584	45.93474	246	60 516	15.68439	49.59839
212	44 944	14.56022	46.04346	247	61 009	15.71623	49.69909
213	45 369	14.59452	46.15192	248	61 504	15.74902	49.79960
214	45 796	14.62874	46.26013	249	62 001	15.77973	49.89990
215	46 225	14.66288	46.36809	250	62 500	15.81139	50.00000
216	46 656	14.69694	46.47580	251	63 001	15.84298	50.09990
217	47 089	14.73092	46.58326	252	63 504	15.87451	50.19960
218	47 524	14.76482	46.69047	253	64 009	15.90597	50.29911
219	47 961	14.79865	46.79744	254	64 516	15.93738	50.39841
220	48 400	14.83240	46.90416	255	65 025	15.96872	50.49752
221	48 841	14.86607	47.01064	256	65 536	16.00000	50.59644
222	49 284	14.89966	47.11688	257	66 049	16.03122	50.69517
223	49 729	14.93318	47.22288	258	66 564	16.06238	50.79370
224	50 176	14.96663	47.32864	259	67 081	16.09348	50.89204
225	50 625	15.00000	47.43416	260	67 600	16.12452	50.99020
226	51 076	15.03330	47.53946	261	68 121	16.15549	51.08816
227	51 529	15.06652	47.64452	262	68 644	16.18641	51.18594
228	51 984	15.09967	47.74935	263	69 169	16.21727	51.28353
229	52 441	15.13275	47.85394	264	69 696	16.24808	51.38093
230	52 900	15.16575	47.95832	265	70 225	16.27882	51.47815
231	53 361	15.19868	48.06246	266	70 756	16.30951	51.57519
232	53 824	15.23155	48.16638	267	71 289	16.34013	51.67204
233	54 289	15.26434	48.27007	268	71 824	16.37071	51.76872
234	54 756	15.29706	48.37355	269	72 361	16.40122	51.86521
235	55 225	15.32971	48.47680	270	72 900	16.43168	51.96152
236	55 696	15.36229	48.57983	271	73 441	16.46208	52.05766
237	56 169	15.39480	48.68265	272	73 984	16.49242	52.15362
238	56 644	15.42725	48.78524	273	74 529	16.52271	52.24940
239	57 121	15.45962	48.88763	274	75 076	16.55295	52.34501
240	57 600	15.49193	48.98979	275	75 625	16.58312	52.44044
241	58 081	15.52417	49.09175	276	76 176	16.61235	52.53570
242	58 564	15.55635	49.19350	277	76 729	16.64332	52.63079
243	59 049	15.58846	49.29503	278	77 284	16.67333	52.72571
244	59 536	15.62050	49.39636	279	77 841	16.70329	52.82045

Appendix II

Table 6 (Continued)

n	n^2	$\sqrt{n}$	$\sqrt{10n}$	n	n^2	$\sqrt{n}$	$\sqrt{10n}$
280	78 400	16.73320	52.91503	315	99 225	17.74824	56.12486
281	78 961	16.76305	53.00943	316	99 856	17.77639	56.21388
282	79 524	16.79286	53.10367	317	100 489	17.80449	56.30275
283	80 089	16.82260	53.19774	318	101 124	17.83255	56.39149
284	80 656	16.85230	53.29165	319	101 761	17.86057	56.48008
285	81 225	16.88194	53.38539	320	102 400	17.88854	56.56854
286	81 796	16.91153	53.47897	321	103 041	17.91647	56.65686
287	82 369	16.94107	53.57238	322	103 684	17.94436	56.74504
288	82 944	16.97056	53.66563	323	104 329	17.97220	56.83309
289	83 521	17.00000	53.75872	324	104 976	18.00000	56.92100
290	84 100	17.02939	53.85165	325	105 625	18.02776	57.00877
291	84 681	17.05872	53.94442	326	106 276	18.05547	57.09641
292	85 264	17.08801	54.03702	327	106 929	18.08314	57.18391
293	85 849	17.11724	54.12947	328	107 584	18.11077	57.27128
294	86 436	17.14643	54.22177	329	108 241	18.13836	57.35852
295	87 025	17.17556	54.31390	330	108 900	18.16590	57.44563
296	87 616	17.20465	54.40588	331	109 561	18.19341	57.53260
297	88 209	17.23369	54.49771	332	110 224	18.22087	57.61944
298	88 804	17.26268	54.58938	333	110 889	18.24829	57.70615
299	89 401	17.29162	54.68089	334	111 556	18.27567	57.79273
300	90 000	17.32051	54.77226	335	112 225	18.30301	57.87918
301	90 601	17.34935	54.86347	336	112 896	18.33030	57.96551
302	91 204	17.37815	54.95453	337	113 569	18.35756	58.05170
303	91 809	17.40690	55.04544	338	114 244	18.38478	58.13777
304	92 416	17.43560	55.13620	339	114 921	18.41195	58.22371
305	93 025	17.46425	55.22681	340	115 600	18.43909	58.30952
306	93 636	17.49286	55.31727	341	116 281	18.46619	58.39521
307	94 249	17.52142	55.40758	342	116 964	18.49324	58.48077
308	94 864	17.54993	55.49775	343	117 649	18.52026	58.56620
309	95 481	17.57840	55.58777	344	118 336	18.54724	58.65151
310	96 100	17.60682	55.67764	345	119 025	18.57418	58.73670
311	96 721	17.63519	55.76737	346	119 716	18.60108	58.82176
312	97 344	17.66352	55.85696	347	120 409	18.62794	58.90671
313	97 969	17.69181	55.94640	348	121 104	18.65476	58.99152
314	98 596	17.72005	56.03570	349	121 801	18.68154	59.07622

Statistical Tables

Table 6 (Continued)

n	n^2	$\sqrt{n}$	$\sqrt{10n}$	n	n^2	$\sqrt{n}$	$\sqrt{10n}$
350	122 500	18.70829	59.16080	385	148 225	19.62142	62.04837
351	123 201	18.73499	59.24525	386	148 996	19.64688	62.12890
352	123 904	18.76166	59.32959	387	149 769	19.67232	62.20932
353	124 609	18.78829	59.41380	388	150 544	19.69772	62.28965
354	125 316	18.81489	59.49790	389	151 321	19.72308	62.36986
355	126 025	18.84144	59.58188	390	152 100	19.74842	62.44998
356	126 736	18.86796	59.66574	391	152 881	19.77372	62.52999
357	127 449	18.89444	59.74948	392	153 664	19.79899	62.60990
358	128 164	18.92089	59.83310	393	154 449	19.82423	62.68971
359	128 881	18.94730	59.91661	394	155 236	19.84943	62.76942
360	129 600	18.97367	60.00000	395	156 025	19.87461	62.84903
361	130 321	19.00000	60.08328	396	156 816	19.89975	62.92853
362	131 044	19.02630	60.16644	397	157 609	19.92486	63.00794
363	131 769	19.05256	60.24948	398	158 404	19.94994	63.08724
364	132 496	19.07878	60.33241	399	159 201	19.97498	63.16645
365	133 225	19.10497	60.41523	400	160 000	20.00000	63.24555
366	133 956	19.13113	60.49793	401	160 801	20.02498	63.32456
367	134 689	19.15724	60.58052	402	161 604	20.04994	63.40347
368	135 424	19.18333	60.66300	403	162 409	20.07486	63.48228
369	136 161	19.20937	60.74537	404	163 216	20.09975	63.56099
370	136 900	19.23538	60.82763	405	164 025	20.12461	63.63961
371	137 641	19.26136	60.90977	406	164 836	20.14944	63.71813
372	138 384	19.28730	60.99180	407	165 649	20.17424	63.79655
373	139 129	19.31321	61.07373	408	166 464	20.19901	63.87488
374	139 876	19.33908	61.15554	409	167 281	20.22375	63.95311
375	140 625	19.36492	61.23724	410	168 100	20.24864	64.03124
376	141 376	19.39072	61.31884	411	168 921	20.27313	64.10928
377	142 129	19.41649	61.40033	412	169 744	20.29778	64.18723
378	142 884	19.44222	61.48170	413	170 569	20.32240	64.26508
379	143 641	19.46792	61.56298	414	171 396	20.34699	64.34283
380	144 400	19.49359	61.64414	415	172 225	20.37155	64.42049
381	145 161	19.51922	61.72520	416	173 056	20.39608	64.49806
382	145 924	19.54482	61.80615	417	173 889	20.42058	64.57554
383	146 689	19.57039	61.88699	418	174 724	20.44505	64.65292
384	147 456	19.59592	61.96773	419	175 561	20.46949	64.73021

Table 6 (Continued)

n	n^2	$\sqrt{n}$	$\sqrt{10n}$	n	n^2	$\sqrt{n}$	$\sqrt{10n}$
420	176 400	20.49390	64.80741	455	207 025	21.33073	67.45369
421	177 241	20.51828	64.88451	456	207 936	21.35416	67.52777
422	178 084	20.54264	64.96153	457	208 849	21.37756	67.60178
423	178 929	20.56696	65.03845	458	209 764	21.40093	67.67570
424	179 776	20.59126	65.11528	459	210 681	21.42429	67.74954
425	180 625	20.61553	65.19202	460	211 600	21.44761	67.82330
426	181 476	20.63977	65.26868	461	212 521	21.47091	67.89698
427	182 329	20.66398	65.34524	462	213 444	21.49419	67.97058
428	183 184	20.68816	65.42171	463	214 369	21.51743	68.04410
429	184 041	20.71232	65.49809	464	215 296	21.54066	68.11755
430	184 900	20.73644	65.57439	465	216 225	21.56386	68.19091
431	185 761	20.76054	65.65059	466	217 156	21.58703	68.26419
432	186 624	20.78461	65.72671	467	218 089	21.61018	68.33740
433	187 489	20.80865	65.80274	468	219 024	21.63331	68.41053
434	188 356	20.83267	65.87868	469	219 961	21.65641	68.48957
435	189 225	20.85665	65.95453	470	220 900	21.67948	68.55655
436	190 096	20.88061	66.03030	471	221 841	21.70253	68.62944
437	190 969	20.90454	66.10598	472	222 784	21.72556	68.70226
438	191 844	20.92845	66.18157	473	223 729	21.74856	68.77500
439	192 721	20.95233	66.25708	474	224 676	21.77154	68.84766
440	193 600	20.97618	66.33250	475	225 625	21.79449	68.92024
441	194 481	21.00000	66.40783	476	226 576	21.81742	68.99275
442	195 364	21.02380	66.48308	477	227 529	21.84033	69.06519
443	196 249	21.04757	66.55825	478	228 484	21.86321	69.13754
444	197 136	21.07131	66.63332	479	229 441	21.88607	69.20983
445	198 025	21.09502	66.70832	480	230 400	21.90890	69.28203
446	198 916	21.11871	66.78323	481	231 361	21.93171	69.35416
447	199 809	21.14237	66.85806	482	232 324	21.95450	69.42622
448	200 704	21.16601	66.93280	483	233 289	21.97726	69.49820
449	201 601	21.18962	67.00746	484	234 256	22.00000	69.57011
450	202 500	21.21320	67.08204	485	235 225	22.02272	69.64194
451	203 401	21.23676	67.15653	486	236 196	22.04541	69.71370
452	204 304	21.26029	67.23095	487	237 169	22.06808	69.78539
453	205 209	21.28380	67.30527	488	238 144	22.09072	69.85700
454	206 116	21.30728	67.37952	489	239 121	22.11334	69.92853

Statistical Tables

Table 6 (Continued)

n	n^2	$\sqrt{n}$	$\sqrt{10n}$	n	n^2	$\sqrt{n}$	$\sqrt{10n}$
490	240 100	22.13594	70.00000	525	275 625	22.91288	72.45688
491	241 081	22.15852	70.07139	526	276 676	22.93469	72.52586
492	242 064	22.18107	70.14271	527	277 729	22.95648	72.59477
493	243 049	22.20360	70.21396	528	278 784	22.97825	72.66361
494	244 036	22.22611	70.28513	529	279 841	23.00000	72.73239
495	245 025	22.24860	70.35624	530	280 900	23.02173	72.80110
496	246 016	22.27106	70.42727	531	281 961	23.04344	72.86975
497	247 009	22.29350	70.49823	532	283 024	23.06513	72.93833
498	248 004	22.31591	70.56912	533	284 089	23.08679	73.00685
499	249 001	22.33831	70.63993	534	285 156	23.10844	73.07530
500	250 000	22.36068	70.71068	535	286 225	23.13007	73.14369
501	251 001	22.38303	70.78135	536	287 296	23.15167	73.21202
502	252 004	22.40536	70.85196	537	288 369	23.17326	73.28028
503	253 009	22.42766	70.92249	538	289 444	23.19483	73.34848
504	254 016	22.44994	70.99296	539	290 521	23.21637	73.41662
505	255 025	22.47221	71.06335	540	291 600	23.23790	73.48469
506	256 036	22.49444	71.13368	541	292 681	23.25941	73.55270
507	257 049	22.51666	71.20393	542	293 764	23.28089	73.62065
508	258 064	22.53886	71.27412	543	294 849	23.30236	73.68853
509	259 081	22.56103	71.34424	544	295 936	23.32381	73.75636
510	260 100	22.58318	71.41428	545	297 025	23.34524	73.82412
511	261 121	22.60531	71.48426	546	298 116	23.36664	73.89181
512	262 144	22.62742	71.55418	547	299 209	23.38803	73.95945
513	263 169	22.64950	71.62402	548	300 304	23.40940	74.02702
514	264 196	22.67157	71.69379	549	301 401	23.43075	74.09453
515	265 225	22.69361	71.76350	550	302 500	23.45208	74.16198
516	266 256	22.71563	71.83314	551	303 601	23.47339	74.22937
517	267 289	22.73763	71.90271	552	304 704	23.49468	74.29670
518	268 324	22.75961	71.97222	553	305 809	23.51595	74.36397
519	269 361	22.78157	72.04165	554	306 916	23.53720	74.43118
520	270 400	22.80351	72.11103	555	308 025	23.55844	74.49832
521	271 441	22.82542	72.18033	556	309 136	23.57965	74.56541
522	272 484	22.84732	72.24957	557	310 249	23.60085	74.63243
523	273 529	22.86919	72.31874	558	311 364	23.62202	74.69940
524	274 576	22.89105	72.38784	559	312 481	23.64318	74.76630

Appendix II

Table 6 (Continued)

n	n^2	$\sqrt{n}$	$\sqrt{10n}$	n	n^2	$\sqrt{n}$	$\sqrt{10n}$
560	313 600	23.66432	74.83315	595	354 025	24.39262	77.13624
561	314 721	23.68544	74.89993	596	355 216	24.41311	77.20104
562	315 844	23.70654	74.96666	597	356 409	24.43358	77.26578
563	316 969	23.72762	75.03333	598	357 604	24.45404	77.33046
564	318 096	23.74868	75.09993	599	358 801	24.47448	77.39509
565	319 225	23.76973	75.16648	600	360 000	24.49490	77.45967
566	320 356	23.79075	75.23297	601	361 201	24.51530	77.52419
567	321 489	23.81176	75.29940	602	362 404	24.53569	77.58866
568	322 624	23.83275	75.36577	603	363 609	24.55606	77.65307
569	323 761	23.85372	75.43209	604	364 816	24.57641	77.71744
570	324 900	23.87467	75.49834	605	366 025	24.59675	77.78175
571	326 041	23.89561	75.56454	606	367 236	24.61707	77.84600
572	327 184	23.91652	75.63068	607	368 449	24.63737	77.91020
573	328 329	23.93742	75.69676	608	369 664	24.65766	77.97435
574	329 476	23.95830	75.76279	609	370 881	24.67793	78.03845
575	330 625	23.97916	75.82875	610	372 100	24.69818	78.10250
576	331 776	24.00000	75.89466	611	373 321	24.71841	78.16649
577	332 929	24.02082	75.96052	612	374 544	24.73863	78.23043
578	334 084	24.04163	76.02631	613	375 769	24.75884	78.29432
579	335 241	24.06242	76.09205	614	376 996	24.77902	78.35815
580	336 400	24.08319	76.15773	615	378 225	24.79919	78.42194
581	337 561	24.10394	76.22336	616	379 456	24.81935	78.48567
582	338 724	24.12468	76.28892	617	380 689	24.83948	78.54935
583	339 889	24.14539	76.35444	618	381 924	24.85961	78.61298
584	341 056	24.16609	76.41989	619	383 161	24.87971	78.67655
585	342 225	24.18677	76.48529	620	384 400	24.89980	78.74008
586	343 396	24.20744	76.55064	621	385 641	24.91987	78.80355
587	344 569	24.22808	76.61593	622	386 884	24.93993	78.86698
588	345 744	24.24871	76.68116	623	388 129	24.95997	78.93035
589	346 921	24.26932	76.74634	624	389 376	24.97999	78.99367
590	348 100	24.28992	76.81146	625	390 625	25.00000	79.05694
591	349 281	24.31049	76.87652	626	391 876	25.01999	79.12016
592	350 464	24.33105	76.94154	627	393 129	25.03997	79.18333
593	351 649	24.35159	77.00649	628	394 384	25.05993	79.24645
594	352 836	24.37212	77.07140	629	395 641	25.07987	79.30952

Statistical Tables

Table 6 (Continued)

n	n^2	$\sqrt{n}$	$\sqrt{10n}$	n	n^2	$\sqrt{n}$	$\sqrt{10n}$
630	396 900	25.09980	79.37254	665	442 225	25.78759	81.54753
631	398 161	25.11971	79.43551	666	443 556	25.80698	81.60882
632	399 424	25.13961	79.49843	667	444 889	25.82634	81.67007
633	400 689	25.15949	79.56130	668	446 224	25.84570	81.73127
634	401 956	25.17936	79.62412	669	447 561	25.86503	81.79242
635	403 225	25.19921	79.68689	670	448 900	25.88436	81.85353
636	404 496	25.21904	79.74961	671	450 241	25.90367	81.91459
637	405 769	25.23886	79.81228	672	451 584	25.92296	81.97561
638	407 044	25.25866	79.87490	673	452 929	25.94224	82.03658
639	408 321	25.27845	79.93748	674	454 276	25.96151	82.09750
640	409 600	25.29822	80.00000	675	455 625	25.98076	82.15838
641	410 881	25.31798	80.06248	676	456 976	26.00000	82.21922
642	412 164	25.33772	80.12490	677	458 329	26.01922	82.28001
643	413 449	25.35744	80.18728	678	459 684	26.03843	82.34076
644	414 736	25.37716	80.24961	679	461 041	26.05763	82.40146
645	416 025	25.39685	80.31189	680	462 400	26.07681	82.46211
646	417 316	25.41653	80.37413	681	463 761	26.09598	82.52272
647	418 609	25.43619	80.43631	682	465 124	26.11513	82.58329
648	419 904	25.45584	80.49845	683	466 489	26.13427	82.64381
649	421 201	25.47548	80.56054	684	467 856	26.15339	82.70429
650	422 500	25.49510	80.62258	685	469 225	26.17250	82.76473
651	423 801	25.51470	80.68457	686	470 596	26.19160	82.82512
652	425 104	25.53429	80.74652	687	471 969	26.21068	82.88546
653	426 409	25.55386	80.80842	688	473 344	26.22975	82.94577
654	427 716	25.57342	80.87027	689	474 721	26.24881	83.00602
655	429 025	25.59297	80.93207	690	476 100	26.26785	83.06624
656	430 336	25.61250	80.99383	691	477 481	26.28688	83.12641
657	431 649	25.63201	81.05554	692	478 864	26.30589	83.18654
658	432 964	25.65151	81.11720	693	480 249	26.32489	83.24662
659	434 281	25.67100	81.17881	694	481 636	26.34388	83.30666
660	435 600	25.69047	81.24038	695	483 025	26.36285	83.36666
661	436 921	25.70992	81.30191	696	484 416	26.38181	83.42661
662	438 244	25.72936	81.36338	697	485 809	26.40076	83.48653
663	439 569	25.74879	81.42481	698	487 204	26.41969	83.54639
664	440 896	25.76820	81.48620	699	488 601	26.43861	83.60622

Appendix II

Table 6 (Continued)

n	n^2	$\sqrt{n}$	$\sqrt{10n}$	n	n^2	$\sqrt{n}$	$\sqrt{10n}$
700	490 000	26.45751	83.66600	735	540 225	27.11088	85.73214
701	491 401	26.47640	83.72574	736	541 696	27.12932	85.79044
702	492 804	26.49528	83.78544	737	543 169	27.14774	85.84870
703	494 209	26.51415	83.84510	738	544 644	27.16616	85.90693
704	495 616	26.53300	83.90471	739	546 121	27.18455	85.96511
705	497 025	26.55184	83.96428	740	547 600	27.20294	86.02325
706	498 436	26.57066	84.02381	741	549 081	27.22132	86.08136
707	499 849	26.58947	84.08329	742	550 564	27.23968	86.13942
708	501 264	26.60827	84.14274	743	552 049	27.25803	86.19745
709	502 681	26.62705	84.20214	744	553 536	27.27636	86.25543
710	504 100	26.64583	84.26150	745	555 025	27.29469	86.31338
711	505 521	26.66458	84.32082	746	556 516	27.31300	86.37129
712	506 944	26.68333	84.38009	747	558 009	27.33130	86.42916
713	508 369	26.70206	84.43933	748	559 504	27.34959	86.48699
714	509 796	26.72078	84.49852	749	561 001	27.36786	86.54479
715	511 225	26.73948	84.55767	750	562 500	27.38613	86.60254
716	512 656	26.75818	84.61678	751	564 001	27.40438	86.66026
717	514 089	26.77686	84.67585	752	565 504	27.42262	86.71793
718	515 524	26.79552	84.73488	753	567 009	27.44085	86.77557
719	516 961	26.81418	84.79387	754	568 516	27.45906	86.83317
720	518 400	26.83282	84.85281	755	570 025	27.47726	86.89074
721	519 841	26.85144	84.91172	756	571 536	27.49545	86.94826
722	521 284	26.87006	84.97058	757	573 049	27.51363	87.00575
723	522 729	26.88866	85.02941	758	574 564	27.53180	87.06320
724	524 176	26.90725	85.08819	759	576 081	27.54995	87.12061
725	525 625	26.92582	85.14693	760	577 600	27.56810	87.17798
726	527 076	26.94439	85.20563	761	579 121	27.58623	87.23531
727	528 529	26.96294	85.26429	762	580 644	27.60435	87.29261
728	529 984	26.98148	85.32292	763	582 169	27.62245	87.34987
729	531 441	27.00000	85.38150	764	583 696	27.64055	87.40709
730	532 900	27.01851	85.44004	765	585 225	27.65863	87.46428
731	534 361	27.03701	85.49854	766	586 756	27.67671	87.52143
732	535 824	27.05550	85.55700	767	588 289	27.69476	87.57854
733	537 289	27.07397	85.61542	768	589 824	27.71281	87.63561
734	538 756	27.09243	85.67380	769	591 361	27.73085	87.69265

Statistical Tables

Table 6 (Continued)

n	n^2	$\sqrt{n}$	$\sqrt{10n}$	n	n^2	$\sqrt{n}$	$\sqrt{10n}$
770	592 900	27.74887	87.74964	805	648 025	28.37252	89.72179
771	594 441	27.76689	87.80661	806	649 636	28.39014	89.77750
772	595 984	27.78489	87.86353	807	651 249	28.40775	89.83318
773	597 529	27.80288	87.92042	808	652 864	28.42534	89.88882
774	599 076	27.82086	87.97727	809	654 481	28.44293	89.94443
775	600 625	27.83882	88.03408	810	656 100	28.46050	90.00000
776	602 176	27.85678	88.09086	811	657 721	28.47806	90.05554
777	603 729	27.87472	88.14760	812	659 344	28.49561	90.11104
778	605 284	27.89265	88.20431	813	660 969	28.51315	90.16651
779	606 841	27.91057	88.26098	814	662 596	28.53069	90.22195
780	608 400	27.92848	88.31761	815	664 225	28.54820	90.27735
781	609 961	27.94638	88.37420	816	665 856	28.56571	90.33272
782	611 524	27.96426	88.43076	817	667 489	28.58321	90.38805
783	613 089	27.98214	88.48729	818	669 124	28.60070	90.44335
784	614 656	28.00000	88.54377	819	670 761	28.61818	90.49862
785	616 225	28.01785	88.60023	820	672 400	28.63564	90.55385
786	617 796	28.03569	88.65664	821	674 041	28.65310	90.60905
787	619 369	28.05352	88.71302	822	675 684	28.67054	90.66422
788	620 944	28.07134	88.76936	823	677 329	28.68798	90.71935
789	622 521	28.08914	88.82567	824	678 976	28.70540	90.77445
790	624 100	28.10694	88.88194	825	680 625	28.72281	90.82951
791	625 681	28.12472	88.93818	826	682 726	28.74022	90.88454
792	627 264	28.14249	88.99428	827	683 929	28.75761	90.93954
793	628 849	28.16026	89.05055	828	685 584	28.77499	90.99451
794	630 436	28.17801	89.10668	829	687 241	28.79236	91.04944
795	632 025	28.19574	89.16277	830	688 900	28.80972	91.10434
796	633 616	28.21347	89.21883	831	690 561	28.82707	91.15920
797	635 209	28.23119	89.27486	832	692 224	28.84441	91.21403
798	636 804	28.24889	89.33085	833	693 889	28.86174	91.26883
799	638 401	28.26659	89.38680	834	695 556	28.87906	91.32360
800	640 000	28.28472	89.44272	835	697 225	28.89637	91.37833
801	641 601	28.30194	89.49860	836	698 896	28.91366	91.43304
802	643 204	28.31960	89.55445	837	700 569	28.93095	91.48770
803	644 809	28.33725	89.61027	838	702 244	28.94823	91.54234
804	646 416	28.35489	89.66605	839	703 921	28.96550	91.59694

Appendix II

Table 6 (Continued)

n	n^2	$\sqrt{n}$	$\sqrt{10n}$	n	n^2	$\sqrt{n}$	$\sqrt{10n}$
840	705 600	28.98275	91.65151	875	765 625	29.58040	93.54143
841	707 281	29.00000	91.70605	876	767 376	29.59730	93.59487
842	708 964	29.01724	91.76056	877	769 129	29.61419	93.64828
843	710 649	29.03446	91.81503	878	770 884	29.63106	93.70165
844	712 336	29.05168	91.86947	879	772 641	29.64793	93.75500
845	714 025	29.06888	91.92388	880	774 400	29.66479	93.80832
846	715 716	29.08608	91.97826	881	776 161	29.68164	93.86160
847	717 409	29.10326	92.03260	882	777 924	29.69848	93.91486
848	719 104	29.12044	92.08692	883	779 689	29.71532	93.96808
849	720 801	29.13760	92.14120	884	781 456	29.73214	94.02127
850	722 500	29.15476	92.19544	885	783 225	29.74895	94.07444
851	724 201	29.17190	92.24966	886	784 996	29.76575	94.12757
852	725 904	29.18904	92.30385	887	786 769	29.78255	94.18068
853	727 609	29.20616	92.35800	888	788 544	29.79933	94.23375
854	729 316	29.22328	92.41212	889	790 321	29.81610	94.28680
855	731 025	29.24038	92.46621	890	792 100	29.83287	94.33981
856	732 736	29.25748	92.52027	891	793 881	29.84962	94.39280
857	734 449	29.27456	92.57429	892	795 664	29.86637	94.44575
858	736 164	29.29164	92.62829	893	797 449	29.88311	94.49868
859	737 881	29.30870	92.68225	894	799 236	29.89983	94.55157
860	739 600	29.32576	92.73618	895	801 025	29.91655	94.60444
861	741 321	29.34280	92.79009	896	802 816	29.93326	94.65728
862	743 044	29.35984	92.84396	897	804 609	29.94996	94.71008
863	744 769	29.37686	92.89779	898	806 404	29.96665	94.76286
864	746 496	29.39388	92.95160	899	808 201	29.98333	94.81561
865	748 225	29.41088	93.00538	900	810 000	30.00000	94.86833
866	749 956	29.42788	93.05912	901	811 801	30.01666	94.92102
867	751 689	29.44486	93.11283	902	813 604	30.03331	94.97368
868	753 424	29.46184	93.16652	903	815 409	30.04996	95.02631
869	755 161	29.47881	93.22017	904	817 216	30.06659	95.07891
870	756 900	29.49576	93.27379	905	819 025	30.08322	95.13149
871	758 641	29.51271	93.32738	906	820 836	30.09983	95.18403
872	760 384	29.52965	93.38094	907	822 649	30.11644	95.23655
873	762 129	29.54657	93.43447	908	824 464	30.13304	95.28903
874	763 876	29.56349	93.48797	909	826 281	30.14963	95.34149

Statistical Tables

Table 6 (Continued)

n	n^2	$\sqrt{n}$	$\sqrt{10n}$	n	n^2	$\sqrt{n}$	$\sqrt{10n}$
910	828 100	30.16621	95.39392	945	893 025	30.74085	97.21111
911	829 921	30.18278	95.44632	946	894 916	30.75711	97.26253
912	831 744	30.19934	95.49869	947	896 809	30.77337	97.31393
913	833 569	30.21589	95.55103	948	898 704	30.78961	97.36529
914	835 396	30.23243	95.60335	949	900 601	30.80584	97.41663
915	837 225	30.24897	95.65563	950	902 500	30.82207	97.46794
916	839 056	30.26549	95.70789	951	904 401	30.83829	97.51923
917	840 889	30.28201	95.76012	952	906 304	30.85450	97.57049
918	842 724	30.29851	95.81232	953	908 209	30.87070	97.62172
919	844 561	30.31501	95.86449	954	910 116	30.88689	97.67292
920	846 400	30.33150	95.91663	955	912 025	30.90307	97.72410
921	848 241	30.34798	95.96874	956	913 936	30.91925	97.77525
922	850 084	30.36445	96.02083	957	915 849	30.93542	97.82638
923	851 929	30.38092	96.07289	958	917 764	30.95158	97.87747
924	853 776	30.39737	96.12492	959	919 681	30.96773	97.92855
925	855 625	30.41381	96.17692	960	921 600	30.98387	97.97959
926	857 476	30.43025	96.22889	961	923 521	31.00000	98.03061
927	859 329	30.44667	96.28084	962	925 444	31.01612	98.08160
928	861 184	30.46309	96.33276	963	927 369	31.03224	98.13256
929	863 041	30.47950	96.38465	964	929 296	31.04835	98.18350
930	864 900	30.49590	96.43651	965	931 225	31.06445	98.23441
931	866 761	30.51229	96.48834	966	933 156	31.08054	98.28530
932	868 624	30.52868	96.54015	967	935 089	31.09662	98.33616
933	870 489	30.54505	96.59193	968	937 024	31.11270	98.38699
934	872 356	30.56141	96.64368	969	938 961	31.12876	98.43780
935	874 225	30.57777	96.69540	970	940 900	31.14482	98.48858
936	876 096	30.59412	96.74709	971	942 841	31.16087	98.53933
937	877 969	30.61046	96.79876	972	944 784	31.17691	98.59006
938	879 844	30.62679	96.85040	973	946 729	31.19295	98.64076
939	881 721	30.64311	96.90201	974	948 676	31.20897	98.69144
940	883 600	30.65942	96.95360	975	950 625	31.22499	98.74209
941	885 481	30.67572	97.00515	976	952 576	31.24100	98.79271
942	887 364	30.69202	97.05668	977	954 529	31.25700	98.84331
943	889 249	30.70831	97.10819	978	956 484	31.27299	98.89388
944	891 136	30.72458	97.15966	979	958 441	31.28898	98.94443

Appendix II

Table 6 (Continued)

n	n^2	$\sqrt{n}$	$\sqrt{10n}$	n	n^2	$\sqrt{n}$	$\sqrt{10n}$
980	960 400	31.30495	98.99495	990	980 100	31.46427	99.49874
981	962 361	31.32092	99.04544	991	982 081	31.48015	99.54898
982	964 324	31.33688	99.09591	992	984 064	31.49603	99.59920
983	966 289	31.35283	99.14636	993	986 049	31.51190	99.64939
984	968 256	31.36877	99.19677	994	988 036	31.52777	99.69955
985	970 225	31.38471	99.24717	995	990 025	31.54362	99.74969
986	972 196	31.40064	99.29753	996	992 016	31.55947	99.79980
987	974 169	31.41656	99.34787	997	994 009	31.57531	99.84989
988	976 144	31.43247	99.39819	998	996 004	31.59114	99.89995
989	978 121	31.44837	99.44848	999	998 001	31.60696	99.94999
				1000	1000 000	31.62278	100.00000

Computed by J. Huang, Department of Statistics, University of Florida.

Statistical Tables

Table 7 Random Numbers

Line	1	2	3	4	5	6	7	8	9	10
1	75029	50152	25648	02523	84300	83093	39852	91276	88988	12439
2	73741	30492	19280	41255	74008	72750	70420	67769	72837	27098
3	07049	98408	27011	76385	15212	03806	85928	81312	14514	55277
4	01033	08705	42934	79257	89138	21506	26797	67223	62165	67981
5	48399	78564	35787	07647	23794	73938	29477	11420	03228	16586
6	70459	73480	06740	79124	14078	72352	07410	93292	93057	18715
7	74770	80185	08181	27417	90866	98444	72870	51219	51481	47916
8	24167	13753	65011	66288	12633	79199	61497	56186	83643	96184
9	24316	80240	62592	53393	57028	61626	56508	84407	97873	27571
10	84565	59254	94435	33322	50014	00180	50954	04099	66005	59141
11	60794	32497	47830	94509	36576	68874	84062	84503	50454	42199
12	99104	14833	97062	48867	19645	78069	91602	46991	57523	22219
13	15604	93654	21487	86036	22827	62637	70378	58539	17827	80108
14	20204	00253	19678	15789	17628	63667	23348	67083	92361	50413
15	71233	73676	00958	42662	47344	00104	74530	46238	06655	23791
16	82846	82954	52107	66054	27358	69664	71760	03577	75622	21536
17	48613	97858	49627	17036	55574	80116	80533	62146	48083	29177
18	42313	91287	66900	79817	76803	42462	63542	99089	22655	44130
19	60879	68102	60700	51281	61386	06782	88214	68246	15552	79093
20	34593	95713	62942	16236	30933	39470	58423	95304	46017	18364
21	96033	10917	01205	08978	43021	77321	76736	64527	96534	98457
22	21932	45476	75464	43497	81807	99369	59945	65349	52588	27386
23	91019	99635	78638	75114	42943	81629	03283	85036	80666	18675
24	86053	48238	14952	55565	98821	92843	67663	70387	13356	46650
25	59700	38346	92770	11506	34101	01051	99390	86884	26788	78768

Table 8 Critical Values for the Wilcoxon Signed-Rank Test

$n = 5(1)50$

One-sided	Two-sided	$n=5$	$n=6$	$n=7$	$n=8$	$n=9$	$n=10$	$n=11$	$n=12$	$n=13$	$n=14$	$n=15$	$n=16$
.05	.10	1	2	4	6	8	11	14	17	21	26	30	36
.025	.05		1	2	4	6	8	11	14	17	21	25	30
.01	.02			0	2	3	5	7	10	13	16	20	24
.005	.01				0	2	3	5	7	10	13	16	19

One-sided	Two-sided	$n=17$	$n=18$	$n=19$	$n=20$	$n=21$	$n=22$	$n=23$	$n=24$	$n=25$	$n=26$	$n=27$	$n=28$
.05	.10	41	47	54	60	68	75	83	92	101	110	120	130
.025	.05	35	40	46	52	59	66	73	81	90	98	107	117
.01	.02	28	33	38	43	49	56	62	69	77	85	93	102
.005	.01	23	28	32	37	43	49	55	61	68	76	84	92

One-sided	Two-sided	$n=29$	$n=30$	$n=31$	$n=32$	$n=33$	$n=34$	$n=35$	$n=36$	$n=37$	$n=38$	$n=39$
.05	.10	141	152	163	175	188	201	214	228	242	256	271
.025	.05	127	137	148	159	171	183	195	208	222	235	250
.01	.02	111	120	130	141	151	162	174	186	198	211	224
.005	.01	100	109	118	128	138	149	160	171	183	195	208

One-sided	Two-sided	$n=40$	$n=41$	$n=42$	$n=43$	$n=44$	$n=45$	$n=46$	$n=47$	$n=48$	$n=49$	$n=50$
.05	.10	287	303	319	336	353	371	389	408	427	446	466
.025	.05	264	279	295	311	327	344	361	379	397	415	434
.01	.02	238	252	267	281	297	313	329	345	362	380	398
.005	.01	221	234	248	262	277	292	307	323	339	356	373

From *Some Rapid Approximate Statistical Procedures* (Revised) by Frank Wilcoxon and Roberta A. Wilcox (Pearl River, N.Y.: Lederle Laboratories, 1964), Table 2. Reproduced by permission of Lederle Laboratories, a division of American Cyanamid Company.

Glossary of Common Statistical Terms

Acceptance Region. Set of values of a test statistic that imply acceptance of the null hypothesis.

Alternative Hypothesis. Hypothesis to be accepted if null hypothesis is rejected.

Analysis of Variance. A procedure for comparing more than two population means. There are many applications of analysis of variance beyond that discussed in the text.

Binomial Experiment. An experiment involving n identical independent trials. (See chapter 4 for an exact description of a binomial experiment.)

Binomial Random Variable. Discrete random variable representing the number of successes y in n identical independent trials. (For an exact definition of a binomial experiment, see chapter 4.)

Central Limit Theorem. Theorem stating that the sampling distribution of the sample mean (or sum) will be approximately normal when certain conditions are satisfied. (See chapter 5.)

Chi-square. Test statistic used to test the null hypothesis of independence for the two classifications of a contingency table. Also has many other statistical applications not discussed in this text.

Circle Chart. A graphical method for describing data; also called a pie chart. (See chapter 2.)

Class Boundary. The dividing point between two cells in a frequency histogram.

Classes. Cells of a frequency histogram.

Class Frequency. The number of observations falling in a class (referring to a frequency histogram).

Confidence Coefficient. The probability that an interval estimate (a confidence interval) will enclose the parameter of interest.

Confidence Interval. Two numbers, computed from sample data, that form an interval estimate for some parameter.

Contingency Table. A two-way table constructed for classifying count data. The entries in the table show the number of observations falling in the cells. The objective of an analysis is to determine whether the two directions of classification are dependent (contingent) upon one another.

Glossary of Common Statistical Terms

Correlation Coefficient. A measure of linear dependence between two random variables.

Degrees of Freedom. A parameter of Student's t, the F, and the chi-square probability distributions. Degrees of freedom measure the quantity of information available in normally distributed data for estimating the population variance σ^2.

Deviation from the Mean. Distance between a sample observation and the sample mean $\bar{y}$.

Discrete Random Variable. A random variable that can assume only a finite number or a countable infinity of values.

Empirical Rule. A rule that describes the variability of data that possess a mound-shaped frequency distribution. (See chapter 3.)

Error of Estimation. The distance between an estimate and the true value of the parameter estimated.

Estimate. A number computed from sample data, used to approximate a population parameter.

F Statistic. Test statistic used to compare variances from two normal populations. Used in the analysis of variance.

Frequency. Number of observations falling in some cell or in some classification category.

Histogram. A graphical method for describing a set of data. (See chapter 2.)

Interval Estimate. Two numbers computed from the sample data. The interval formed by the numbers should enclose some parameter of interest. An interval estimate is usually called a confidence interval.

Least Squares. A method of curve fitting that selects as the best-fitting curve the one that minimizes the sum of the squares of deviations of the data points from the fitted curve. (See chapter 10.)

Level of Significance. Refers to the outcome of a specific statistical test of an hypothesis. The level of significance of the test is the probability of drawing a value of the test statistic that is as contradictory, or more contradictory, to the null hypothesis than the value observed.

Linear Correlation. Dependence between two random variables.

Lower Confidence Limit. The smaller of the two numbers that form a confidence interval.

Mean. The average of a set of measurements. The symbols $\bar{y}$ and μ denote the means of a sample and a population, respectively.

Median. The middle measurement when a set is ordered according to numerical value. (See chapter 3.)

Mode. The measurement in a set that occurs with greatest frequency.

Nonparametric Methods. Usually refer to statistical tests of hypotheses about population probability distributions—but not about specific parameters of the distributions. (See chapter 15.)

Normal Distribution. A bell-shaped probability distribution. The curve possesses a specific mathematical formula. (See References.)

Null Hypothesis. The hypothesis under test in a statistical test of an hypothesis.

Paired-Difference Test. A statistical test for the comparison of two population means. The test is based on paired observations, one from each of the two populations.

Parameter of a Population. A numerical descriptive measure of a population.

Parametric Methods. Statistical methods for estimating parameters or testing hypotheses about population parameters.

Percentiles. See chapter 3 for definition.

Pie Chart. A graphical method for describing data; also called a circle chart. (See chapter 2.)

Point Estimate. *See* Estimate.

Population. The set of measurements, existing or conceptual, that is of interest to the experimenter. Samples are selected from the population.

Probability. As a practical matter, we think of the probability of an event as a measure of one's belief that the event will occur when the experiment is conducted once. The exact definition, giving a quantitative measure of this belief, is subject to debate. The relative frequency concept is most widely accepted.

Probability Distribution, Continuous. A smooth curve that gives the theoretical frequency distribution for the continuous random variable. An area under the curve over an interval is proportional to the probability that the random variable will fall in the interval.

Probability Distribution, Discrete. A listing, a mathematical formula, or a histogram that gives the probability associated with each value of the random variable.

p **Value.** Level of significance of a statistical test.

Quartiles. See chapter 3 for definition.

Random Sample. A sample of n measurements selected in such a way that every different sample of n elements in the population has an equal probability of being selected.

Random Variable. A random variable is associated with an experiment. Its values are numerical events that cannot be predicted with certainty.

Range of a Set of Measurements. The difference between the largest and smallest members of the set.

Rank Correlation Coefficient. A rank correlation coefficient is a coefficient of linear correlation between two random variables that is based on the ranks of the measurements, not their actual values. *See also* Correlation Coefficient.

Regression Line. Line fit to data points, using the method of least squares.

Rejection Region. Set of values of a test statistic that indicates rejection of the null hypothesis.

Relative Frequency. Class frequency divided by the total number of measurements.

Glossary of Common Statistical Terms

Sample. A subset of measurements selected from a population.

Sampling Distribution. The probability distribution for a sample statistic.

Significance Level. *See* Level of Significance.

Sign Test. A nonparametric statistical test used to compare two populations. (See chapter 15.)

Spearman's Rank Correlation Coefficient. One of several correlation coefficients based on the ranks of the two random variables.

Standard Deviation. A measure of data variation. (See chapter 3.)

Standardized Normal Distribution. A normal distribution with mean and standard deviation equal to 0 and 1, respectively. The standardized normal variable is denoted by the symbol z.

Statistical Test. A procedure for making an inference about one or more population parameters by using information from sample data. The procedure is based on the concept of proof by contradiction.

Student's t. A test statistic used for small-sample tests of means.

Student's t Distribution. A particular symmetric mound-shaped distribution that possesses more spread than the standard normal probability distribution.

Test Statistic. A function of the sample measurements, used as a decision maker in a test of an hypothesis.

Type I Error. Rejecting the null hypothesis when it is true.

Type II Error. Accepting the null hypothesis when it is false and the alternative hypothesis is true.

Upper Confidence Limit. The larger of the two numbers that form a confidence interval.

Variance. A measure of data variation. (See chapter 3.)

z Score. A standardized score formed by subtracting the mean and dividing by the standard deviation.

z Statistic. A standardized normal random variable that is frequently used as a test statistic.

Selected Answers

CHAPTER 1

1.1. (a) The population is the set of weights of all shrimp on the diet.
(b) The sample is the set of weights of the 100 shrimp selected from the pond.
(c) A single weight that typifies the collection of weights contained in the population, for example, the "average" or mean weight.
(d) A measure of reliability is needed so that you will know how much faith you can place in your inference.

1.2. (a) The population is the set of radioactivity levels that could be measured at any point in the suspect area.
(b) The sample is the set of two hundred radioactivity level measurements obtained within the area.
(c) The smallest and the largest radioactivity level measurement in the population; also, the "average" or mean activity level.
(d) same as answer to 1.1(d)

1.3. (a) The population is the set of the numbers of children in all households that receive welfare support in the city.
(b) The sample is the set of numbers of children corresponding to the 400 households selected from the welfare roles.
(c) The characteristic of interest would be a number that typifies the number of children per welfare household—for example, the "average" or mean number per household.
(d) same as answer to 1.1(d)

CHAPTER 2

2.5.

Racial-Ethnic Group	Number	Percentage
Indian	793	36.7
Japanese	591	27.3
Chinese	435	20.1
Filipino	343	15.9

2.9., 2.11.

Class Boundaries	f_i	f_i/n
531.5–721.5	11	.23
721.5–911.5	10	.21
911.5–1101.5	6	.13
1101.5–1291.5	5	.10
1291.5–1481.5	4	.08
1481.5–1671.5	1	.02
1671.5–1861.5	3	.06
1861.5–2051.5	4	.08
2051.5–2241.5	1	.02
2241.5–2431.5	3	.06
2431.5–2621.5	0	.00

2.10., 2.12.

Class Boundaries	f_i	f_i/n
18.5–27.5	2	.04
27.5–36.5	6	.12
36.5–45.5	5	.10
45.5–54.5	4	.08
54.5–63.5	2	.04
63.5–72.5	6	.12
72.5–81.5	4	.08
81.5–90.5	6	.12
90.5–99.5	8	.16
99.5–108.5	7	.14

2.13. The range in patient waiting times is 6 to 108 minutes. To construct a relative frequency histogram with 9 subintervals, we use a class interval width of

$$\frac{108.5 - 5.5}{9} = \frac{103}{9} \approx 11.4$$

The class boundaries, the class frequencies (f_i), and relative frequencies (f_i/n) are shown in the table.

Class	Class Boundaries	f_i	f_i/n
1	5.5–16.9	9	.18
2	16.9–28.3	11	.22
3	28.3–39.7	10	.20
4	39.7–51.1	10	.20
5	51.1–62.5	3	.06
6	62.5–73.9	4	.08
7	73.9–85.3	2	.04
8	85.3–96.7	0	.00
9	96.7–108.1	1	.02
Total		50	1.00

CHAPTER 3

3.3. 50% of all workers earn more than $8.48 per hour; 50% earn less than $8.48.

3.4. (a) 20.32 (b) 19.5 (c)

Class Boundaries	f_i	f_i/n
5.5–10.5	4	.08
10.5–15.5	12	.24
15.5–20.5	11	.22
20.5–25.5	11	.22
25.5–30.5	6	.12
30.5–35.5	4	.08
35.5–40.5	0	.00
40.5–45.5	2	.04

(d) no

3.6. $\bar{y} = 4.0$; $s^2 = 9.0$; $s = 3.0$

3.8. Yes; $y = 83.4$ is more than four standard deviations away from the mean.

3.9. Approximately 68% will lie between .7 and 1.9. Approximately 95% will lie between .1 and 2.5. All or nearly all will lie between $-.5$ and 3.1.

3.11. (b) $\Sigma (y - \bar{y})^2 = 9.20$
(c) $s^2 = 2.30$; $s = 1.52$
(d) $\bar{y} - 2s = -.44$; $\bar{y} + 2s = 5.64$

3.12. (b) $\Sigma (y - \bar{y})^2 = 5.5$
(c) $s^2 = 1.1$; $s = 1.05$
(d) $\bar{y} - 2s = -.6$; $\bar{y} + 2s = 3.6$

3.13. (b) $\Sigma (y - \bar{y})^2 = 20.4$
(c) $s^2 = 2.2667$; $s = 1.51$
(d) $\bar{y} - 2s = -.62$; $\bar{y} + 2s = 5.42$

3.14. (a) $s^2 = 70.0996$; $s = 8.37$
(b)

(k)	$\bar{y} - ks$ to $\bar{y} + ks$	Percentage in Interval
1	11.95 to 28.69	70%
2	3.58 to 37.06	96%
3	-4.79 to 45.43	100%

Selected Answers

3.16. (a)

Class	Class Boundaries	f_i	f_i/n
1	1.5–2.5	1	.0143
2	2.5–3.5	1	.0143
3	3.5–4.5	3	.0429
4	4.5–5.5	5	.0714
5	5.5–6.5	5	.0714
6	6.5–7.5	12	.1714
7	7.5–8.5	18	.2571
8	8.5–9.5	15	.2143
9	9.5–10.5	6	.0857
10	10.5–11.5	3	.0429
11	11.5–12.5	0	.0000
12	12.5–13.5	1	.0143
Total		70	1.0000

(b) $\bar{y} = 7.73$
(c) $\Sigma (y - \bar{y})^2 = 271.8429$; $s^2 = 3.9398$; $s = 1.98$

Interval	Percentage	Empirical Rule
$\bar{y} \pm s$: 5.75–9.71	71	68%
$\bar{y} \pm 2s$: 3.77–11.69	96	95%
$\bar{y} \pm 3s$: 1.79–13.67	100	almost all

3.17. (a) $s \approx \text{range}/4 = 1$ (b) $s = 1.23$

3.18. (a) $s \approx \text{range}/4 = 2.0$
(b) $s = 2.02$

3.19. (a) $s \approx \text{range}/4 = 1.25$
(b) $s = 1.51$

3.20. (a) $s \approx 9.50$ (b) $s = 8.37$

3.22. (a) .348 (b) $s \approx .07$
(c) $s = .095$

3.23. $\bar{y} = 3.6$; $s = 2.07$

3.24. $\bar{y} = 2.14$; $s = 1.77$

3.26. according to the Empirical Rule, 68%

3.27. (a) $\bar{y} = 56.9$; median = 56.0; according to definition, there are two modes, 45 and 56.
(b) $s^2 = 305.4333$; $s = 17.48$

3.28. A measurement of 20% lies exactly one standard deviation ($s = 6\%$) away from the mean. $\bar{y} = 14\%$. The proportion of cities giving percentages of unlisted numbers exceeding 20% would probably be in the neighborhood of 16%.

3.29. (a) $\bar{y} = 167.6$; $s^2 = 249.3$; $s = 15.79$
(b) $\bar{y} = 129.43$; $s^2 = 583.9524$; $s = 24.17$

3.30. $\bar{y} = 2.1$; $s^2 = 2.6211$; $s = 1.62$; all the measurements fall within the interval $\bar{y} \pm 2s$, or -1.14 to 5.34.

3.31. (a) $s \approx \text{range}/4 = 1.0$
(b) $s^2 = 1.3132$; $s = 1.15$
(c) $\bar{y} = 1.05$; 95% of the measurements fall within the interval $\bar{y} \pm 2s$, or -1.25 to 3.35.

3.32. Approximately 68% of the measurements will lie between 76° and 90°; approximately 95% will lie between 69° and 97°; all or nearly all will lie between 62° and 104°.

3.33. $\bar{y} = 102.2$; $s^2 = 141.7$; $s = 11.90$

3.34. $\mu \approx 45$

3.35.

Class	Class Boundaries	f_i	f_i/n
1	81.9–91.1	7	.2187
2	91.1–100.3	11	.3437
3	100.3–109.5	5	.1562
4	109.5–118.7	6	.1875
5	118.7–127.9	1	.0312
6	127.9–137.1	2	.0625
Total		32	1.0000

3.36. (a) $\bar{y} = 101.59$; $s = 13.33$

(b) $s \approx 13.75$
(c) 88.26 to 114.92 (72%); 74.93 to 128.25 (94%); 61.60 to 141.58 (100%)

CHAPTER 4

4.1. yes, assuming only two possible outcomes

4.2. no

4.3. yes; estimate will be biased

4.4. .0016; .0256; .0272

4.5. .729; .972

4.8. .143; not enough evidence to prove effectiveness of new drug

4.9. $\mu = 500$; $\sigma = 15.81$; 452.57 to 547.43

4.10. $\mu = 6400$; $\sigma = 35.78$; yes, $y = 6200$ is more than five standard deviations away from the mean.

4.11. (a) .4032 (b) .4713

4.12. (a) .2580 (b) .3849

4.13. (a) .4015 (b) .2794

4.14. (a) .4947 (b) .2406

4.15. (a) .0542 (b) .1857

4.16. .0401

4.17. .8729

4.18. $z_0 = 0$

4.19. $z_0 = 1.96$

4.20. $z_0 = 2.37$

4.21. $z_0 = 1.645$

4.22. $z_0 = 1.96$

4.23. (a) .5 (b) .1056 (c) .9699 (d) .8664 (e) .3413

4.24. (a) .0475 (b) approx. 0 (c) .5934

4.25. .0475

4.26. .8664

4.27. .0336

4.29. .188

4.30. .062

4.31. (a) discrete (b) discrete (c) continuous (d) discrete (e) continuous (f) continuous

4.32. (b) no

4.33. not binomial

4.35. No, $y = 1373$ is more than 15 standard deviations above the mean

4.36. (a) trials may not be independent (b) p = proportion of infected plants (c) $\mu = 200$; $\sigma = 10$; no, $y = 242$ is more than four standard deviations from the mean.

4.38. (a) .59 (b) $1 - .59 = .41$

4.39. (a) 0 (b) $p < .99$

4.40. $z = (y - \mu)/\sigma$

4.41. (a) .4332 (b) .4641

4.42. (a) .4938 (b) .4332

4.43. (a) .2881 (b) .5762

4.44. (a) .95 (b) .9902

4.45. (a) .5119 (b) .0267

4.46. $z = .524$

4.47. $z = 1.645$

4.48. $z = 1.96$

4.49. $z = -.50$

4.50. $z = .75$

4.51. .4649

4.52. .1335

4.53. .0764

4.54. approximately 70 minutes

4.55. approximately 66; approximately 52

4.56. $\mu = 6$; $\sigma = 2.19$

CHAPTER 5

5.2. The process is not likely to approximate random sampling. For example, if the sampling is conducted late in the day, a disproportionately large number of discarded homeowners might be those without children and those who might have negative views on expenditures for public schools.

5.5. According to the Central Limit Theorem, the sampling distribution of $\bar{y}$ will be approximately normal, with $\mu_{\bar{y}} = \mu = 55$ and $\sigma_{\bar{y}} = \sigma/\sqrt{n} = 2.0$.

5.6. According to the Central Limit Theorem, the sampling distribution of $\bar{y}$ will be approximately normal, with $\mu_{\bar{y}} = \mu = 192$ and $\sigma_{\bar{y}} = \sigma/\sqrt{n} = 5.55$.

5.7. see table 5.3

5.9. (a) .2389
(b) .5222
(c) .0764

5.10. (a) .0548
(b) If the mean oxygen content really were 6.0 ppm, the probability that a sample mean would exceed 6.5 ppm is very small (.0548). This suggests that the mean oxygen content is not 6.0 ppm. The data suggest that it is larger than 6.0 ppm.

5.11. According to the Central Limit Theorem, the sampling distribution of $\hat{p}$ will be approximately normal, with $\mu_{\hat{p}} = p = .7$ and $\sigma_{\hat{p}} = \sqrt{pq/n} = .014$.

5.12. about 1.00

5.13. .8464

5.14. (a) .5684   (b) .5684

5.15. (a) .2033   (b) .0668

5.16. (a) The sampling of five insurance claims satisfies the five characteristics of a binomial experiment.
(b) .0023; the probability of observing one or fewer fraudulent claims in a sample of 100, when the true percentage of fraudulent claims is 10%, is very small. Consequently, you must conclude that either you have observed a rare event or that the true percentage of claims that are fraudulent is not 10%. It appears that the percentage of fraudulent claims is less than 10%.

5.17. .2076

5.18. (a) .3085   (b) .6826   (c) .1587

5.19. If the store were open 6 days per week, the mean of the four-week (24 days) shrinkage would be 24($320) = $7680. The standard deviation would be $\sigma\sqrt{n} = \$80\sqrt{24} = \$391.92$. The sampling distribution of the total shrinkage would be approximately normal (see the Central Limit Theorem for sums).

5.20. .0768

5.21. .0793

5.22. $\mu = 3500$; $\sigma = 32.4$

5.23. No; $y = 2600$ lies 27.8 standard deviations below $\mu = 3500$. This is so improbable it suggests that μ is less than 3500 and hence that the proportion of incorrect returns this year is less than 70%.

5.24. .0110

5.25. A highly improbable result has occurred, assuming $p = .05$. This might lead us to believe that the fraction defective exceeds .05.

5.26. $P(y \geq 35) = .1685$

5.27. (a) .0125

5.28. .1038

5.29. .3844

5.30. Approx. 0; because the observed percentage defective (.9) lies so many standard deviations away from the expected (or mean) value (.5), we conclude that either we have observed a rare event or that the manufacturer's claim is incorrect. In fact, the data suggest that the percentage of defectives exceeds .5 percent.

CHAPTER 6

6.1. (a) the set of all registered voters in the state
(b) Divide the state into a series of dis-

tricts. A random sample of districts could be selected; and within the selected districts, a sample of voting precincts could be selected. The voting records within the sampled precincts could be used to obtain information about the percentage (or proportion) of registered voters who have voted at least once over the past two years.

6.2. estimation; the binomial parameter p

6.3. (a) the perceptual set of all lifetimes for all fuses of that particular type manufactured by the company
(b) testing an hypothesis about a population mean

6.4. Obtain a sample of fuses from different suppliers or outlets where the fuses are sold. One important thing would be to avoid (if possible) selecting the sample of fuses from the same shipment (lot or batch) from the manufacturer.

6.6. point estimate; $\bar{y} = 12.2$; bound = .75

6.7. point estimate: $\bar{y} = 63.1$; bound = 3.95

6.9. $\bar{y} = 6.3$; bound = .65

6.10. $\bar{y} = 39.8$; bound = 4.86

6.11. $\bar{y} = 160$; bound = 9.53

6.12. $n = 400$, that is, 290 additional values

6.13. $13.00 \pm 1.645 \, (2.20/\sqrt{70})$, or 12.57 to 13.43

6.14. $13.00 \pm 2.58 \, (2.20/\sqrt{70})$, or 12.32 to 13.68

6.15. $6.5 \pm 1.96 \, (2.6/\sqrt{42})$, or 5.71 to 7.29

6.16. $6.5 \pm 2.58 \, (2.6/\sqrt{42})$, or 5.46 to 7.54

6.17. $3300 \pm 1.96 \, (500/\sqrt{30})$, or 3121.08 to 3478.92

6.18.

n	Confidence Interval	Width
30	3121.08 to 3478.92	357.84
60	3173.48 to 3426.52	253.04
90	3196.70 to 3403.30	206.60
120	3210.54 to 3389.46	178.92

In general, the width of the confidence interval is inversely related to the $\sqrt{n}$. For example, if the sample size is quadrupled (say from 30 to 120), the width is decreased by $\sqrt{4} = 2$.

6.19. $3.27 \pm 1.96 \, (.23/\sqrt{100})$, or 3.22 to 3.32

6.20. $.18 \pm 1.96 \, (.08/\sqrt{100})$, or .16 to .20

6.21. point estimate: $\hat{p} = .28$

6.22. bound: .0669

6.23. $\hat{p} = .175$; bound = .054

6.24. $\hat{p} = .675$; bound = .148

6.25. $\hat{p} = .047$; bound = .013

6.26. (a) random sample
(b) bound = .016

6.27. (a) bound = .024

6.28. (a) bound = .2

6.29. $\hat{p} = .29$; bound = .03

6.30. $.68 \pm 1.645\sqrt{(.68)(.32)/1000}$, or .66 to .70

6.31. $.26 \pm 1.96\sqrt{(.26)(.74)/50}$, or .14 to .38

6.32. .10 to .42

6.33. $\hat{p} = .25$. The value may be an overestimate when solicited.

6.34. $.07 \pm 1.96\sqrt{(.07)(.93)/1200}$ or .06 to .08

6.35. $.69 \pm 1.96\sqrt{(.69)(.31)/400}$, or .64 to .74

6.36. $.26 \pm 1.96\sqrt{(.26)(.74)/1000}$, or .23 to .29

Selected Answers

6.37. (b) no, sample size unknown

6.38. Parameter: a numerical descriptive measure of the population (e.g., μ). Statistic: a numerical descriptive measure of the sample (e.g., $\bar{y}$).

6.39. No; then the population parameters of interest are known and there is no need to estimate them.

6.40. (a) Point estimate: a single number computed from the sample, which is used to estimate the parameter of interest. Interval estimate: the sample data are used to compute two numbers; and the interval formed by the two numbers is an interval estimate of the parameter of interest.
(b) estimation and hypothesis testing
(c) bound on error
(d) confidence coefficient and the width of the confidence interval

6.41. $\hat{p} = .10$; bound $= .03$

6.42. $120 \pm 1.96\,(5.50/\sqrt{100})$, or 118.92 to 121.08

6.43. $\bar{y} = 3.50$; bound $= .27$

6.44. 3.24 to 3.76

6.45. $\bar{y} = 25.83$; bound $= 3.22$

6.46. Central Limit Theorem

6.47. 22.67 to 28.99

6.48. $3250 \pm 2.58\,(420/\sqrt{70})$, or 3120.49 to 3379.51

6.49. $3.2 \pm 1.645\,(0.3/\sqrt{50})$, or 3.13 to 3.27

6.50. (a) no (b) no (c) yes

6.51. $.55 \pm 1.96\sqrt{(.55)(.45)/200}$, or .48 to .62

6.52. $.38 \pm 1.96\sqrt{(.38)(.62)/870}$, or .35 to .41

6.53. $\hat{p} = .48$; bound $= .08$

6.54. $.19 \pm 1.645\sqrt{(.19)(.81)/300}$, or .15 to .23

6.55. $.10 \pm 2.58\sqrt{(.1)(.9)/400}$, or .06 to .14

6.56. $.52 \pm 2.58\sqrt{(.52)(.48)/1508}$, or .49 to .55

6.57. $.12 \pm 1.96\sqrt{(.12)(.88)/1508}$, or .10 to .14

6.58. $.54 \pm 1.96\sqrt{(.54)(.46)/1540}$, or .52 to .56

CHAPTER 7

Note: Abbreviations used are R.R. = rejection region; T.S. = test statistic.

7.1. $z = \dfrac{63.7 - 68}{14.2/\sqrt{50}} = -2.14$. Since $-2.14 < -1.645$, reject H_0.

7.2. rejection region: $|z| > 2.33$; insufficient evidence to reject H_0

7.3. $H_0: \mu = 525$.
$H_a: \mu > 525$.
T.S.: $z = \dfrac{\bar{y} - \mu_0}{\sigma/\sqrt{n}}$.
R.R.: For $\alpha = .05$, reject H_0 when $z > 1.645$.

7.4. $H_0: \mu = 525$.
$H_a: \mu \neq 525$.
T.S.: $z = \dfrac{\bar{y} - \mu_0}{\sigma/\sqrt{n}}$.
R.R.: For $\alpha = .01$, reject H_0 if $|z| > 2.58$.

7.5. $H_0: \mu = 0$.
$H_a: \mu > 0$.
T.S.: $z = \dfrac{\bar{y} - \mu_0}{\sigma/\sqrt{n}}$.
R.R.: For $\alpha = .05$, reject H_0 if $z > 1.645$.

7.6. (a) $z = \dfrac{10.3}{4.6/\sqrt{35}} = 13.25$; reject H_0
(b) Not with this information. One would have to study separate groups of overweight persons simultaneously, one group using the fixed diet only and another group with the diet plus the weight-reducing agent.

7.7. $H_0: \mu = 8.2$.
$H_a: \mu < 8.2$.

Selected Answers

7.8. T.S.: $z = \dfrac{7.6 - 8.2}{1.8/\sqrt{50}} = -2.36$.
R.R.: For $\alpha = .05$, reject H_0 if $z < -1.645$. Reject H_0.

7.9. $H_0: \mu = .3$.
$H_a: \mu > .3$.
T.S.: $z = \dfrac{.7 - .3}{.4/\sqrt{60}} = 7.75$.
R.R.: For $\alpha = .05$, reject H_0 if $z > 1.645$. Reject H_0.

7.10. $H_0: p = .50$.
$H_a: p \neq .50$.

7.11. For $\hat{p} = 110/200 = .55$, $z = 1.41$. Insufficient evidence to reject H_0.

7.12. For $\hat{p} = .45$, $z = 2.04$. Reject H_0.

7.13. For $\hat{p} = .56$, $z = 3.79$. Reject H_0.

7.14. $H_0: p = .001$.
$H_a: p > .001$.
T.S.: $z = \dfrac{\hat{p} - p_0}{\sigma_{\hat{p}}} = \dfrac{.002 - .001}{.000316} = 3.16$.
R.R.: For $\alpha = .05$, reject H_0 if $z > 1.645$. Reject H_0.

7.15. $H_0: p = .05$.
$H_a: p > .05$.
T.S.: $z = \dfrac{\hat{p} - p_0}{\sigma_{\hat{p}}} = \dfrac{.08 - .05}{.01} = 3.08$.
R.R.: For $\alpha = .05$, reject H_0 if $z > 1.645$. Reject H_0.

7.16. For $\hat{p} = .12$, $z = 0.67$. Insufficient evidence to reject H_0.

7.17. For $y = 18$ defectives, $z = 2.67$. Reject H_0.

7.18. (a) $p = .0314$
(b) $p = 2(.0314) = .0628$

7.19. (a) .0102 (b) .0204

7.20. $z = 1.74$; level of significance: .0409

7.21. $H_0: p = .5$.
$H_a: p \neq .5$.
T.S.: $z = \dfrac{\hat{p} - .5}{\sigma_{\hat{p}}}$, where $\sigma_{\hat{p}} = \sqrt{(.5)(.5)/100} = .05$.

R.R.: For $\alpha = .05$, reject H_0 if $|z| > 1.96$.

7.22. (a) For $\hat{p} = .61$, $z = 2.20$. Reject H_0.
(b) $\alpha = .05$ is the probability of incorrectly rejecting H_0.

7.23. .0278

7.24. For $\hat{p} = .45$, $z = -1.73$. Insufficient evidence to reject H_0.

7.25. $H_0: p = .5$.
$H_a: p > .5$.
T.S.: $z = 4.67$.
Level of significance: less than .001.

7.26. $z = 18.91$; reject H_0

7.27. less than .001

7.28. $H_0: \mu = 1.6$.
$H_a: \mu \neq 1.6$.
T.S.: $z = \dfrac{\bar{y} - \mu_0}{s/\sqrt{n}} = \dfrac{2.2 - 1.6}{.57/\sqrt{36}} = 6.32$.
R.R.: For $\alpha = .05$, reject H_0 if $|z| > 1.96$. Reject H_0.

7.29. $z = \dfrac{4.05 - 4}{.12/\sqrt{50}} = 2.95$; reject H_0

7.30. $z = \dfrac{6.2 - 5}{5.2/\sqrt{36}} = 1.38$; insufficient evidence to reject H_0

7.31. $z = \dfrac{78.3 - 80}{2.9/\sqrt{30}} = -3.21$; reject H_0

7.32. $z = \dfrac{.025 - .027}{.003/\sqrt{50}} = -4.71$; reject H_0

7.33. For $\hat{p} = .25$, $z = 1.25$; level of significance: .1056.

7.34. Large sample sizes are required to run the z test for tests related to μ or p.

7.35. For $\hat{p} = .69$, $z = 3.59$. Reject H_0.

7.36. For $\hat{p} = .07$, $z = 1.30$. Insufficient evidence to reject H_0.

7.37. $z = 5.60$; reject H_0

7.38. $z = -2.71$; reject H_0

Selected Answers

7.39. $z = -8.1$; reject H_0

7.40. $z = -2.74$; reject H_0

7.41. H_0: $p = .001$.
H_a: $p > .001$.
T.S.: $z = 2.12$.
R.R.: For $\alpha = .05$, reject H_0 if $z > 1.645$. Reject H_0.

7.42. .017

7.43. $z = -2.86$; reject H_0

7.44. .0042

CHAPTER 8

Note: R.R. = rejection region; T.S. = test statistic.

8.1. (a) $\mu_{\bar{y}_1 - \bar{y}_2} = \mu_1 - \mu_2 = 200$
(b) $\sigma_{\bar{y}_1 - \bar{y}_2} = \sqrt{\dfrac{\sigma_1^2}{n_1} + \dfrac{\sigma_2^2}{n_2}} = 5.76$

8.2. 95% of the sample differences will be within $2(6.01) = 12.02$ of $\mu_1 - \mu_2 = 200$.

8.3. H_0: $\mu_1 - \mu_2 = 0$.
H_a: $\mu_1 - \mu_2 > 0$.
T.S.: $z = 2.24$.
R.R.: For $\alpha = .05$, reject H_0 if $z > 1.645$. Reject H_0.

8.4. .0125

8.5. 5.76 to 12.24

8.6. 4.73 to 13.27

8.7. For H_a: $\mu_1 - \mu_2 > 0$, $z = 1.77$. Reject H_0.

8.8. $z = 24.75$; reject H_0

8.9. (a) $\mu_{\hat{p}_1 - \hat{p}_2} = p_1 - p_2 \doteq .2$
(b) $\sigma_{\hat{p}_1 - \hat{p}_2} = \sqrt{\dfrac{p_1 q_1}{n_1} + \dfrac{p_2 q_2}{n_2}} = .02$

8.10. For $\hat{p}_1 = .36$ and $\hat{p}_2 = .25$, $-.14$ to $.36$.

8.11. $z = 1.12$; insufficient evidence to reject H_0

8.12. For $\hat{p}_1 = .58$ and $\hat{p}_2 = .46$, $z = 1.70$. Level of significance is $p = .0446$.

8.13. $-.034$ to $.124$

8.14. $-.01$ to $.21$

8.15. $z = -10.69$; reject H_0

8.16. $z = 8.46$; reject H_0

8.18. For $\hat{p}_1 = .32$ and $\hat{p}_2 = .40$, $z = -1.67$. Insufficient evidence to reject H_0.

8.19. $(.69 - .40) \pm 1.96(.07)$, or $.15$ to $.43$.

8.20. $z = 3.96$; reject H_0

8.21. -3.98 to 5.38

8.22. $z = 3.70$; $p < .001$

8.23. $z = 3.33$; reject H_0

8.25. $(260 - 294) \pm 2.58(14.81)$, or -72.22 to 4.22

8.26. $z = -1.45$; insufficient evidence to reject H_0

8.27. $z = 3.78$; reject H_0

8.28. $z = 3.60$; reject H_0

8.29. H_a: $\mu_1 - \mu_2 > 0$; $z = -1.72$. Reject H_0.

8.30. 633.94 to 4006.06

8.31. $z = -2.81$; $p = .005$

8.32. $z = 2.54$; reject H_0

8.33. $z = 3.36$; reject H_0

8.34. $z = 1.36$; insufficient evidence to reject H_0

8.35. H_0: $\mu_1 - \mu_2 = 0$.
H_a: $\mu_1 - \mu_2 \neq 0$.
T.S.: $z = -4.03$
$p < .002$.

8.36. $-.18 \pm .09$, or $-.27$ to $-.09$

CHAPTER 9

Note: R.R. = rejection region; T.S. = test statistic.

9.1. The distribution of the test statistic $\dfrac{\bar{y} - \mu_0}{s/\sqrt{n}}$ for $n < 30$ is no longer normal but possesses a t distribution.

Selected Answers

9.2. (a) Reject H_0 if $t < -1.761$.
(b) Reject H_0 if $|t| > 2.074$.
(c) Reject H_0 if $t > 2.015$.

9.3. (a) Reject H_0 if $t < -2.624$.
(b) Reject H_0 if $|t| > 2.819$.
(c) Reject H_0 if $t > 3.365$.

9.4. T.S.: $t = 1.64$.
R.R.: Reject H_0 if $t > 1.74$. Insufficient evidence to reject H_0.

9.5. H_0: $\mu = 650$.
H_a: $\mu < 650$.
T.S.: $t = \dfrac{638.6 - 650}{12.74/\sqrt{5}} = -2.00$.
R.R.: Reject H_0 if $t < -2.132$.
Insufficient evidence to reject H_0.

9.6. T.S.: $t = 5.99$.
R.R.: Reject H_0 if $t > 1.746$.
Reject H_0.

9.7. $t = -1.25$; insufficient evidence to reject H_0

9.8. 1.55 to 2.85

9.9. $38.6 \pm 2.776 (2.55)$, or 31.52 to 45.68

9.10. 3.56 to 3.91

9.11. 3.55 to 4.97

9.12. (a) Reject H_0 if $|t| > 2.064$.
(b) Reject H_0 if $t > 2.624$.
(c) Reject H_0 if $t < -1.86$.

9.13. T.S.: $t = -2.67$.
R.R.: Reject H_0 if $t < -1.703$.

9.14. $.005 < p = \text{value} < .01$

9.15. $t = -18.51$; reject H_0

9.16. T.S.: $t = 1.61$.
R.R.: Reject H_0 if $|t| > 2.306$.
Insufficient evidence to reject H_0.

9.17. T.S.: $t = -4.54$.
R.R.: Reject H_0 if $|t| > 1.96$.
Reject H_0.

9.18. less than .01

9.19. $6.2 \pm 2.878(3.75)\sqrt{\dfrac{1}{10} + \dfrac{1}{10}}$, or 1.37 to 11.03

9.20. $-212.67 \pm 1.812(19.90)\sqrt{\dfrac{1}{6} + \dfrac{1}{6}}$, or -233.49 to -191.85

9.21. $-6.30 \pm 2.101(4.20)\sqrt{\dfrac{1}{10} + \dfrac{1}{10}}$, or -10.25 to -2.35

9.22. $t = -6.96$; reject H_0

9.23. $t = 1.08$; insufficient evidence to reject H_0

9.24. greater than .10

9.25. $t = .58$; insufficient evidence to reject H_0

9.26. 21.90 to 26.48

9.27. $2.28 \pm 1.895(0.25)/\sqrt{8}$, or 2.11 to 2.45

9.28. 1.90 to 2.22

9.29. 8.90 to 9.04

9.30. (a) $t = -2.16$; $p = .056$
(b) $t = 1.689$; $p = .122$

9.31. -57.42 to 48.26

9.32. $t = -2.14$; insufficient evidence to reject H_0

9.33. 1.92 to 10.48

9.34. $t = -1.49$; insufficient evidence to reject H_0

9.35. $t = -.17$; insufficient evidence to reject H_0

9.36. $t = 6.90$; reject H_0

9.37. $p < .01$

9.38. $t = 2.46$; insufficient evidence to reject H_0

9.39. -1.41 to 1.21

9.40. $3.83 \pm 2.132 (0.68) \sqrt{\dfrac{1}{3} + \dfrac{1}{3}}$, or 2.65 to 5.01

9.41. $-.34$ to 5.54

9.42. $t = -3.33$; reject H_0

CHAPTER 10

10.3. $y = 7.8$

10.7. $\hat{y} = .50 + 1.70x$

Selected Answers

10.8. (a) $\hat{y} = .05 + 1.35x$
(b) 8.15

10.9. $\hat{y} = 2.47 + 1.63x$

10.10. (a) $\hat{y} = 4.70 + 1.97x$
(b) 73.66

10.11. $\hat{y} = -191.27 + .28x$; $\hat{y} = 816.73$

10.13. $t = 6.41$; reject H_0

10.14. (b) $\hat{y} = 62.81 + 4.39x$

10.15. $t = 4.68$; reject H_0

10.17. $\hat{\rho} = .995$

10.18. $\hat{\rho} = .924$

10.19. $t = 6.41$; reject H_0

10.20. $\hat{\rho} = .856$

10.21. (a) $\hat{y} = 31.33 - 7.33x$
(b) $s^2 = 1.2$
(c) $t = -12.96$; reject H_0
(d) $\hat{y} = 14.47$

10.22. (a) $\hat{\rho} = -.9715$
(b) $t = -12.96$; reject H_0

10.23. (b) $\hat{y} = 11.82 + 1.36x$
(c) $s^2 = 1.42$
(d) $t = 5.71$; reject H_0
(e) $\hat{y} = 18.62$

10.24. (a) $\hat{\rho} = .907$
(b) $t = 5.71$; reject H_0

10.25. (b) $\hat{y} = 1.00 + 9.39x$
(c) $t = 11.39$; reject H_0
(d) $\hat{y} = 21.66$

10.26. $\hat{\rho} = .99$; yes

10.27. (b) $\hat{y} = 11.24 + 1.31x$
(c) $t = 3.46$; reject H_0
(d) $\hat{\rho} = .77$
(e) $\hat{y} = 50.54$

10.28. $\hat{\rho} = .63$

10.29. $\hat{\rho} = .67$

10.30. (a) 2.23 to 6.55
(c) 1.24 to 7.54

10.31. (a) $\hat{y} = 9.275 + .505x$
(b) .29981 to .71019
(c) .9754

CHAPTER 11

11.1. (a) 8
(b) Yes; reject H_0: $\mu_d = 0$ in favor of H_a: $\mu_d \neq 0$;
$t = -2.76$ exceeds $t_{.025} = 2.306$.
(c) -4.99 to -1.21

11.2. (a) Reject H_0: $\mu_d = 0$ in favor of H_a: $\mu_d > 0$; $t = 5.45$ exceeds 1.833.
(b) 1.58 to 3.82

11.3. Will likely lose information; for explanation, see section 11.2.

11.4. (a) No; do not reject H_0: $\mu_d = 0$; $t = 1.83$ is less than $t_{.05} = 2.015$.
(b) $-.14$ to 2.90

11.5. (a) No; do not reject H_0: $\mu_d = 0$; $t = -1.84$ does not exceed .895 in absolute value
(b) -8.71 to 1.21

11.6. $n \approx 324$

11.7. $n \approx 25$

11.8. $n \approx 396$

11.9. $n \approx 100$

11.12. paired-difference experiment

11.13. No; pairing only increases the information in the experiment if the experimental units within pairs are more homogeneous than those between pairs; it occurs when variation within pairs is less than variation between pairs.

11.14. (a) No; do not reject H_0: $\mu_d = 0$; $t = 1.67$ is less than $t_{.025} = 2.571$.
(b) $-.04$ to .20
(c) $-.13$ to .29
(d) No, because the two observations in each individual pair are not independent.
(e) If no pairing occurred in the design, that is, if you have collected independent random samples, then a paired-difference analysis will result in a loss of information (because you lose degrees of freedom for your Student's t).

11.15. (b) Reject H_0: $\mu_d = 0$ in favor of H_a: $\mu_d \neq 0$; $t = 4.92$ exceeds 2.447.

11.16. (b) Reject H_0: $\mu_d = 0$ in favor of H_a: $\mu_d > 0$; $t = 4.09$ exceeds 2.821.

11.17. $n \approx 81$

11.18. $n = 324$

11.19. $n = 81$

11.20. $n \approx 144$

11.21. (b) Do not reject H_0: $\mu_d = 0$; $t = 2.39$ does not exceed 2.571.

11.22. The design enables you to eliminate variation in the durability difference measurements caused by environmental differences from one location to another.

11.23. (a) No; do not reject H_0: $\mu_d = 0$; $t = 2.12$ is less than $t_{.025} = 2.306$. (b) $-.01$ to $.21$

11.24. $n = 157$

11.25. $n = 1111$

11.26. $n = 2500$

11.27. $n = 703$

11.28. $n = 100$

11.29. $n = 144$

11.30. $n \approx 61$ pairs

11.31. $n_1 = n_2 \approx 15$

CHAPTER 12

12.1. (a) 2.91
(b) 3.71
(c) 2.42
(d) 4.28
(e) 3.03

12.2. (a) 2.215
(b) 2.48
(c) 1.96
(d) 2.70
(e) 2.76

12.3. No; do not reject H_0: $\sigma_1^2 = \sigma_2^2$; $F = 1.716$ is less than $F_{.05} = 3.68$.

12.4. No; do not reject H_0: $\sigma_A^2 = \sigma_B^2$; $F = 3.154$ is less than $F_{.05} = 3.79$.

12.5. (a) H_0: $\sigma_1^2 = \sigma_2^2$; H_a: $\sigma_1^2 > \sigma_2^2$
(b) H_a: $\sigma_1^2 \neq \sigma_2^2$

12.6. Do not reject H_0: $\sigma_1^2 = \sigma_2^2$; $F = 2.73$ does not exceed 2.98.

12.7. Do not reject H_0: $\sigma_1^2 = \sigma_2^2$; $F = 2.70$ does not exceed 5.35.

12.8. Reject H_0: $\sigma_1^2 = \sigma_2^2$ in favor of H_a: $\sigma_1^2 > \sigma_2^2$; $F = 2.27$ exceeds 2.12.

12.9. (a) Reject H_0: $\sigma_1^2 = \sigma_2^2$ in favor of H_a: $\sigma_1^2 \neq \sigma_2^2$ $F = 2.27$ exceeds 2.12.
(b) Do not reject H_0: $\sigma_1^2 = \sigma_2^2$ in favor of H_a: $\sigma_1^2 \neq \sigma_2^2$; $F = 2.27$ is less than 2.94.

12.10. Do not reject H_0: $\sigma_1^2 = \sigma_2^2$ $F = 3.57$ does not exceed the critical value, with $v_1 = 9$, $v_2 = 9$, $\alpha = .01$, of $F_{.01} = 5.35$.

12.11. Reject H_0: $\sigma_1^2 = \sigma_2^2$ in favor of H_a: $\sigma_1^2 \neq \sigma_2^2$ since $F = 3.57$ exceeds the critical value, for $\alpha = .10$, of $F = 3.18$.

12.12. (a) Reject H_0: $\sigma_1^2 = \sigma_2^2$ in favor of H_a: $\sigma_1^2 \neq \sigma_2^2$ since $F = 2.20$ exceeds the critical value, for $\alpha = .10$, $v_1 \approx 15$, $v_2 = 23$, of $F \approx 2.13$.

12.13. Reject H_0: $\sigma_1^2 = \sigma_2^2$ in favor of H_a: $\sigma_1^2 \neq \sigma_2^2$ since $F = 3.13$ exceeds the critical value, for $\alpha = .10$, $v_1 = v_2 = 12$, $F = 2.69$.

12.14. H_0: $\sigma_1^2 = \sigma_2^2$; H_a: $\sigma_1^2 > \sigma_2^2$

12.15. Yes; reject H_0: $\sigma_1^2 = \sigma_2^2$ in favor of H_a: $\sigma_1^2 > \sigma_2^2$; $F = 3.619$ is greater than $F_{.05} = 3.18$.

12.16. (a) The variances of the two populations should be equal.
(b) No; do not reject H_0: $\sigma_A^2 = \sigma_B^2$, $F = 5.897$ is less than $F_{.05} = 6.39$.

CHAPTER 13

13.2. $F = 8.72$; reject H_0

13.3. .86 to 2.88

13.4. $F = 3.34$; reject H_0

13.5. $F = 2.51$; insufficient evidence to reject H_0

Selected Answers

13.6. $F = 6.05$; $.01 < p\text{-value} < .05$

13.7. -4.03 to 22.03

13.8. $F = 22.86$; reject H_0

13.9. (b) 1.732
(c) $F = 6.055$; $p\text{-value} = .022$

13.10. $F = 17.69$; reject H_0

13.11. $-.37$ to $-.13$

13.12. $F = 14.73$; reject H_0

13.13. (b) $F = 1.799$; $p\text{-value} = .225$

13.15. (a) $F = 14.6$; reject H_0
(b) Machine 2

13.16. (a) 6.63 to 13.81
(b) -1.41 to 5.77

13.17. (a) preparation B
(b) $F = 20.60$; reject H_0

13.18. $F = 64.50$; reject H_0

13.20. (b),(c) $F = 2.64$; do not reject H_0
(d) locations 1 and 4

13.21. 1.01 to 7.77

CHAPTER 14

14.2. (a) 2   (b) 4

14.3. (b)

	I	II	Total
Favorable	11	11	22
Unfavorable	9	9	18
Total	20	20	40

14.4. $\chi^2 = 6.46$; reject H_0

14.5. $\chi^2 = 6.09$; reject H_0

14.6. (b) $.025 < p\text{-value} < .05$

14.7. $\chi^2 = 32.64$; reject H_0

14.8. $\chi^2 = 7.38$; $p\text{-value} > .10$

14.9. $\chi^2 = 17.43$; reject H_0

14.10. $\chi^2 = 7.86$; reject H_0

14.11. $\chi^2 = 4.00$; insufficient evidence to reject H_0

14.12. $\chi^2 = 9.11$; reject H_0

14.13. $\chi^2 = 1.43$; insufficient evidence to reject H_0

14.14. $\chi^2 = 6.20$; insufficient evidence to reject H_0

14.15. $\chi^2 = 4.36$; insufficient evidence to reject H_0

14.16. $\chi^2 = 32.96$; $p\text{-value} < .005$

14.17. (a) .323, .248, .167, .167

(b) OBSERVED VALUES TITLE 14.17
```
 42.    61.   20.   12.   135.
 88.   185.  100.   60.   433.
130.   246.  120.   72.   568.
```
EXPECTED VALUES
```
 30.9   58.5   28.5   17.1
 99.1  187.5   91.5   54.9
```
OBS. MINUS EXP.
```
 11.1    2.5   -8.5   -5.1
-11.1   -2.5    8.5    5.1
```
CHI SQUARE BY CELL
```
  4.0    0.1    2.5    1.5
  1.2    0.0    0.8    0.5
```
CHI SQUARE = 10.71

3 DEGREE(S) OF FREEDOM

14.18. $.01 < p\text{-value} < .025$

14.19. (a) .10, .14, .19
(b) $\chi^2 = 3.31$; insufficient evidence to reject H_0

14.22.

OBSERVED VALUES TITLE 14.22
```
181.  137.  318.
101.  111.  212.
282.  248.  530.
```
EXPECTED VALUES
```
168.2  148.8
112.8   99.2
```
OBS. MINUS EXP.
```
 11.8  -11.8
-11.8   11.8
```
CHI SQUARE = 4.40

1 DEGREE(S) OF FREEDOM

Selected Answers

CHAPTER 15

Note: R.R. = rejection region; T. S. = test statistic.

15.1. H_0: $p = .5$.
H_a: $p \neq .5$.
T.S.: $z = \dfrac{29 - 25}{3.54} = 1.13$.
R.R.: For $\alpha = .05$, reject H_0 if $|z| > 1.96$. Insufficient evidence to reject H_0.

15.2. .258.

15.3. For $y = 15$ and $n = 20$, $z = 2.236$. Reject H_0.

15.4. For $y = 4$ and $n = 12$, $z = -1.15$. Insufficient evidence to reject H_0.

15.5. For $y = 1$ and $n = 11$, $z = -2.71$. Reject H_0.

15.6. H_0: $p = .5$.
H_a: $p \neq .5$.
T.S.: $z = \dfrac{1 - 5}{\sqrt{2.50}} = -2.53$.
R.R.: For $\alpha = .05$, reject H_0 if $|z| > 1.96$. Reject H_0.

15.7. For $y = 11$ and $n = 14$, $z = 2.14$. Reject H_0.

15.8. $T = 4$; reject H_0

15.9.

WILCOXON SIGNED RANK TEST

FILES
  APP 1—APPROACH 1
  APP 2—APPROACH 2

APP 1	APP 2	DIFFERENCE
3.00000	2.00000	−1.00000
4.00000	2.00000	−2.00000
3.00000	5.00000	2.00000
5.00000	4.00000	−1.00000
4.00000	3.00000	−1.00000
3.00000	2.00000	−1.00000
3.00000	4.00000	1.00000
4.00000	3.00000	−1.00000
3.00000	2.00000	−1.00000
4.00000	2.00000	−2.00000
5.00000	4.00000	−1.00000
3.00000	4.00000	1.00000
2.00000	1.00000	−1.00000
3.00000	2.00000	−1.00000
5.00000	3.00000	−2.00000
5.00000	4.00000	−1.00000
5.00000	3.00000	−2.00000
2.00000	3.00000	1.00000
4.00000	2.00000	−2.00000
4.00000	3.00000	−1.00000

SUM OF POSITIVE SIGNED RANKS: 40.00000
SUM OF NEGATIVE SIGNED RANKS: 170.00000
NUMBER OF NONZERO DIFFERENCES: 20

T = 40.00000
PROBABILITY < = 0.02

15.10. $T = 4$; reject H_0

15.11. $z = 2.71$; reject H_0

15.12.

WILCOXON SIGNED RANK TEST

FILES
  PL—PLACEBO
  BEN—BENZEDRINE

PL	BEN	DIFFERENCE
250.00000	258.00000	8.00000
271.00000	285.00000	14.00000
243.00000	245.00000	2.00000
252.00000	250.00000	−2.00000
266.00000	268.00000	2.00000
272.00000	278.00000	6.00000
293.00000	280.00000	−13.00000
296.00000	305.00000	9.00000
301.00000	319.00000	18.00000
298.00000	308.00000	10.00000
310.00000	320.00000	10.00000
286.00000	293.00000	7.00000
306.00000	305.00000	−1.00000
309.00000	313.00000	4.00000

SUM OF POSITIVE SIGNED RANKS: 89.00000
SUM OF NEGATIVE SIGNED RANKS: 16.00000
NUMBER OF NONZERO DIFFERENCES: 14

$T = 16.00000$
PROBABILITY $< = 0.02$

15.13.

WILCOXON RANK SUM TEST

FILES
  1ST: DRUG—POST-PRE (DRUG)
  2ND: CTL—POST-PRE (CONTROL)

	DRUG	CTL
	55.70000	62.60000
	52.50000	56.0000
	57.90000	56.40000
	58.00000	63.60000
	55.20000	71.90000
	55.80000	66.20000
	44.10000	69.50000
	45.70000	60.30000
	54.00000	66.00000
	34.20000	72.90000
NO. OBSERVATIONS:	10	10
MEDIAN:	54.60000	64.80000
SUM OF RANKS:	59.0000	151.0000
AVERAGE RANK:	5.9000	15.1000

$Z = -3.48$

15.15. $z = 1.66$; insufficient evidence to reject H_0.

15.16.

```
UNPAIRED T TEST

  FILES
    1ST: C1—CONTROL
    2ND: TR1—TREATED
```

	C1	TR1
	9.00000	59.00000
	12.00000	44.00000
	36.00000	63.00000
	77.50000	87.50000
	−7.50000	30.50000
	32.50000	57.50000
	−17.00000	51.00000
	36.00000	−9.00000
	42.00000	75.50000
	65.00000	28.00000
	30.50000	82.00000
	45.50000	65.00000

	1ST	2ND
NO. OF OBSERVATIONS	12.	12.
MEAN	30.12500	52.83333
STANDARD DEVIATION	27.50878	26.79496
STANDARD ERROR	7.94110	7.73504
COEFFICIENT OF VARIATION	91.31545	50.71601

RATIO OF MEANS (2ND/1ST)	1.75380
DIFFERENCE OF MEANS (2ND-1ST)	22.70833
STANDARD ERROR OF DIFFERENCE	11.08566
95% CONFIDENCE INTERVAL FOR DIFFERENCE OF MEANS	[−0.28333, 45.69999]
RATIO OF VARIANCES (2ND/1ST)	0.94878
T STATISTIC (EQUAL VARIANCES)	2.04844
DEGREES OF FREEDOM	22
PROBABILITY	0.05263

15.17. $z = -1.85$; p-value $= .0644$

15.18. $z = -.63$; insufficient evidence to reject H_0.

15.19. $z = 2.11$; reject H_0

15.20. (a) $\hat{\rho}_s = .55$

(b) $z = 1.83$; insufficient evidence to reject H_0

15.21. (a) $\hat{\rho}_s = .95$

(b) $z = 2.84$; reject H_0

15.22. $z = 1.90$; insufficient evidence to reject H_0

15.23. $z = -1.73$; insufficient evidence to reject H_0

15.24. (a) $T = 84$

(b) $z = -3.81$; reject H_0

Index

Acceptance region, 166
   how to select, 169–170
Additivity of probabilities, 76–78
Alpha, 169
Alternative hypothesis, 165
Analysis of variance, 319–339
   for comparing more than two means, 321–328
   logic behind, 319–321
   and test of hypothesis, 325–326
Applications of statistics, 5–8
Arithmetic mean, 38–40

Bar chart, 15–18
Beta, 169
Biased samples, 392–394
Binomial experiment, 82
Binomial parameter, 150
   confidence interval for, 155–156
   difference between two, 192–193, 200–205
   estimator of, 150
   point estimation of, 150–152
   and sample size for estimating, 291–293
   statistical test for, 176–178
Binomial probability distribution, 81–89
   and approximating normal curve, 91, 125
   mean of, 87
   standard deviation of, 87
Binomial random variable, 83–89, 124–130
   mean of, 87, 125
   probability distribution for, 84
   standard deviation of, 87, 125
Bound on error of estimation, 143, 151, 195, 201, 279, 289–291

Central Limit Theorem, 116–123, 191
   for means, 116–121, 191
   significance of, 121–122
   for sums, 121
Central tendency, measures of, 38–45
Characteristic, 2
Chi-square distribution, 344–345
   degrees of freedom for, 345
   tabulated values of, 345
Chi-square statistic, 344–345
Chi-square test of independence, 346
Circle chart, 14–15
Class boundaries, 20–21
Classes, 20
   number of, 23–24

Class frequency, 21
Class width, 20
Coefficient of determination, 268
Coefficient of linear correlation, 266–269.
   See also Correlation coefficient
Collection of data, 2
Comparing more than two population means, 321–328
Comparing two methods of classification, 342–348
Comparing two (or more) populations, 362–386
Comparing two population means, 194–198, 229–232, 234–236, 280–286
Comparing two population proportions, 200–205
Comparing two population variances, 303–316
Comparisons, 189–214, 303–316, 321–328
Complement of event, 77
Conceptual population, 3, 115
Conditional probability, 78–80
Confidence coefficient, 147
Confidence interval, 147–149
   based on paired-difference design, 285
   for binomial parameter p, 155–156
   for difference between two binomial parameters, 201
   for difference between two population means, 195, 234–236
   large-sample, 149, 156, 195, 201
   for population mean, 146–149, 227–228
   for slope, 261–262
   small-sample, 227–228, 234–236
   summary of, 403–409
Confidence limits, 148
Contingency table, 341–359
Continuous random variable, 80, 91
Correlation, 247–277
Correlation coefficient, 266–269, 379
   interpretation of, 267–268
   linear, 266–269
   of population, 266
   rank, 379
   statistical test for, 269, 381–382
Curvilinear relationship, 250–251

Decision making, 111, 166, 168
Degrees of freedom, 220
   for chi-square, 345

Degrees of freedom (*continued*)
  for F, 304
  for Student's t, 220
Descriptive statistics
  graphical methods for, 13–31
  numerical methods, for, 37–68
Design of an experiment, 279–300
  noise-reducing, 280–286
  paired-difference, 280–286
  randomized block, 286
  volume-increasing, 281, 286
Deviation, 51
  from mean, 51
  and regression line, 255–257
  standard, 53–59
Difference between two binomial parameters, 192–193, 200–205
  confidence interval for, 201
  mean of, 192–193
  point estimation of, 201
  standard deviation of, 192–193
  test for, 203
Difference between two means, 191–192, 194–198, 280–286
  confidence interval for, 195, 234–236, 285
  estimating, 195
  mean of, 191–192
  point estimation of, 195
  standard deviation of, 191–192
  test for, 197, 229–232, 283–284
Discrete random variable, 80
Disraeli, 389
Distribution, probability. *See* Probability distribution
Distribution, skewed, 42
Dot diagram, 50–51

Empirical Rule, 53–55, 57–59, 116
Enumerative data, 341
Error
  bound on, 143, 151, 195, 201
  of estimation, 142, 151, 195, 201
  probability of type I, 169
  probability of type II, 169
  type I, 169
  type II, 169
Estimate
  interval, 140, 149, 155–156, 195, 201, 227, 235, 261
  point, 139, 143, 150, 195, 201
Estimation, 111, 137–156. *See also* Interval estimation; Point estimation
  for binomial parameter, 150–152, 155–156
  for difference between two binomial parameters, 201
  for difference between two means, 195, 234–236, 285
  error of, 142, 151, 195, 201
  goodness of, 143, 147, 151
  for population mean, 141–145, 146–149, 227–228
  for slope, 261–262
Events, 74–75
  complement of, 77
  independent, 79–80
  mutually exclusive, 76
Expected cell count, 343
Experiment, 74
  binomial, 82
  design of, 279–300

F distribution, 304–307
  degrees of freedom for, 304
  tabulated values of, 304–306
F statistic, 305–306, 324
F test, 307
  in analysis of variance, 324–326
Freehand regression line, 249–254
Frequency, 16
Frequency distribution, 18–24, 37, 53
Frequency histogram, 18–24
  and area under, 28
  classes for, 20
  construction of, 21–24
  interpretation of, 28–31
  and number of classes, 23–24
  and probability, 80–81
Frequency polygon, 27–28
Frequency table, 21

Gauss, Karl Friedrich, 91
Glossary of common statistical terms, 435–438
Gosset, W. S., 219
Graphical distortions, 391–392
Graphical methods, 13–31
Group frequency, 16

Histogram, 18–24
  classes for, 20
  construction of, 21–24
  interpretation of, 28–31
Hypothesis
  alternative, 165
  null, 165–166
  research, 165–166
  summary of, in tests, 403–409
  test of. *See* Test of hypothesis

Independence of two classifications, 342–348
Independent events, 79–80
Independent samples, 80
Inference, 2–3, 73–74, 111, 137–156, 165–181, 189–214, 260–264, 280–286, 304–310, 342–348. *See also* Estimation; Test of hypothesis
  goodness of, 8, 73–74
  large-sample, 137–156, 165–181, 189–214, 381–382
  small-sample, 217–245
Intercept. *See* y-intercept
Interval estimate, 139–140

Index

Interval estimation, 139–140. *See also* Confidence interval
  of a binomial parameter p, 155–156
  of difference between two population means, 195, 234–236, 285
  of difference between two population proportions, 201
  of a population mean, 146–149, 227–228
  of slope, 261–262

Large-sample confidence interval
  for binomial parameter, 156
  for difference between two binomial parameters, 201
  for difference between two means, 195
  for population mean, 149
Large-sample statistical test, 165–181, 197, 203, 381–383. *See also* Test of hypothesis
Least squares. *See* Method of least squares
Least squares estimates, 257
Least squares prediction equation, 255–258
Level of significance, 179–182
Limits, confidence, 148
Linear correlation, coefficient of, 266–269
Linear equation, 251–254
  slope of, 251–253
  y-intercept for, 251–253
Linear relationship, 250–251
Lower boundary, of classes, 20–21
Lower confidence limit, 148
Lying with statistics, 389–396

Mann-Whitney U test, 374
Mean,
  arithmetic, 38–40
  of binomial probability distribution, 87
  of binomial random variable, 87, 125
  and Central Limit Theorem, 116–121
  inferences concerning, 141–145, 146–149, 167–174, 217–226
  of $(\hat{p}_1 - \hat{p}_2)$, 192–193
  of population, 40
  of sample, 40
  of sample means, 117, 122, 141
  of sample proportions, 128–129, 151
  and sample size for estimating, 288–291
  of sample sums, 121
  of $(\bar{y}_1 - \bar{y}_2)$, 191–192
Measurement, 2
Measures of central tendency, 38–45
  mean, 38–40
  median, 40–43, 49–50
  mode, 43–45
Measures of variability, 38, 47–59, 60–62, 64–66
  percentile, 49
  quartile, 49–50
  range, 47–49
  standard deviation, 53–59
  variance, 50–53

Median, 40–43, 49–50
Method of least squares, 254, 255–258
  and estimate of slope, 256–257
  and estimate of y-intercept, 256–257
Mode, 43–45
Mound-shaped distribution, 53–55
Multiple regression, 270–272
  and computer solution, 271–272
  and prediction equation, 270
Mutually exclusive events, 76

Noise-reducing experimental design, 280–286
Nonparametric methods, 361–386
Nonparametric tests, 361–386
  sign test, 362–367
  Spearman rank correlation test, 378–382
  Wilcoxon rank-sum test, 373–376
  Wilcoxon signed-rank test, 369–372
Normal approximation to binomial, 91, 125
Normal curve, 91
  area under, 91
Normal probability distribution, 91–100
  and area of, 91
  and Central Limit Theorem, 116–123
  standardized, 96–100
  tabulation of areas for, 93–95
Null hypothesis, 165–166
Numerical descriptive measures, 37–68, 111

Objective of statistics, 2, 4, 138
Observed cell count, 343
One-tailed statistical test, 171
Ordinal data, 361
Outcome, 74

Paired-difference confidence interval, 285
Paired-difference experiment, 280–286
Paired-difference test, 283–284
Parameters, population, 38, 111, 138, 266
Parametric tests. *See* Test of hypothesis
Percentiles, 49
p hat, 128
Pie chart, 14–15
Point estimate, 139, 143, 150, 195, 201
Point estimation, 139
  of binomial parameter, 150–152
  for difference between two binomial parameters, 201
  for difference between two means, 195
  of population mean, 141–145
Pooled estimate of variance, 229–230
Population, 3
  conceptual, 3, 115
Population parameters, 38
  binomial parameter p, 150
  correlation coefficient, 266
  mean, 40
  standard deviation, 53
  variance, 52
Predicted value of y, 255

Prediction equation, 248
  and method of least squares, 255–258
  for multiple regression, 270–272
  and regression line, 248–254
Probability, 73–100
  additivity of, 76–78
  and area under histogram, 28
  binomial, 84–89
  calculation of, relative frequency approach, 75
  conditional, 78–80
  of event, 75
  and inference, 73–74
  and level of significance, 179–180
  of normal random variable, 91–95
  properties of, 78
  as relative frequency, 74–75
  of type I error, 169
  of type II error, 169
  unconditional, 79
Probability distribution, 73–100
  binomial, 81–89
  chi-square, 344–345
  F, 304–307
  normal, 91–100
  for random variable, 80, 84, 91
  and relation to relative frequency histogram, 80
  standardized, 96–100
  Student's t, 219–220
Proportion, 127–130, 150–152
  comparisons of, 200–205
  interval estimation for, 155–156
  mean for, 128–129, 151
  point estimation of, 128, 150–152
  sampling distribution of, 124–130, 151, 193
  standard deviation for, 128–129, 151
  statistical test for, 176–178
p value. See Significance level

Quartiles, 49–50

Randomized block design, 286
Random number table, use of, 113–115
Random sample, 112–113
  importance of, 112
Random sampling, 111–115
Random variable, 77, 80–81
  binomial, 83–89, 124–130
  continuous, 80, 91
  discrete, 80
  normal, 91
  probability distribution for, 80, 84, 91
  standard normal, 96
Range, 20, 47–49
Range approximation to s, 64–66, 290
Rank correlation coefficient, 379
Regression, 247–277
  and method of least squares, 255–258
  multiple, 270–272

Regression line, 249–254, 255–258
  deviations from, 255–256
  and estimates for slope and intercept, 256–257
  and inferences, 260–264
  slope of, 251–253
  y-intercept for, 251–253
Rejection region, 166
  how to select, 169–170
  one-tailed, 171
  two-tailed, 171–172
Relative frequency, 21
Relative frequency concept of probability, 74–75
Relative frequency histogram, 22–23
  and area under, 28, 80
  classes for, 20
  construction of, 22–23
  interpretation of, 28–31
  and probability, 80–81
Reliability of inference, 8, 73–74, 169, 179–182
Research hypothesis, 165
rho, 266

Sample, 3
  independent, 80
  mean of, 40
  random, 112, 113
  standard deviation of, 53
  variance of, 52
Sample size, determination of, 288–294
Sampling, 2, 111–130
  random, 111–115
Sampling distribution, 111–130
  for differences, 190–194
  for $\hat{p}$, 128–130
  for $(\hat{p}_1 - \hat{p}_2)$, 192–193
  for sums, 121
  for $\bar{y}$, 116–123, 141–142
  for $(\bar{y}_1 - \bar{y}_2)$, 191–192
Scatter diagram, 249–254
Short method for calculating variance, 60–62
Sigma (sum), 40
Significance level, of statistical test, 179–182
Sign test, 362–367
Skewed distribution, 42
Slope, 251–253
  confidence interval for, 261–262
  estimate of, 256–257
  inferences about, 260–264
  statistical test for, 260–264
Small-sample confidence interval
  for difference between two population means, 234–236
  for population mean, 227–228
Small-sample statistical test, 217–226, 229–232
Spearman rank correlation coefficient, 379
  test with, 381–382
Standard deviation, 53–59, 60–62, 64–66

Index

of binomial probability distribution, 87
of binomial random variable, 87, 125
calculation of, using shortcut method, 60–62
check on, 64–66
of $(\hat{p}_1 - \hat{p}_2)$, 192–193
population, 53
range approximation for, 64–66
sample, 53
of sample means, 117, 122, 141
of sample proportions, 128–129, 151
of sample sums, 121
of $(\bar{y}_1 - \bar{y}_2)$, 191–192
Standard error
of mean, 142
of difference, 194
Standard normal distribution, 96–100
Standard normal random variable, 96–100
Statistical inference. *See* Inference
Statistics, 1–4, 38, 111
descriptive, 13–31, 37–68
inferential. *See* Inference
Student's t distribution, 219–220
assumptions for, 220
degrees of freedom for, 220
tabulated values for, 220–222
Student's t statistic, 219, 223, 232, 261, 269, 283
Student's t test, 223–224, 232, 261, 269, 283, 321–323
Summary of useful statistical tests, 403–409
Summation notation, 40
Sum of squares of deviations, in regression, 255–257
Sum of squares for error, 256
Sums, and Central Limit Theorem, 121
S_{xx}, 257
S_{xy}, 257
S_{yy}, 261–262

Tabulated areas. *See also* Appendix II
of chi-square distribution, 345
of F distribution, 304–306
of normal distribution, 93–95
of t distribution, 220–222
t distribution. *See* Student's t distribution
Test of hypothesis, 138, 165–181
acceptance region for, 166
alternative hypothesis for, 165–166
for binomial parameter, 176–178
for comparing more than two population means, 321–328
for comparing two means, 197, 229–232
for comparing two (or more) populations, 362–386
for comparing two proportions, 203
for comparing two variances, 304–310
concerning the independence of two methods of classification, 342–348
for correlation coefficient, 269
elements of, 165–166

large-sample, 165–181, 197, 203, 381–382
nonparametric, 362–386
null hypothesis for, 165–166
one-tailed, 171
paired-difference, 283–284
for population mean, 167–174, 217–226
for rank correlation, 381–382
rejection region for, 166
research hypothesis for, 165–166
significance level of, 179–182
for slope of a line, 260–264
small-sample, 217–226, 229–232
summary of, 174, 403–409
test statistic for, 166, 223, 306, 344
two-tailed, 171–172
and type I error, 169
and type II error, 169
using signs, 362–367
Test statistic, 166
t statistic. *See* Student's t statistic
t test. *See* Student's t test
Two-tailed statistical test, 171–172
Two-way classification of data, 342
Type I error, 169
Type II error, 169

Unconditional probability, 79
Upper confidence limit, 148

Variability
among sample means, 116–123, 324–325
measures of, 38, 47–59, 60–62, 64–66
within populations, 319–321, 323–325
Variable, 77, 248
Variance, 50–53, 59
analysis of, 319–339
calculation of, using short method, 60–62
comparison of, 303–316
inferences concerning, 304–310
pooled estimate of, 229–230
population, 52
sample, 52
Volume-increasing experimental design, 281, 286

Wells, H. G., 391
Wilcoxon rank-sum test, 373–376
Wilcoxon signed-rank test, 369–372

y bar, 40
y-intercept, 251–253
estimate of, 256–257

z score, 96
z statistic, 93–100, 173, 197, 364, 367, 374, 381
z test, 167–174, 197, 203, 364–365, 366–367, 374, 381–382